WETTING AND WETTABILITY

Edited by **Mahmood Aliofkhazraei**

Wetting and Wettability

http://dx.doi.org/10.5772/62031
Edited by Mahmood Aliofkhazraei

Contributors

Rita Khanna, Kyo-Han Kim, Kim In-Hye, Nikola Slepičková Kasálková, Petr Slepicka, Zdeňka Kolská, Vaclav Svorcik, Javier Narciso, Huixia Wang, Franco Ferrero, Monica Periolatto, Abdelmonem Mohamed Amer, Aleksandr Kryshtal, Sergei Dukarov, Vladimir Sukhov, Ion N. Mihailescu, Liviu Duta, Andrei Popescu, Irina Zgura, Ligia Frunza, Marius Enăchescu, Antoniu Moldovan, Ping Kuang, Kristen Constant, Yuekun Lai

CBS, Edition **2018**

Published by InTech

Janeza Trdine 9, 51000 Rijeka, Croatia

Publishing Process Manager Sandra Bakic

Technical Editor SPi Global

Cover Designer InTech Design Team

Additional hard copies can be obtained from orders@intechopen.com

Wetting and Wettability, Edited by Mahmood Aliofkhazraei
p. cm.
ISBN 978-953-51-2215-9

INTECH

open science | open minds

Contents

Preface

Each substance has an end at a point or place. This point or place is the surface and it is not an actual end point; but it is a point to distinguish two states or physical concepts. On the surface of a liquid, the molecules have fewer neighbors in comparison with the bulk volume. As a result the energy interaction shows its self in the surface tension, a force which is being applied to decrease the area of the free surface of the liquid. Traditionally, the surface tension can be assumed as a force in the unit of the length which can be counted by the unit of Newton on squared meter, or energy on the units of the surface. The surface tension, implies the interface between liquid and vapor, which is an example of the surface tensions. The equilibrium between these surface tensions, decides that a droplet on a solid surface, would have a droplet form or will change to layer form. As we saw in the recently published papers, the lexicology of the word "wettability" has not been clearly paraphrased.

This book collects new developments in wetting and wettability science. I like to express my gratitude to all of the contributors for their high quality manuscripts. I hope open access format of this book will help all researchers and that they will benefit from this collection.

Dr. Mahmood Aliofkhazraei
Tarbiat Modares University
Iran
www.aliofkhazraei.com

Chapter 1

Water Vapor Adsorption and Soil Wetting

Abdelmonem Mohamed Ahmed Amer

Additional information is available at the end of the chapter

http://dx.doi.org/10.5772/60953

Abstract

Soil water management and irrigation practices largely depend on a timely and accurate characterization of temporal and spatial soil moisture dynamics in the root zone. Consequently, measurements and detailed information about soil water sorption, water content, behavior, and potential are required. In that concern, water vapor adsorption is an important phenomenon in arid and semi-arid regions, as well as in dry periods of tropical soils. Therefore, quantifying adsorption is important for agricultural water management, surface energy balance studies, ecological studies, and remote sensing investigations (changes in surface soil moisture content will affect land surface properties such as albedo, emissivity, and thermal inertia). The vapor pressure and isothermal adsorption of water vapor can be used to predict soil moisture adsorption capacity (*Wa*), specific surface area, and hydro-physical properties of arid soils such as in Egypt and in the tropical soils in Ecuador. Theory of adsorption of water vapor on soil particles is developed among the mono-molecular and poly-molecular adsorption with respect to Brunauer, Emmett, and Teller (BET) theory. Data of soil-water adsorption (W%) at different relative vapor pressures (P/Po) can be obtained for the soils, where the W% values are increased with increasing P/Po in general. The highest values of water adsorption capacity (*Wa*), specific surface area (S), and other hygro-physical properties such as adsorbed layers and maximum hygroscopic water are observed in the clay depths of soil profiles, while the lowest values can be found in coarse textured soils (sandy and sandy loam soils profiles). Two equations were assumed: (1) to predict P/Po at water adsorption capacity (*Wa*) and (2) to apply *Wa* in prediction of soil moisture retention, i.e., ψ (W) function at pF < 4.5.

Keywords: Water adsorption capacity, vapor pressure isotherm, soil hydro-physical properties, specific surface area, poly-molecular adsorption, soil wetting

1. Introduction

The depletion in irrigation water in arid and semi-arid regions, as well as in some tropical zones, may be due to the discharge scarcity of water resources in dry periods at which the high temperature and dry weather supports to more evapotranspiration. In that concern, water vapor is adsorbed from the atmosphere by a thin layer of top soil where the amounts of adsorbed water can be considerable, up to 70% of daily evaporation depending on water vapor amount, pressure, and temperature [20, 17, 1].

Adsorbed water on soil surface layer can be caused not only by vapor adsorption but also by dew deposition. Dew deposition is a phenomenon recorded for most soil and climate types [16]. It occurs with decreasing the temperature, particularly during the night, when dew point is reached and it results in a discernible wetting of the surface.

The vapor pressure and isothermal adsorption of water vapor is used to predict soil moisture adsorption capacity (*Wa*) and the specific surface area (S). The specific surface area is closely related to the physic-chemical soil properties, which refer to the absence or presence of internal pores. It can be used to evaluate quality, fertility, chemical reactions, and mineralogical composition of soils and subsequently, its nutrition elements status for crop production.

2. Water vapor pressure

According to the kinetic theory, molecules in a liquid are in continuous motion reflecting their thermal energy. Occasionally, one or another of the molecules absorbs sufficient momentum to leap out of the liquid into the atmosphere above it. The relative movement rates of molecules depend upon the concentration of vapor in the atmosphere relative to its concentration at a state of equilibrium (i.e., when the movement in both directions of water and air is equal). An atmosphere that is at equilibrium with a body of pure water that is at atmospheric pressure is considered to be saturated with water vapor; the partial pressure of the vapor in such an atmosphere is called the saturation (or equilibrium) vapor pressure. The vapor pressure at equilibrium with any body of water depends upon the physical condition of the water (pressure and temperature) and its chemical condition (solutes), but does not depend on the absolute or relative quantity of liquid or gas in the system [11]. The saturation vapor pressure increases with increasing temperature. If the temperature range is not too wide, the dependence of saturation vapor pressure on temperature is expressible by the equation:

$$ln\ Po = a - b / T \tag{1}$$

where *ln Po* is the logarithm to the base "e" of the saturation vapor pressure Po, *T* is the absolute temperature, and *a* and *b* are constants.

3. Atmospheric humidity

Air humidity is expressed by either absolute humidity, which is known by estimation of the amount of water vapor already existing in unit volumes of air, or relative humidity, which is estimated as a percentage of the amount water vapor already existing in unit volumes of air and the amount of water required for the steam to satisfy this volume of air at the same temperature. Generally, the water vapor percentage change from one place to another is less significant in desert areas due to the lack of water, as well as in the polar regions where it is much colder and there is less evaporation and less possibility of the air to bear the water vapor; while water vapor is higher in the air in warm, rainy, and tropical regions [7]. On the other hand, the water vapor density decreases with height greater than the density of the main constituent gases of the air, i.e., when rising from 1.5 to 2 km above sea level, water vapor density is less twice than that of atmospheric layer in contact with the surface of the earth, and vanish at altitudes of more than 10–15 km [8]. In general, at all times and at different temperatures a part of water vapor remains in a gaseous phase even with predominance of condensation in the air. Therefore, water vapor is considered as one of the air gases component, and has a significant pressure like the rest of the gases in the atmosphere. This pressure (P) is linked to the quantity water vapor (Q) by the relationship [6]:

$$Q = \frac{1.06P}{(1+\alpha\ T)} \tag{2}$$

where α represents volumetric expansion coefficient of air ($\alpha = 0.004$), and T is air temperature. From the relationship, it is clear that Q = P when the temperature rises to 15°C, and thus relative humidity can be expressed in terms water vapor pressure rather than mass (quantity), i.e, relative humidity = actual pressure (P)/pressure saturated (P_0).The difference between saturated vapor pressure and actual vapor pressure is represented in the lack of air humidity, which is known as lack of saturation (or saturation deficit). The lack of saturation is equal to zero at the dew point, which means that the air could not carry other amounts of water vapor. The dew point can be obtained when the temperature is reduced to the degree at which the condensation of water vapor begins, and then the saturation pressure is reached. Thus, relative humidity can be expressed as a percentage of the saturated water vapor pressure at the dew point (i.e., at the temperature of condensation) to the saturated water vapor pressure at normal temperature. Thus, it is clear that the saturation of the air with water vapor is directly influenced by air temperature degree.

4. Water vapor and soil humidity

Water vapor either reaches the soil from the atmosphere or is formed in the soil by the evaporation of water. The migration of water vapor in soil depends not only on the difference of vapor pressure in different sites, but also on the capacity of soil particle surfaces to attract and absorb the molecules of vapor.

On the other hand, the relative humidity values in soil can be transformed to water potential values ψ(Pa) according to Equation (3) [7, 14]:

$$\Psi = \frac{RT}{M} \ln \frac{P}{Po} \tag{3}$$

where R is the gas constant (8.31 J mol^{-1} K^{-1}), T is the temperature of the air (K), M is the molecular weight of water (0.018015 kg mol^{-1}), and P and Po are the actual and saturated vapor pressures (Pa), respectively. The term (P/Po) x100 is equal to relative humidity (H).

However, soil water contains solutes—mainly electrolytic salts—in highly variable concentrations. Since the vapor pressure of electrolytic solutions is lower than that of pure water, accordingly, soil water also has a lower vapor pressure even when the soil is saturated. In an unsaturated soil, the capillary and adsorptive effects further lower the potential and the vapor pressure.

Adsorption is an interfacial phenomenon resulting from the differential forces of attraction or repulsion occurring among molecules or ions of different phases at their exposed contact surfaces. Consequently, various types of adsorption can occur according to the different phases, such as the adsorption of gasses upon solids, gasses upon liquid surfaces, or liquids upon solids.

A distinction should be made between **adsorption**, being a surface attachment or repulsion, and its complementary term **absorption**, which refers to cases in which one phase penetrates or permeates another. For instance, some of the molecules in the gaseous phase (water vapor) may strike the surface of the liquid (soil moisture) and be absorbed in it. Actually, it is often difficult to separate the phenomenon of adsorption from that of absorption, particularly in the case of highly porous systems, and hence the non-committal term sorption is frequently employed.

On contact of soil particles with water vapor, electro-molecular force of interactions are formed that strongly attract dipoles of water to the surface of mineral particles; and the greater the unit surface area of the particles, the larger the number of water molecules in a bonded state.

5. Soil moisture films and adsorption capacity

The electro-molecular force is very strong, as much as several hundreds of mega-Pascal (MPa) for the first layer of bonded (adsorbed) molecules of water at the surface of particles. In the layers of water, the 1–3 rows of molecules that are closest to the mineral particle surface are firmly immobile adsorbed water or the so-called **adsorbed films**. Beyond these layers, the interaction forces attenuate rapidly and reach zero at a distance of approximately 0.5 mm from the surface of the mineral particles; and hence, more moisture layers will envelope the particles and form the loosely bonded water or **wetting films**.

Amer [8] studied the molecules layers of the adsorbed films and proved that the three layers of adsorbed water can be expressed in term of soil moisture adsorption capacity (Wa) in the following form:

$$Wa = Wm + 2Wme \tag{4}$$

where ***Wm*** is the moisture content of the soil when water vapor is adsorbed in the monolayer, and ***Wme*** is the external mono-adsorbed layer of soil moisture content.

The *Wm* and *Wme* are used also to determine total, external, and internal specific surface areas (*S*, *Se*, and *Si*) by applying the BET method (Brunauer et al. [10]) as modified and described by Farrar [12] and Globus [13].

The property of moisture adsorption capacity (*Wa*) can be introduced as the critical limit between adsorbed and absorbed wetting films (pellicles) of soil moisture content, as well as correspond to capillary condensation [7]. However, the *Wa* values can be also derived from the soil moisture tension curve by applying the corresponding values of log (ψc) where ψc is the capillary condensation attitude [9].

The moisture adsorption capacity (*Wa*), maximum hygroscopic water (*MH*), specific surface area (*S*), and particle size composition are the most important indices characterizing the hygro-physical, physicochemical, and heat properties of soil. Moreover, these parameters are inter-related to each other; therefore, the value of a parameter of them can be obtained from the data of the other parameters.

6. Adsorption isotherms and BET theory

The relationship between vapor pressure and moisture content is difficult to be deduced by means of thermodynamics but can be obtained experimentally or from theories of soil moisture involving the molecular structure of the water films, whereas, soil sample is maintained in equilibrium with an atmosphere of water vapor, a thin film of moisture is adsorbed on the external surface of the soil particles. So the water vapor adsorption isotherms (at 25°C) on dried soil particles can be determined gravimetrically using different saturated salt solutions that have specific relative water vapour pressure (P/Po) value for each solution. Table 1 shows the values of P/Po for some saturated solutions of salts and also the corresponded chemical potential or its equivalent pressure at 25°C [7].

The vapor adsorption isotherms, specific surface area, moisture adsorption capacity, and hygroscopic parameters of soil have been found to be correlated with soil physical properties such as texture, cohesion, clay percentage, clay minerals, cation exchange capacity, water retention, and permeability. Figure 1 shows the water vapor adsorption isotherms for sandy soils, located at Mláky II near Sekule Bratislava (southwest Slovakia).

Salt	P/Po	ψ			
		Atm.	pF	KPa	Joule/gm
$K_2Cr_2O_7$	0.98	27	4.44	2780	2.78
K_2SO_4	0.97	43	4.62	4370	4.37
KNO_3	0.92	107	5.06	10830	10.83
$ZnSO_4.7H_2O$	0.87	188	5.28	19010	19.01
KCl	0.84	233	5.38	23580	23.58
KBr	0.81	292	5.46	29550	29.55
NH_4Cl	0.77	348	5.56	35820	35.82
NaCl	0.75	386	5.60	39120	39.12
$NaNO_3$	0.74	413	5.62	41840	41.84
NH_4NO_3	0.62	653	5.82	66200	66.20
$Ca(NO_3)_2.4H_2O$	0.50	942	5.98	95530	95.53
$K_2CO_3.2H_2O$	0.43	1155	6.07	117000	117.00
$KCH_3COO.1.5H_2O$	0.22	2031	6.32	205780	205.78
$LiCl.H_2O$	0.12	2781	6.47	292000	292.00

Table 1. Values of chemical potential (ψ) of some saturated solutions of salts and its equivalent pressures to relative vapor pressure at 20 °C.

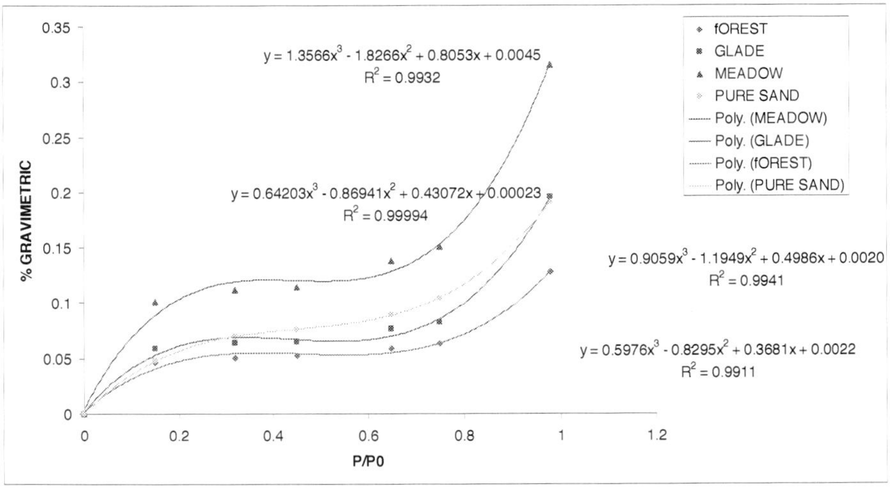

Figure 1. Unpublished data of the relationship between gravimetric adsorbed water (W%) and vapor pressure P/Po in sandy soils (94% sand in average) with plant cover.

On the other hand, Figure 2 shows the adsorption isotherms for the clay minerals. However, a number of theories have been proposed to explain the observed relation between the vapor pressure and moisture content of a soil. The most widely used theory is from Brunauer, Emmett, and Teller [11], whereas they derived what has come to be known as the BET equation based on multilayer adsorption theory. In the BET theory, the explanation proposed for sigmoid type isotherm (Figure 1) is that the adsorption is in multi-molecular layers on the surface rather than a mono-molecular one. Farrar [12] and Amer [9, 8] used the water vapor adsorption isotherm method by applying BET theory based on the assumption that the isotherm is made up of monolayer physical adsorption combined with capillary condensation as follows:

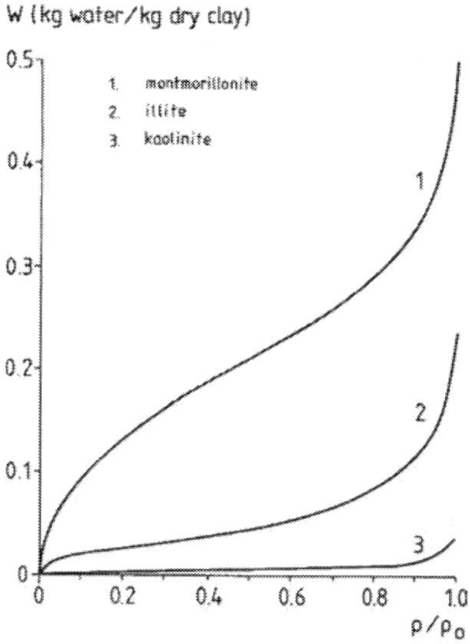

Figure 2. Adsorption isotherms for water on different clay minerals (reprinted from Verhoefe, et al., 2006 [20] upon Orchiston, 1954 [19]), P /Po denote the soil relative humidity.

$$\frac{\frac{P}{Po}}{V\left(1-\frac{P}{Po}\right)}=\frac{1}{VmC}+\frac{C-1}{VmC}\cdot\frac{P}{Po} \tag{5}$$

where V is the volume of gas adsorbed at pressure P, V_m is the volume of a single layer of adsorbed molecules over the entire surface of the adsorbent (soil particles), P_o is the gas pressure required for monolayer saturation at the temperature of the experiment, and C is a constant for the particular gas, adsorbent, and temperature;

$$C = exp\frac{E1 - El}{RT} \tag{6}$$

whereas, E_1 is adsorption heat of the water adsorbed layer, E_L is condensation adsorption heat.

At values of *P* that are far from P_s and C > 1, adsorption leads to the formation of a monomolecular layer. As *P* approaches P_s, the number of free active centers that are always present on the adsorbent surface decreases and the multiplicity of complexes increases. At $P = P_s$, the condensation of vapor occurs.

7. Prediction of adsorbed layers (Wm & Wme) and adsorption capacity (Wa)

In order to estimate *Wm* and *Wme*, and then calculate the soil moisture adsorption capacity (*Wa*), the relation between relative vapor pressure (*P*/*Po*) and moisture content (*W*%*)* should be experimental, obtained by maintaining a soil sample in isothermal equilibrium with an atmosphere of water vapour.

The following linear form of the BET equation can be applied using the gravimetric of a single layer of adsorbed molecules over the entire surface of the soil particles:

$$\frac{P}{W(Po - P)} = \frac{1}{WmC} + \frac{C-1}{WmC} \cdot \frac{P}{Po} \tag{7}$$

where *Wm* is the moisture content when the soil surface is completely covered by a monomolecular layer of water. *C* is a function of the state of the first adsorbed uni-molecular layer of water and soil particles surface. *W* is adsorbed soil moisture content (%) equilibrated with *P*/*Po*, whereas P and *Po* are the actual and saturated water vapor pressures. By plotting *P*/*W* (*Po*–*P*) as ordinate versus *P*/*Po* at the segment 0–0.42 of the adsorption isotherm as abscissa, a straight line would be obtained. The intercept on the y-axis is then $1/W_mC$ and the slope is $C - 1/W_mC$. Hence W_m and C can be determined.

To determine W_{me} the BET Equation (7) can be developed with some assumptions to the next form:

$$W = \frac{Wme}{\left(1 - Ke\frac{P}{Po}\right)} \cdot \frac{Ce}{\left(Ce + \frac{Po}{P} - Ke\right)} + Wi \tag{8}$$

At high relative water vapor pressures, it can be assumed that the amount $\frac{Ce}{\left(Ce + \frac{Po}{P} - Ke\right)}$ is equal unit, and then Equation (8) becomes:

http://dx.doi.org/10.5772/60953

$$W = \frac{W_{me}}{\left(1 - K_e \frac{P}{P_0}\right)} + W_i \tag{9}$$

where the suffixes (e) and (i) refer to the external and internal surfaces, respectively. The values of K_e in the indicated P/*P*o range were stated by Farrar [12] as 0.9 ±0.01; but practically, it seems that K_e is an arbitrary coefficient ranged from 0.70 to 0.90 [4].

Equation (9) can be represented in the linear equation y = mx + c, where y = *W*, m = W_{me}, x = 1 / 1 – $K_e P/Po$, and c = *W*i, so W_{me} can be obtained graphically as the intercept on the y – axis.

From W_m *and* W_{me} the soil water adsorption capacity (Wa) can be calculated (Equation 4).

8. Soil specific surface area

The specific surface of the adsorbent (soil) can be calculated by determining the number of molecules (volumetrically or gravimetrically) and multiplying this by the cross-sectional area of the molecules. Assuming that a single water molecule occupies some constant area on the sorbent surface (usually taken as 10.8°A^2), the total specific surface area (*S*) of the soil is then calculated as *S* = 36.16 W_m m^2/g, and accordingly, the external specific surface is estimated as *Se* = 36.16 W_{me} m^2/g.

However, the internal specific surface area (*Si*) may be calculated by the difference between *S* and *Se*.

In general, high clay content in soil means increasing the specific surface areas, hygroscopic water, soil moisture content and retention, and water adsorption capacity (Wa) (Figure 3).

Figure 3. Clay soil samples were taken at depth >90 cm from the INIAP research station, Pichilingue, Ecuador ~containing ferrous and ferric minerals with the high values of specific surface area and water adsorption capacity (Final Report by Amer, 2014 [2] - SENESCYT, Prometeo Project, Ecuador).

Hygroscopic water exists as a very thin film at the solid-liquid interfaces of the soil particles. At the maximum hygroscopic water (*MH*), the surface of soil particles is almost completely covered with individual molecules of water. However, it is known that the maximum hygroscopic water (*MH*) is determined practically at P/Po = 0.98.

9. Absorption water in relation to P/Po and pF scale

Equation (7) can be developed to predict the vapor pressure (*P*/*Po*) at water adsorption capacity (*Wa*) as follows:

$$\frac{\frac{P}{Po}}{w\left(1-\frac{P}{Po}\right)} = A + \mathrm{B}\frac{P}{Po} \tag{10}$$

where, A = $1/W_mC$, and B = $C\text{-}1/W_mC$

From Equations 7 and 10;

$$W\left(\frac{Po-P}{Po}\right) = \frac{\frac{P}{Po}}{A+B\frac{P}{Po}} \tag{11}$$

Then at Wa:

$$\left(\frac{P}{Po}\right)wa = \frac{P}{Wa\left(Po-P\right)B} - \frac{A}{B}\text{and } P = Wa\left(Po-P\right)\left(APo+BP\right) \tag{12}$$

Relative vapor pressure $[P/Po]_{wa}$ at *Wa* is ranged between 0.42 to 0.51, indicating that at this range the soil moisture reach water *adsorption* capacity. Above this range (P/Po > 0.51) the *absorption* process is prevailing. Hence, to predict the soil absorbed water, Equation (7) can be rearranged to be available for application along the range of high relative vapor pressure (0.50 < P/Po > 1.0):

$$w = \frac{\frac{P}{Po}}{\left[1-\frac{P}{Po}\right]\left[\frac{1}{WmC}+\frac{C-1}{WmC}.\frac{P}{Po}\right]} \tag{13}$$

By applying the intercept ($1/W_m$) and the slope ($C\text{-}1/W_m$), we can obtain W% at P/Po values.

On the other hand, soil water potential (ψ) values corresponded to the total range (0.0 < P/Po > 1.0) can be calculated using Equation 3 in the following form (pF equation):

$$pF = 6.502 + log\left[2 - log\ H\right]$$

where pF = soil moisture potential, expressed as the common logarithm of the suction (ψ) in cm of water; H is the relative humidity (H= *P/Po* x 100); and *P/Po* is the relative water vapor pressure [*P* being the actual water vapor pressure on the soil particles and *Po* being the saturation vapor pressure of water at 20°C]. These parameters can be obtained by applying different appropriate salt solutions, whereas, the soil moisture potential (ψ) values which resulting from equilibrating soil samples with salt solutions that have different relative humidity (H). The ψ values expressed in pF at adsorption and absorption processes are in the range pF = 4.5–7.0.

10. Application of Wa for prediction of soil wetting

The increase in soil water suction is associated with a decreasing thickness of the hydration envelopes covering the soil particle surfaces and vice versa. The following equation was advanced by Nerpin & Chudnovski [17] to describe soil wetness (h) within limited suction ranges:

$$\psi = B / W^{\wedge} n \tag{14}$$

where ψ is matric or capillary-sorption potential, *B* and *n* are constants.

Taking moisture adsorption capacity (Wa) in consideration, the concept of the last relation (14) can be developed to predict the soil wetting expressed in moisture potential or matric suction (ψ) at pF < 4.5 as follows:

$$\psi = \frac{\psi a}{\left(\mathrm{Wi} - \dfrac{\mathrm{Wc}}{\mathrm{Wme}}\right)^{\mathrm{n}}} \tag{15}$$

or in log form;

$$log\psi = \log\psi_a - n\log\left(Wi - \frac{Wc}{Wme}\right) \tag{16}$$

where ψi and ψa are capillary-sorption potentials at soil water content (*Wi*) and moisture adsorption capacity (*Wa*), respectively. The *Wc* is called boundary moisture films.

By plotting log ψi against log (Wi – Wc/Wme), we obtain log ψa as intercept and *n* as the slope.

Substituting the coefficients parameters log ψa and *n* in the equation 16, the log ψi then could be obtained at arbitrary soil water content (*Wi*). From the relationship between log ψi and *Wi* we obtain the integrated soil moisture (retention) curve (Figure 4) or ψ(W) function.

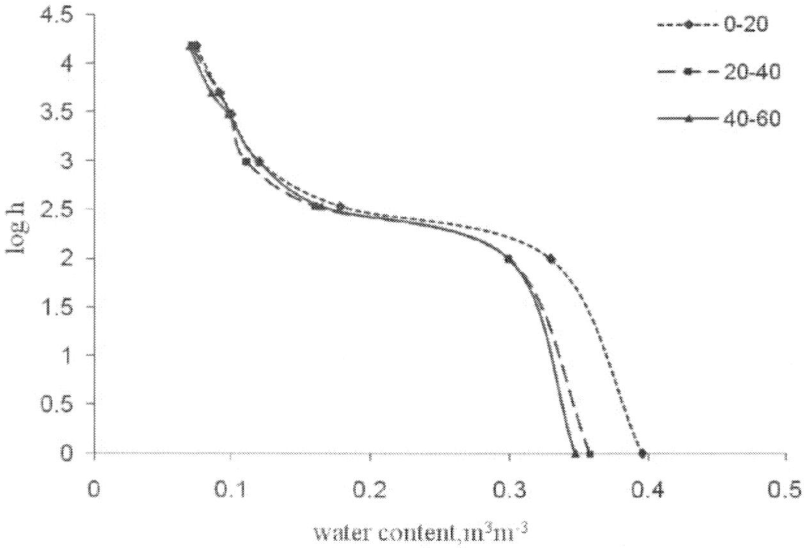

Figure 4. Moisture retention curves for three depths (cm) in Nubaria calcareous sandy loam soil (west Nile Delta, Egypt).

The moisture potential (ψi) can be expressed in log h (= pF), and called soil matric suction or pressure head (h), which is expressed as potential per weight (m) in SI units. In this case, the pF term is useful to apply as pF = log h, when h is expressed in cm (Figure 4).

Author details

Abdelmonem Mohamed Ahmed Amer

Address all correspondence to: amer_abdel@hotmail.com

Soil Science Department, Faculty of Agriculture, Menoufia University, Shebin El-Kom, Egypt

References

[1] Agam N., Berliner P: Diurnal water content changes in the bare soil of a coastal desert. 2004. J. Hydrometeor., 5, 922–933.

[2] Amer A.M: Moisture dynamics and available water capacity in root zone as influenced by swelling pressure and water table in tropical soils. 2014: Final Report, submitted to SENESCYT, Prometeo Project, Ecuador.

[3] Amer A.M: Prediction of hydraulic conductivity and sorptivity in soils at steady state infiltration. 2011. Archives of Agronomy and Soil Science. Published Online July 2011. http://www.tandfonline.com/loi/gags20 http://dx.doi.org/10.1080/03650340.2011.572877.

[4] Amer A.M: Moisture adsorption capacity and surface area as deduced from vapor pressure isotherms in relation to hygroscopic water of soils. 2009. Biologia, 64: 516–521.

[5] Amer A.M., Logsdon S.D., Davis D: Prediction of hydraulic conductivity in unsaturated soils. 2009. Soil Sci., 174, 9: 508–515.

[6] Amer, A.M: Soil Hydro-physics & Agricultural Irrigation and Drainage."2nd Part, *Water Requirements and Irrigation & Drainage*". 2004. El-Dar Al-Arabia Publishing Foundation, Cairo, Egypt, I.S.B.N 977-258-195-7. (in Arabic).

[7] Amer A.M: Soil hydro-physics. 1st Part, 2003. Al-Dar Alarabia Publishing Foundation, Cairo, Egypt, ISBN 9 77-258-179-5. (in Arabic).

[8] Amer A.M: Surface area measurements as related to water vapour adsorption in arid soils of Egypt. 1993. pp. 619–627. In: Proceedings of the. IV Int. Conf. Desert development, Mexico City, Mexico.

[9] Amer A.M: An approach towards estimation of the external specific surface from soil moisture characteristics curve. 1986. Int. R. T. Conf. Micro-Irrigation, Budapest, Hungary.

[10] Brunauer S., Emmett P.H., Teller E: Adsorption of gases in multi-molecular layers. 1938. J. Am. Chem. Soc., 60: 309.

[11] El-Fiky Y.S: Studies on hydro-physical and physicochemical properties of new reclaimed soils in Egypt. 2002. Ph.D. Thesis, Soil Sci. Dept., Faculty of Agric. Menoufia Univ., Egypt.

[12] Farrar D.M: The use of vapor pressure and moisture content measurements to deduce the internal and external surface area of soil particles. 1963. J. Soil Sci., 14: 303–321.

[13] Globus A.M: On specific soil surface area computing by one point on the water vapor sorption isotherm. 1996. Eurasian Soil Sci., 28: 154–155.

[14] Goebel M., Bachmann J., Woche S. K., Fischer W. R., Horton R: Water potential and aggregate size effects on contact angle and surface energy. 2004. Soil Sci. Soc. Am. J., 68: 383–393.

[15] Hillel D: "Particle sizes and specific surface" in Environmental soil physics. 1998. Academic press, NY.

[16] Jacobs A. F. G., Heusinkveld B. G., Berkowicz S. M: Dew deposition and drying in a desert system: A simple s imulation model. 1999. J. Arid Environ., 42, 211–222.

[17] Kosmas C., Marathianou M., Gerontidis S., Detsis V., M., Tsara M., Poesen J: Parameters affecting water vapour adsorption by the soil under semi-arid climatic conditions. 2001 Agric. Water Manage., 48, 61–78.

[18] Nerpin V., Chudnovski A.F: Energy and mass-transfer in plant-soil-air system. 1975. Hydro-Meteo Izdat., Leningrad. (in Russian).

[19] Orchiston, H. D: Adsorption of water vapour: 1. Soils at 25°C. 1954. Soil Sci., 76, 453–465.

[20] Verhoef A., Diaz-Espejo A., Knight J. R., Villagarcia L., Fernandez J. E: Adsorption of water vapor by bare soil in an olive grove in Southern Spain. 2006. Journal of Hydrometeorology, 7, 1011–1026.

Chapter 2

Wetting Properties at Nanometer Scale

Antoniu Moldovan and Marius Enachescu

Additional information is available at the end of the chapter

http://dx.doi.org/10.5772/60886

Abstract

The proposed chapter reviews a series of experimental techniques which enable the accurate quantitative study of wetting properties. The introductive part presents some of the many phenomena and processes influenced by wetting, underlining the importance of understanding the fundamental science involved. A few historical considerations about the quantitative study of wetting and related phenomena are given. Next, some of the "classical" techniques employed for studies at the macroscopic scale are presented. The importance of studies of such phenomena at micro- and nanometer level is underlined, as a consequence of the enormous influence that micro- and nanodevices play in our day to day activities, and examples of quantitative studies, involving various measurement techniques, are given from literature. A description of the basic phenomena related to polarization forces in Scanning Polarization Force Microscopy (SPFM) technique is given, followed by experimental details concerning the actual implementation of the technique. Examples of applications of SPFM are given from literature (from the spreading of liquid crystals on solid substrates to studies of corrosion at nanometer level). Particularly, it is emphasized how this versatile technique was successfully used for direct measurements of contact angles for liquid micro- and nano-droplets, enabling the calculation of the dependence of surface potential energy between the surfaces, the spreading coefficient and the disjoining pressure for micro- and nano-droplets.

Keywords: wetting, contact angle, surface potential energy, disjoining pressure, micro- and nano-droplets, scanning polarization force microscopy

1. Introduction

Wetting, capillarity and adhesion phenomena are so common in our daily lives that most of them go unnoticed. Every painted object is a result of wetting; the oleophobic screens of our smartphones and tablets are a result of wetting-related technology [1]; the integrated circuits that are everywhere have been exposed during fabrication to several steps of wetting processes; the oil inside the engines of our cars wets the moving parts and keeps them running; fruits have natural waxes on their surface to protect them from over- or dehydration [2], etc. Understanding these phenomena and the physical quantities that are involved is extremely important if we want to enhance the desired effects and diminish or eliminate the undesired effects [3,4].

Wetting, capillarity, and adhesion have been subjects of systematic study for at least the last two centuries, the Young's equation being formulated at the beginning of the 19th century. However, basic studies of capillarity, adhesion, and wetting can be traced back to antiquity: Hero of Alexandria (10–70 AD) studied liquids and surface tension, and described in his book "Pneumatics" a series of inventions based on capillary effects, and Pliny the Elder (23–79 AD) wrote in his studies about his interpretation related to the glassy wakes of ships. During the Renaissance, Leonardo da Vinci (1452–1519) analyzed capillarity and envisioned that capillary networks fed mountain water streams. In the 18th century, Benjamin Franklin (1706–1790) was investigating the ability of oil in suppressing waves. Bridging two centuries, Pierre-Simon de Laplace (1749–1827) introduced the term Laplace pressure by investigating the concept of meniscus and offering a theoretical description of the meniscus. A younger polymath but almost during the same period, Thomas Young (1773–1829) described the wetting of solids by fluids [5].

In technology, the implementation of many real-life applications is based on effects related to surface tension, e.g., lab-on-a-chip technology [6], inkjet printing [7], and superhydrophobic surfaces [1]. In biology, surface tension is involved in many basic functions, such as weight support and propulsion at the water surface [5], in natural strategies for water-repellency [2], in the functioning of lungs [8,9], etc. The dynamics of raindrops, groundwater flows, oil spill dynamics, chemical leaching, the water-repellency of soils, and disease transmission via droplet exhalation represent a few of the surface-tension-related phenomena and applications in geophysics and environmental science [5].

2. Techniques for surface energy and wetting investigation from macro- to nanoscale

Most of the techniques involved in the determination of wetting properties are based on measurements of the contact angle. The contact angle is a convenient parameter which describes the ability of a liquid to wet a surface in a gaseous environment, and it is related to the interfacial energies by the well-known Young's equation:

$$\gamma_{lv}\cos\theta = \gamma_{sv} - \gamma_{sl}, \tag{1}$$

where γ_{sv}, γ_{sl}, and γ_{lv} are the surface energies corresponding to the solid–vapor, solid–liquid, and liquid–vapor interfaces, respectively, and θ is the contact angle. Figure 1 shows various degrees of wetting as quantified by the contact angle. Small contact angles (<< 90°) correspond to high wettability, while large contact angles (>> 90°) correspond to low wettability. For water, if the contact angle is smaller than 90°, the solid surface is considered hydrophilic and if the contact angle is larger than 90°, the solid surface is considered hydrophobic. Usually, bare metallic and ceramic surfaces are highly hydrophilic while many polymers exhibit hydrophobic surfaces. Contact angles can be static and dynamic angles. Static contact angles are measured when the droplet is standing on the surface and its three-phase boundary is not moving. When the three-phase boundary is moving, dynamic contact angles can be measured, and they are referred to as advancing and receding contact angles. Contact angle hysteresis is the difference between the advancing and receding contact angles. Advancing and receding contact angles give the maximum and minimum values the static contact angle can have on the surface. Contact angle measurements are used in fields ranging from coatings to printing and to oil recovery.

At the macroscopic scale, the contact angle can be measured by conventional optical techniques, whereas at the micro- and nanoscale, it is only accessible either by more advanced optical techniques, by indirect measurements, or by direct measurements with scanning probe techniques, as will be shown in the following sections.

Scanning probe techniques are also able to measure interaction forces and energies directly, similar to the surface forces apparatus, which will be reviewed in the following section.

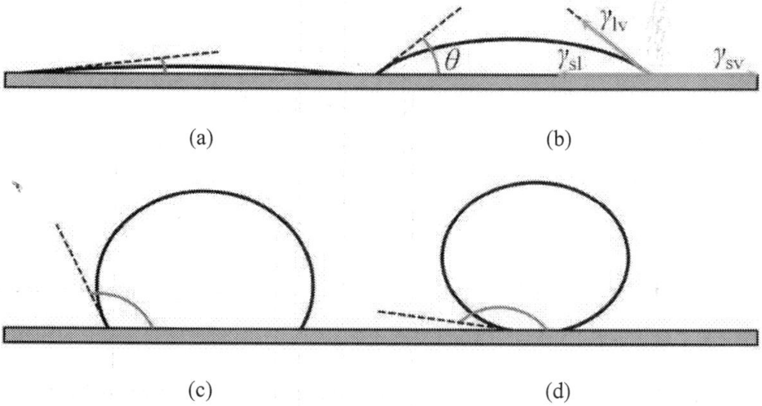

Figure 1. Contact angles for various degrees of wetting: a) almost total wetting, $\theta \sim 0°$; b) good wetting, $\theta < 90°$; c) poor wetting, $\theta > 90°$; d) almost no wetting, $\theta \sim 180°$

2.1. Surface forces apparatus

The surface forces apparatus (SFA), introduced by Tabor, Winterton, and Israelachvili in the 1970s [10-12], enables the determination of the distance dependence of forces between surfaces, with resolution down to 10 nN. The forces are usually measured directly by the deformation of a spring attached to one of the surfaces or by means of capacitive sensors. The surfaces are held close to each other in crossed cylinder geometry (Figure 2). The distance between the surfaces, measured by interferometry with a resolution of ~0.1 nm, is usually controlled by means of piezoelectric positioners. The crossed cylinders usually have the same radius, and this geometry is equivalent – from the point of view of the distance dependence of the forces – to a sphere of the same radius in the vicinity of a plane [13]. The samples are usually deposited (sputtered, adsorbed, etc.) in the form of thin layers on mica sheets, which are bent and fixed to transparent semi-cylindrical lenses. Mica is usually employed because it is transparent and can easily be cleaved to obtain atomically flat surfaces. Initially, the technique was developed for measurements in vacuum or air, and was subsequently extended for liquid and controlled gas environment.

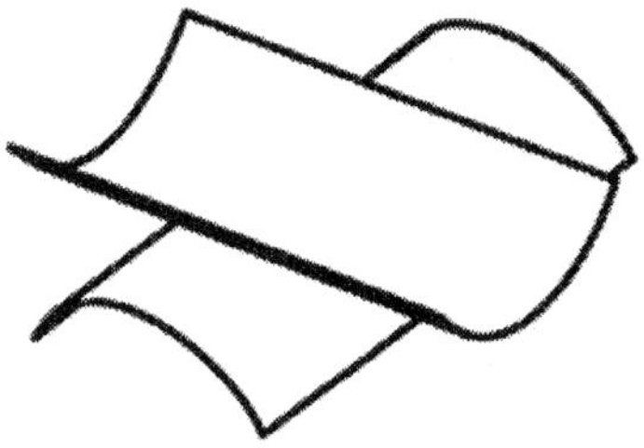

Figure 2. Crossed cylinder geometry of the samples in the surface forces apparatus

One of the samples is attached to a spring cantilever system with known elastic constant and the other sample is attached to the piezoelectric positioning system. The samples are first brought into close vicinity (< 1μm) by a coarse positioning mechanism, usually a stepper motor or a micrometer screw. The distance between the surfaces is measured by an optical technique using fringes of equal chromatic order (FECO), in a white light interferometer. The separation between the surfaces, their shape as well as the refractive index of the material between them, can be calculated from the positions and shapes of the fringes. The piezoelectric positioner moves one of the samples by a controlled amount. The deformation of the spring holding the other sample can be determined from the optically measured distance between the samples, and thus the force can be calculated.

This technique has allowed the accurate determination of fundamental interactions between surfaces: van de Waals, electric double layer, adhesion, capillary, solvation, hydration, steric, hydrophobic, etc. It has also been extended to measuring lateral forces (shear and friction) and forces in the dynamic regime. The SFA technique can now be used to measure both normal and lateral forces between surfaces in liquids with a distance resolution of less than 1Å, and it can be involved even in measuring forces in electrochemical and biological systems [14]. Additionally, SFA can be combined with complementary techniques, such as AFM [15], X-ray scattering [16], IR spectroscopy, and fluorescence microscopy [17], in order to be able to perform different measurements at the same time on the same sample.

2.2. Wilhelmy method

In the Wilhelmy method, the contact angle is determined indirectly by measuring the force exerted on a solid sample of simple shape (plate, rod, wire, etc.), which is brought in contact with a liquid (Figure 3). The Wilhelmy method finds nowadays wide use in the preparation and monitoring of Langmuir–Blodgett films.

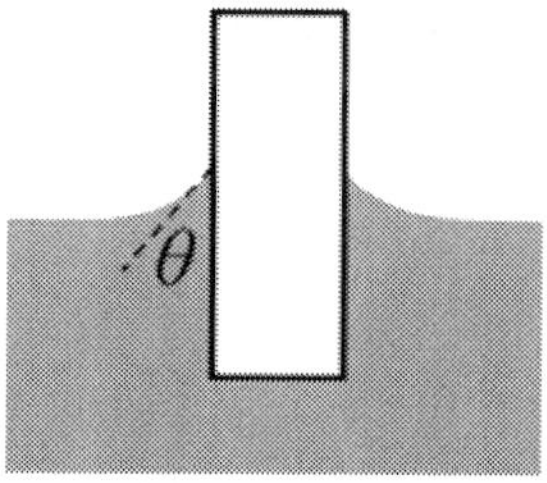

Figure 3. Formation of the liquid meniscus when the sample is submerged in a wetting liquid

The force is usually detected by a sensitive balance which measures the apparent weight of the sample as it is inserted into and retracted out of the liquid (Figure 4a). The apparent weight (G_{app}) at any moment is the sum of the actual weight (G), the wetting force and the buoyancy:

$$G_{app} = G + \gamma_{lv} L cos\theta - \rho V g, \tag{2}$$

where γ_{lv} is the surface tension of the liquid, L is the length of the contact line, θ is the contact angle, ρ is the density of the liquid, V is the submerged volume of the sample, and g is the gravitational acceleration.

The force detected by the balance during a full cycle of insert–retract (submersion cycle) has the general shape described in Figure 4b, and it represents an excellent choice for measuring the dynamic contact angles on any homogeneous, regularly shaped sample. During the initial approach (A), the measured force is actually the weight of the sample. When the sample touches the liquid (B), the surface tension of the liquid creates a relatively sharp increase or decrease of the force – depending whether the liquid wets the sample or not, respectively (Figures 3 and 4 show the case of a wetting liquid). Upon further submersion (C), the increasing buoyancy causes the apparent weight to decrease. In this region the contact angle is the advancing contact angle. While the sample is retracted (D) the contact angle is the receding contact angle. The offset between the C and D portions of the graph is due to the contact angle hysteresis, which is mainly an effect of the contact line being pinned by defects and irregularities of the sample surface [18-20].

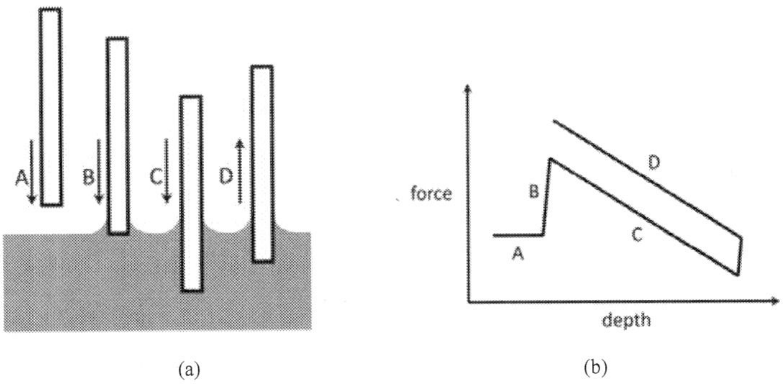

Figure 4. (a) Submersion cycle with the Wilhelmy balance technique; b) the apparent weight of the sample measured by the Wilhelmy balance during a submersion cycle

An adaptation of this technique for measurements at the nanoscale is described in the work of Yazdanpanah [21]. The samples are nanowires with diameters of up to 500 nm, which extend from the apex of AFM tips – so-called nanoneedle probes. The wires, made of Ag_2Ga alloy, are grown at room temperature from droplets of melted gallium, on the surface of silver-coated AFM tips. The forces are conveniently determined with the AFM through force-distance spectroscopy measurements. The buoyancy can be neglected due to the reduced volume of the wires, so the corresponding submersion and retraction portions of the force curves are horizontal (Figure 5). As the diameters of the wires are accessible from SEM measurements, the contact angle values can readily be determined from the force measurements. In the described experiments, for most of the studied liquids, the authors observed only very small contact angle hysteresis, which they attributed to the low dimensions of the defects which would pin the contact line and to the low values of the liquids' surface tensions.

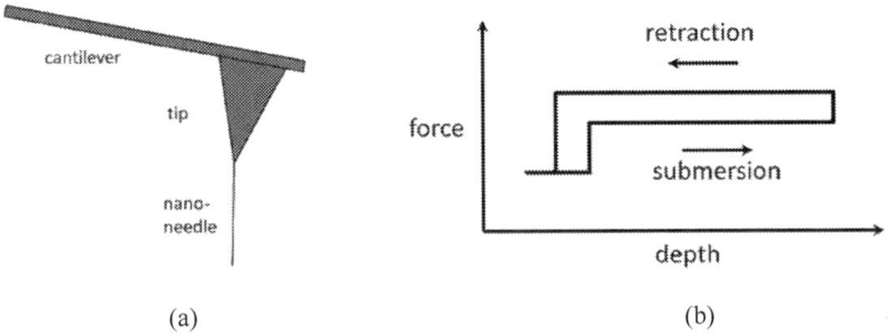

Figure 5. (a) Schematic representation of nanoneedle tips; b) The wetting force measured with the Wilhelmy balance technique at the nanoscale, with nanoneedle tips

http://dx.doi.org/10.5772/60886

2.3. Optical methods

2.3.1. Sessile drop

One of the most widely used techniques for studies of wetting properties is the sessile drop technique, in which the contact angle is directly measured from the profile of a liquid drop. A general setup consists of a horizontal sample stage illuminated from behind, a liquid dispensing syringe or pipette and a visualization system (Figure 6). Early setups used telescopes fitted with a goniometer eyepiece (protractor) to measure visually the contact angle [22]. Modern setups make use of digital video cameras, which allow measurements in the dynamic range and recording data for postprocessing.

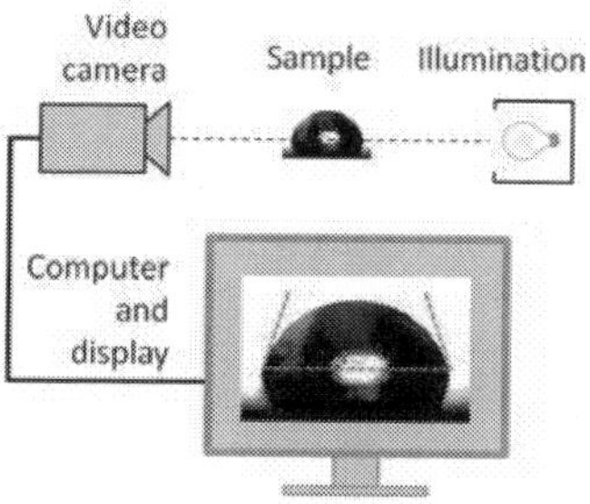

Figure 6. Schematic representation of the contact angle measurement setup for sessile drop technique

If someone intends to characterize solid surface energies, the Sessile Drop method is highly appropriate. Thus, using a liquid having a known surface energy and generating a drop of it on the solid surface to be investigated, from the shape of the drop, the specific contact angle, and the known surface energy of the liquid, one can have enough parameters to calculate the surface energy of the solid sample. In such experiments a specific probing liquid is used, and for a trustable determination of the solid surface energy several different probing liquids should be used.

This method has the advantage of a relatively simple design and operation. It allows a huge variety of samples to be studied: from "standard" materials (polymers, metals, glass, textiles, etc.) [23,24], to films of colloid particles [25-27], layers of bacteria [28,29], or even viruses [30].

2.3.2. Interference microscopy

Interference microscopy allows the determination of the three-dimensional shape of drops on transparent or reflective substrates, and the subsequent calculation of the contact angle. In interference microscopy the contact angle value is calculated using the fringe patterns formed by the interfering beams reflected from the solid–liquid and the liquid–vapor interfaces. In the work described in reference [31], droplet profiles were reconstructed by analyzing the intensity profiles along the interference pattern (Newton rings) created when observing the droplets from above with an interference microscope. The positions of the intensity maxima and minima were used for the reconstruction of the droplet profiles. This method is only suited for contact angles less than ~30°, and for contact angles lower than 15° it offers high precision.

Fischer and Ovryn describe a method for the reconstruction of droplet profiles for higher values of contact angles, from phase images recorded using sufficiently high numerical aperture objectives in a confocal interference microscope [32]. The method uses a scalar model based on geometrical optics principles, in order to determine the differential optical path length through the droplets. However, this method requires a prior estimation of the shape of the droplets in order to determine the actual exact shape of the droplets, as it needs to be incorporated in an iterative algorithm.

2.4. Scanning probe techniques

2.4.1. Atomic Force Microscopy (AFM)

In atomic force microscopy, a sharp tip is raster-scanned in the vicinity of the sample surface, in the region of the van der Waals forces. A force-sensing mechanism (in most cases a "light lever", as shown in Figure 7), coupled to a feedback loop, which controls the separation between tip and sample, helps maintain a constant interaction force during scanning, allowing the tip to follow the topography of the surface.

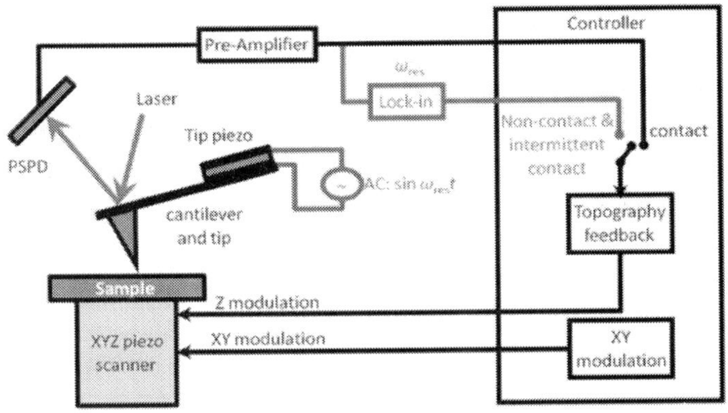

Figure 7. The common light lever technique and the signals in contact and noncontact AFM

Techniques of this family have been employed in studies of wetting phenomena. Yu and coworkers report about the direct measurement of macroscopic contact angles for millimeter size glycerol droplets [33]. The contact angles are determined directly from topography images taken at the edge of the droplets, in noncontact mode. Another study, by Checco and coworkers, reports about noncontact AFM measurements of the topography of micron-sized alkane droplets [34]. However, the true noncontact regime is difficult to maintain in AFM measurements due to the very low separation between the tip and the liquid surface (< 5 nm), which increases the risk of the tip being captured by the liquid [35,36]. Jung and Bhushan describe a method for the indirect measurement of the contact angle for micro- and nanodroplets of the liquid glycerol mixed with a small amount of rhodamine [37]. A modified AFM tip (NADIS – nanodispenser) is employed for the controlled deposition of micro- and nanodroplets. The contact angle is calculated after measuring the exact volume of the deposited liquid (from the change in resonant frequency of the cantilever before and after releasing the droplet), its height (from force-distance curves taken in the center of the droplet) and diameter (from AFM topography images of the rhodamine trace left on the substrate after evaporation).

2.4.2. Scanning Polarization Force Microscopy (SPFM)

In order to overcome the difficulties of maintaining a noncontact regime in AFM measurements, Scanning Polarization Force Microscopy (SPFM) has been proposed and implemented at the Lawrence Berkeley National Laboratory [38]. The technique is based on the measurement and control of electric polarization forces, which appear when a conductive AFM tip is electrically biased with respect to the sample surface. As the range of electric interaction is larger than the range of van der Waals interaction, the tip can follow the topography contour at a larger distance (typically 10–20 nm) than "classical" noncontact AFM. This makes SPFM a very powerful and versatile technique, which is able to image a large variety of wetting-related processes: from molecularly thin films of water [38], capillary phenomena and adsorption of water on surfaces [39-41], droplets of ionic solutions [36], wetting and corrosion by sulfuric acid [42,43], liquid crystals [44], lubricants [45,46], to bio-membranes and lyposomes [47,48].

The following section describes in detail the SPFM principle and method.

3. SPFM principle and method

3.1. Polarization phenomena and forces

In scanning polarization force microscopy an electrical bias is applied to a conductive AFM tip, which leads to an accumulation of electric charge on the tip. As the physical dimensions of the tip apex are very small (radius on the order of few tens of nanometers), the electric field in its vicinity is greatly enhanced. When the tip is brought in the vicinity of the sample, the strong electric field causes a local accumulation of electric charge of opposed sign, which can be conveniently modeled by the image charge, as shown in Figure 8. The two opposed charges create an attractive electrostatic polarization force with magnitude in the nanonewton range for tip-sample distances of few tens of nanometers and tip bias of a few volts [49]. This force can be detected by the usual AFM light lever technique and is kept constant by the topography feedback loop while scanning the sample, in order to obtain a constant force image.

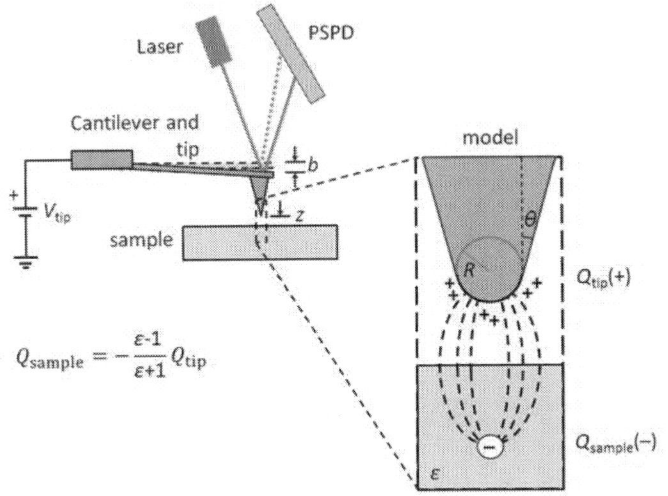

Figure 8. Tip charge and induced image charge with a positive bias applied to the AFM tip

The electrostatic force acting on the tip can be written as [49,50]:

$$F = -\frac{1}{2}\frac{\partial C}{\partial z}\left(V_{\text{tip}} - V_{\text{CPD}}\right)^2 \cong -4\pi\varepsilon_0 \frac{\varepsilon - 1}{\varepsilon + 1} f\left(\frac{R}{z}\right)\left(V_{\text{tip}} - V_{\text{CPD}}\right)^2, \tag{3}$$

where C is the tip-sample capacitance and V_{CPD} is the local contact potential difference. The function $f(R/z)$ is characteristic of the tip-sample geometry. Although it is basically impossible to calculate an exact analytical expression of this function, some approximations can provide useful results.

If the tip and cantilever are approximated by a sphere plus a flat plate, the resulting contributions to the capacitance are

$$C_{\text{tip}}\left(z\right) \cong 2\pi\varepsilon_0 R(\ln\frac{R}{z} + 2) \text{ and } C_{\text{lever}}\left(z\right) \cong \frac{\varepsilon_0 S}{z + D}, \tag{4}$$

where S is the area of the plate corresponding to the tip and D is the separation between the model sphere and the plate [49]. For typical values of these parameters found in practice (e.g., $R = 100$ nm, $z = 10$ nm, $S = 1000$ μm², $D = 5$ μm), it is found that the lever capacitance exceeds the tip capacitance by two orders of magnitude. However, the corresponding forces, $\partial C_{\text{tip}}/\partial z$ and $\partial C_{\text{lever}}/\partial z$, are comparable. Moreover, because $z \ll D$, $\partial C_{\text{lever}}/\partial z$ is almost insensitive to variations of z, while $\partial C_{\text{tip}}/\partial z$ follows the topography.

From equation (4), in the range $z < R$, the electrostatic force can be approximated by the following function [49]:

http://dx.doi.org/10.5772/60886

$$F \cong \left(\frac{AR}{z} + B\right)\left(V_{\text{tip}} - V_{\text{CPD}}\right)^2, \tag{5}$$

where A and B are constants on the order of 10^{-11} N/V^2 each. It can be seen that in this region the force has a $1/z$ dependence.

4. Comparison of SPFM with AFM

As the separation between tip and sample is larger in SPFM than in standard contact and noncontact AFM operations, the risk of the tip touching the surface is minimized. This is especially useful in studies of soft and liquid samples, where contact AFM cannot be considered at all for topography measurements and noncontact AFM is generally difficult to operate.

However, this leads to a loss of lateral resolution compared to standard AFM. In SPM, generally, the lateral resolution cannot be less than the tip-sample separation, and is also influenced by the dimensions of the tip apex. So in most of the SPFM experiments the lateral resolution is on the order of few tens of nanometers. The vertical resolution does not suffer from this loss and is limited by the noise level of the system, usually better than one nanometer.

4.1. DC-SPFM

In DC-SPFM the bias applied to the tip is constant, and the electric polarization force has the form described by equation (3). It can be seen that besides topographical influence (accounted for by the function *f*), the force is dependent on variations of the local dielectric constant of the sample. For constant force topography imaging this will reflect in an apparent topography contrast for adjacent areas of the sample having the same height and different dielectric constants. This is perfectly illustrated in the studies of Hu and coworkers [36,38] where the first layers of water adsorbed on the hydrophilic surface of freshly cleaved mica were imaged in DC-SPFM in increasing relative humidity. It was found that two distinct phases exist, and the topography contrast indicated a difference in dielectric constant between the two. Moreover, the shape of one of the phases' domains suggests a crystalline structure for this phase.

4.2. AC-SPFM

In AC-SPFM, the bias applied to the tip has an AC and a DC component: $V_{\text{tip}} \sin \omega t + V_{\text{DC}}$. Equation (3) will give

$$F \cong -4\pi\varepsilon_0 \frac{\varepsilon - 1}{\varepsilon + 1} f\left(\frac{R}{z}\right)\left(V_{\text{tip}}\mathbf{sin}\omega t + V_{\text{DC}} - V_{\text{CPD}}\right)^2 =$$

$$\begin{aligned} &= -4\pi\varepsilon_0 \frac{\varepsilon - 1}{\varepsilon + 1} f\left(\frac{R}{z}\right)\left[-\frac{1}{2}V_{\text{tip}}^2\mathbf{cos}2\omega t + 2V_{\text{tip}}\left(V_{\text{DC}} - V_{\text{CPD}}\right)\mathbf{sin}\omega t + \frac{1}{2}V_{\text{tip}}^2 + \left(V_{\text{DC}} - V_{\text{CPD}}\right)^2\right] = \\ &= F(2\omega) + F(1\omega) + F(0\omega) \end{aligned} \tag{6}$$

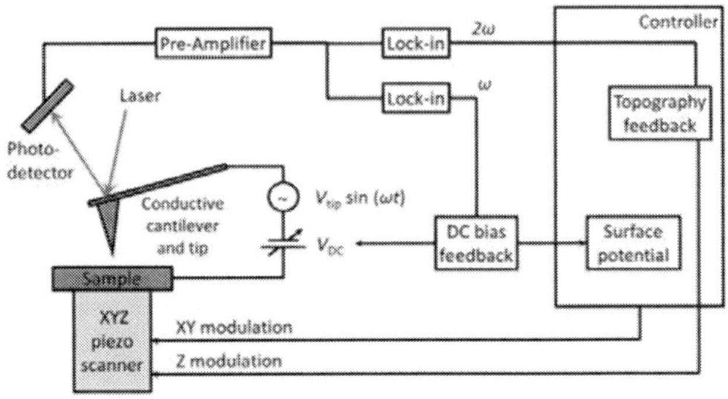

Figure 9. Schematic view of the SPFM setup

The polarization force has three components with different time dependencies: 2ω, 1ω, and a DC component. The 2ω and 1ω components can be separated with the use of lock-in amplifiers. The amplitude of the 2ω component is used for constant force topography feedback. The amplitude of the 1ω component can be used to determine V_{CPD}: a feedback loop which controls the value of V_{DC} is set to null the amplitude of the 1ω component, which, as seen in equation (6), happens when V_{DC} equals V_{CPD}. Figure 9 shows a common setup for SPFM experiments.

4.3. Forces and energies in wetting at nanoscale

For macroscopic drops, where most of the liquid is outside the range of long-range forces, the contact angle is defined as the angle at which the liquid surface meets the substrate, measured through the liquid (Figure 10a). For micron and nanometer size droplets, where most of the liquid is affected by the long range forces, the liquid meniscus does not meet the substrate at a well-defined angle (Figure 10b). The effective contact angle, further called microscopic contact angle, is determined by the slope of the droplet profile at the inflexion points [43].

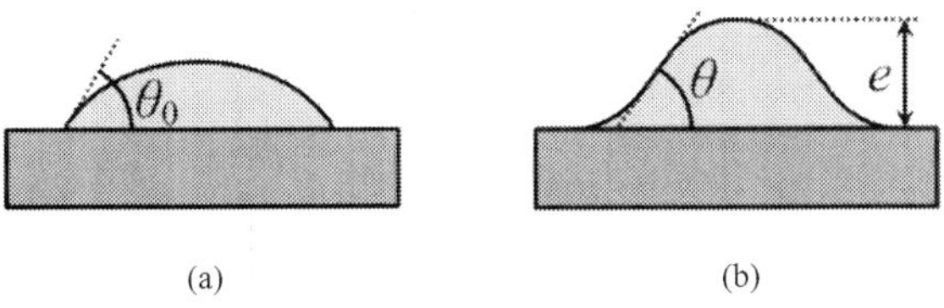

Figure 10. Contact angle for a) macroscopic droplets; b) microscopic and nanometer size droplets

In order to study the wetting behavior of nanodroplets, following a similar approach of de Gennes [49,51,52], the effect of long-range forces will be analyzed. The free energy for a circularly symmetric drop defined by the shape $z(r)$ can be written as an integral over the area covered by the drop:

http://dx.doi.org/10.5772/60886

$$G = G_0 + \int_{drop} 2\pi r \, \mathbf{d}r \left[-S + \frac{\gamma_{lv}}{2}\left(\frac{\mathbf{d}z}{\mathbf{d}r}\right)^2 + P(z) - \frac{\mu_{vapor} - \mu_{liq}}{v_{mol}} \cdot z \right] \quad (7)$$

The first term under the integral that characterizes the wetting properties of a surface by a given liquid is the spreading coefficient, $S = \gamma_{sv} - \gamma_{sl} - \gamma_{lv}$, where γ_{sv}, γ_{sl}, and γ_{lv} are the surface energies corresponding to the solid–vapor, solid–liquid and liquid–vapor interfaces, respectively. Valid for the case of shallow droplets is the second term, which is due to the excess surface caused by the curvature of the drop. The third term $P(z)$ is the surface potential energy between the surfaces. The supersaturation in terms of chemical potentials of the vapor and liquid is described by the last term; v_{mol} is the molecular volume of the liquid [51].

Considering a constant volume of droplet, $V = \int z(r)\ 2\pi r\ dr$, and assuming that the droplets have spherical cap shape [52], the minimization of the free energy G in equation (7) leads to the relation between the microscopic contact angle θ, the surface potential $P(e)$, and the disjoining pressure defined as $\Pi(e) = -\frac{dP}{de} = -P'(e)$ [53]:

$$\theta^2 = \theta_0^2 + \frac{2}{\gamma}\left[P(e) + e\Pi(e)\right], \quad (8)$$

where e is the height of the droplet, θ_0 is the macroscopic contact angle, $P(0)$ is the spreading coefficient S which characterizes the wetting properties of a surface by a given liquid at short ranges. If $S > 0$ the liquid will completely wet the surface; if $S < 0$ a contact angle will exist, determined by Young's equation $\gamma_{lv}\cos\theta = \gamma_{sv} - \gamma_{sl}$. Thus, the contact angle θ is influenced by the disjoining pressure and depends on the interfacial energies. The disjoining pressure is related to the spreading coefficient S by

$$S = \int_0^\infty \Pi(x)\,dx = P(e)_{e\to 0} \quad (9)$$

By rearranging the terms of equation (8) it can be obtained that

$$P(e) - eP'(e) = \left(\theta^2 - \theta_0^2\right)\frac{\gamma}{2} \quad (10)$$

Thus, using the dependence described in equation (10), the potential energy $P(e)$ between the surfaces can be quantitatively determined after measuring the dependence of contact angle on droplet height in the case of spherical shaped droplets of small height. The ability of SPFM to image the topography of liquid samples makes it an ideal candidate for such determinations, as described in the following section.

5. Application of SPFM techniques for nanometer scale wetting investigations

This section presents results regarding the characterization of wetting properties at nanoscale, obtained in the Center for Surface Science and NanoTechnology (CSSNT) at the University "Politehnica" of Bucharest. The results were reported in references [54-56].

5.1. Wetting properties of glycerol and sulfuric acid on highly oriented pyrolitic graphite and aluminum

In this study, glycerol and sulfuric acid were chosen for the formation of liquid droplets on highly oriented pyrolytic graphite (HOPG); on aluminum-covered mica, the droplets were formed by glycerol [54].The HOPG substrates were freshly cleaved prior to the experiments. The aluminum film was deposited on freshly cleaved mica by thermal evaporation in vacuum. The substrates were verified by optical microscopy prior to the deposition of liquid droplets. In order to create the droplets, an evaporation–condensation technique was chosen: a few milliliters of glycerol were heated in a Berzelius glass to ~100°C. The substrates were held upside down at ~5 mm above the liquid until their surface achieved a "foggy" appearance, which indicated the presence of microscopic droplets. Optical microscopy (Figure 11a and c) confirmed the presence of small droplets [54].

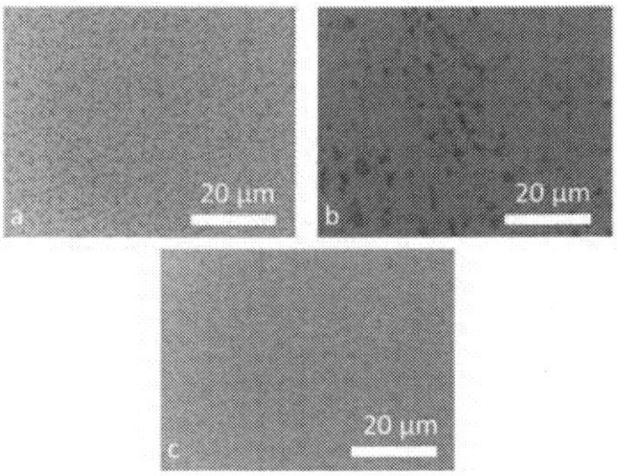

Figure 11. Optical microscopy images of the samples after the deposition of the droplets: a) glycerol on HOPG; b) sulfuric acid on HOPG; c) glycerol on aluminum-covered mica

Sulfuric acid droplets (50% vol.) were created on the substrates by casting a macroscopic drop (~50 µl) and then absorbing the liquid at one corner of the substrate with the aid of a lens-cleaning tissue, avoiding contact between the tissue and the substrate surface. The substrates appeared dry at a first inspection with the naked eye. However, optical microscopy images revealed the presence of liquid droplets of various sizes and shapes (Figure 11b) [54]. The droplet deposition took place at room temperature and ~50% relative humidity. The images shown in Figure 11 are typical for the distribution of glycerol and H_2SO_4 droplets on the substrates.

http://dx.doi.org/10.5772/60886

Figure 12 shows typical SPFM-AC topography images of the samples after the deposition of the droplets on HOPG and aluminum. The images show that the dispersion of the droplets depends on liquid type and substrate. The deposited droplets have shapes close to spherical caps, which allows us to use the theoretical model of de Gennes [51] in order to determine the surface potential energy *P*(*e*) between glycerol/H_2SO_4 and HOPG and between glycerol and aluminum.

Figure 12. Typical SPFM-AC images of the samples after the deposition of the droplets: a) glycerol on HOPG, 5 µm × 5 µm; b) sulfuric acid on HOPG, 3 µm × 3 µm; c) glycerol on aluminum-covered mica, 5 µm × 5 µm

Topography profiles of some of the droplets are shown in Figure 13, for glycerol and sulfuric acid on HOPG, and for glycerol on aluminum. The profiles are plotted along segments that pass through the point of maximum height for each chosen droplet [54].

For micro- and nanodroplets, as can be seen in Figure 13 and as discussed in Section 3.e, the liquid meniscus does not meet the solid surface at a precise angle and the graph of the line-cut profile has two inflection points (one on each side of the peak). The contact angle is calculated by measuring the slope (first derivative) of the line-cut profile at these inflection points [35]. Contact angle values corresponding to the droplets were plotted as a function of droplet height. Figures 14 and 15 show the results for glycerol and H_2SO_4 on HOPG and aluminum-covered mica [54]. A decrease of contact angle with droplet height is observed for all cases, which indicates that the surface potential *P*(*e*) is negative, i.e., the interaction forces between surfaces are hydrophobic or attractive.

Using relation $\theta^2=\theta_0^2+\frac{2}{\gamma}[P(e)+e\Pi(e)]$ and the dependence of contact angle on droplet height (Figures 14 and 15), we calculated the dependence of surface potential energy *P*(*e*) between the surfaces for glycerol/H_2SO_4 on HOPG and for glycerol on aluminum-covered mica. The results are shown in Figures 16 and 17. The macroscopic contact angle θ_0 was determined as an asymptotic fitting parameter for the microscopic contact angle dependencies. In a related experiment we measured the macroscopic contact angle by optical methods and the results are in good agreement with the value found in literature [33].

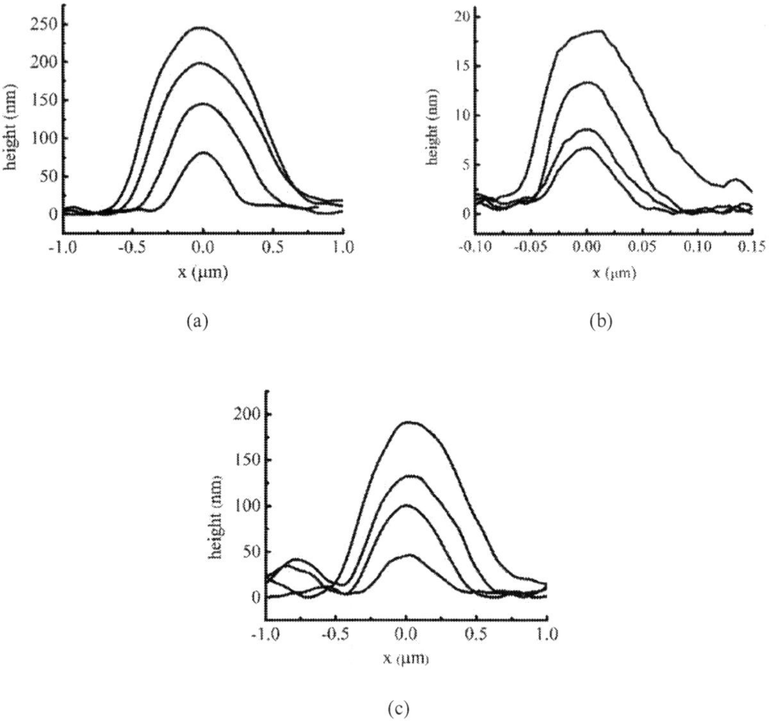

Figure 13. Line-cut profiles of droplets of: a) glycerol on HOPG; b) sulfuric acid on HOPG; c) glycerol on aluminum

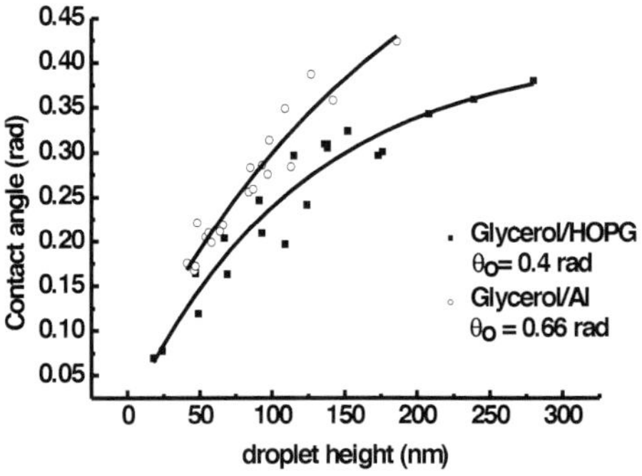

Figure 14. Dependence of contact angle on droplet height for glycerol deposited on HOPG and on aluminum-covered mica

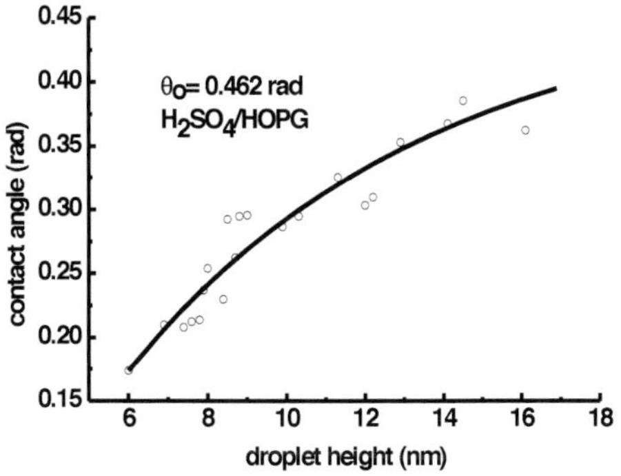

Figure 15. Dependence of contact angle on droplet height for H_2SO_4 deposited on HOPG

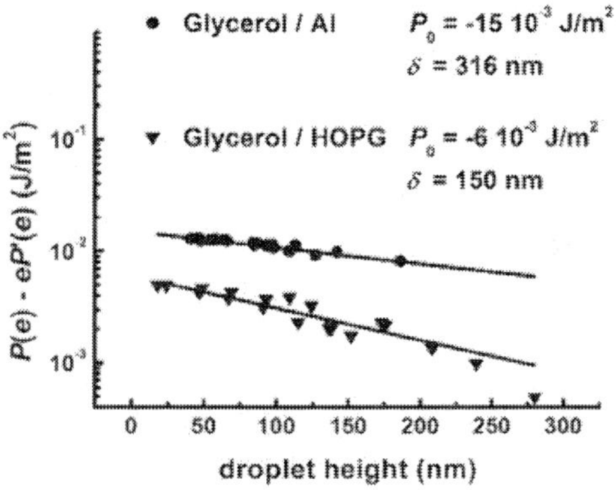

Figure 16. Semilog plot of $P(e) - eP'(e)$ vs. e for glycerol deposited on HOPG and on aluminum-covered mica

On HOPG and aluminum, glycerol and H_2SO_4 form droplets whose shapes are close to spherical caps. The contact angle varies with height as shown in Figures 14 and 15. The decrease of the contact angle with decreasing droplet height indicates that the surface potential $P(e)$ is negative and an exponential dependence $P(e)=P_0\exp(-e/\delta)$ [52] with distance gives a good fit (Figures 16 and 17). From the fitting parameters, P_0 and δ, we determine the dependence of the potential energy on droplet height [54]:

$$P(e) = -6 \cdot 10^{-3} \exp(-e / 150 \text{ nm}) \text{J} / \text{m}^2 \text{ for glycerol on HOPG}$$

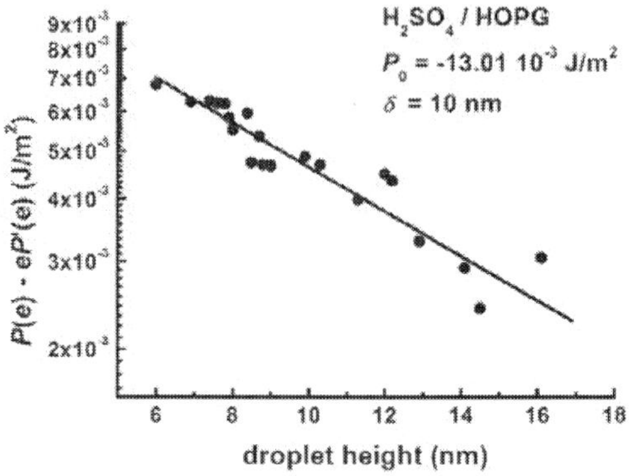

Figure 17. Semilog plot of $P(e) - eP'(e)$ vs. e for H_2SO_4 on HOPG

$$P(e) = -15 \cdot 10^{-3} \exp(-e / 316 \text{ nm}) \text{J} / \text{m}^2 \text{ for glycerol on Al}$$

$$P(e) = -13 \cdot 10^{-3} \exp(-e / 10 \text{ nm}) \text{J} / \text{m}^2 \text{ for } H_2SO_4 \text{ on HOPG}$$

The exponential dependence of surface potential energy $P(e)$ in all cases indicates attractive interaction between the liquid–substrate interfaces, having a decay length $\delta = 150$ nm for glycerol on HOPG, $\delta = 316$ nm for glycerol on aluminum, and $\delta = 10$ nm for H_2SO_4 on HOPG, which dominate over the characteristic length of the van der Waals interactions. In our calculations we have used the values of 73 mJ/m^2 and 64 mJ/m^2 for the surface tensions of H_2SO_4 solution and glycerol, respectively.

The strength of the potential at $e = 0$ nm gives the spreading coefficient S for each liquid on the respective substrate [54]:

$$S = P(0) = -6 \cdot 10^{-3} \text{J} / \text{m}^2 \text{ for glycerol on HOPG}$$

$$S = P(0) = -15 \cdot 10^{-3} \text{J} / \text{m}^2 \text{ for glycerol on Al}$$

$$S = P(0) = -13 \cdot 10^{-3} \text{J} / \text{m}^2 \text{ for } H_2SO_4 \text{ on HOPG}$$

In the case of glycerol on HOPG and on aluminum, the values for spreading coefficient indicate a very weak hydrophobic interaction by comparison with surface tension of glycerol; the value for spreading coefficient indicates a value ten times (on HOPG), respectively, four times (on Al) smaller compared to the surface tension of glycerol, and in the case of H_2SO_4 on HOPG the value for spreading coefficient indicates a value five times smaller compared to the surface tension of H_2SO_4. These potential energies give negative disjoining pressures Π of ~0.4 atm at e close to zero for glycerol on HOPG, 0.47 atm for glycerol on aluminum and ~13 atm for H_2SO_4 on HOPG. In the case of H_2SO_4 on HOPG the strength of the force appears to be thirty times bigger than that for glycerol on HOPG or aluminum.

5.2. Wetting properties of glycerol on mica and stainless steel

Glycerol was chosen for the formation of liquid droplets in these experiments and the substrates were mica and stainless steel [55].

Mica substrates were freshly cleaved prior to the experiments. Stainless steel substrates were polished using various grades of abrasive paper (gradually increasing the grit) and finally using slurry of alumina particles approximately 20 nm in size, dispersed on a felt disc. They were then sonicated for several cycles in methanol and deionized water for five minutes per cycle.

Glycerol droplets were created on the substrates by condensation, as described in section 4.a. The presence of microscopic droplets was confirmed by optical microscopy inspection (Figure 18).

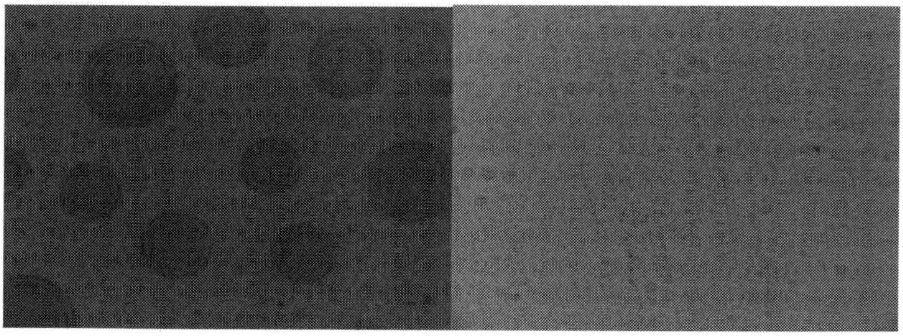

(a) (b)

Figure 18. Optical microscopy images of the substrates after the deposition of the droplets: a) glycerol on mica; b) glycerol on stainless steel. Field of view ~80 μm × 64 μm

Optical images shown in Figure 18 are typical for the distribution of glycerol droplets on the substrates in these experiments. On mica, glycerol tends to form many small droplets and few large droplets, while on stainless steel it only forms many small droplets. This could be attributed to the higher roughness of the steel surface, which prevents the droplets form migrating on the surface and merging into larger droplets [55].

Figure 19 shows typical SPFM-AC topography images of the samples after the deposition of the droplets. These images confirm the general aspect of the droplets from the optical microscopy analysis.

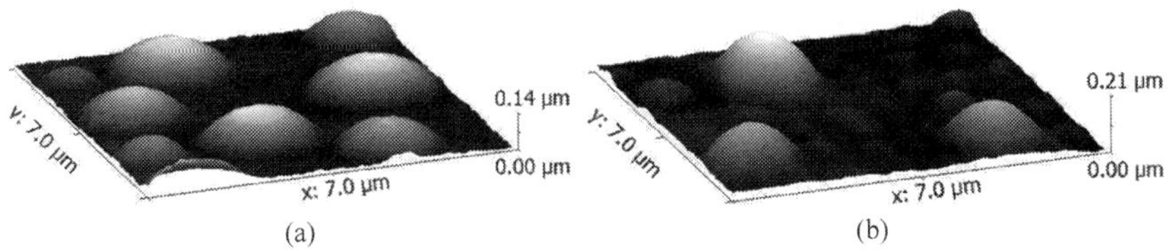

Figure 19. Typical SPFM-AC images of the deposited droplets of glycerol on: a) mica, b) stainless steel; field of view 7 µm × 7 µm

Droplet profiles were extracted from the SPFM topography images of the samples. Some of the profiles are plotted in Figure 20 [55].

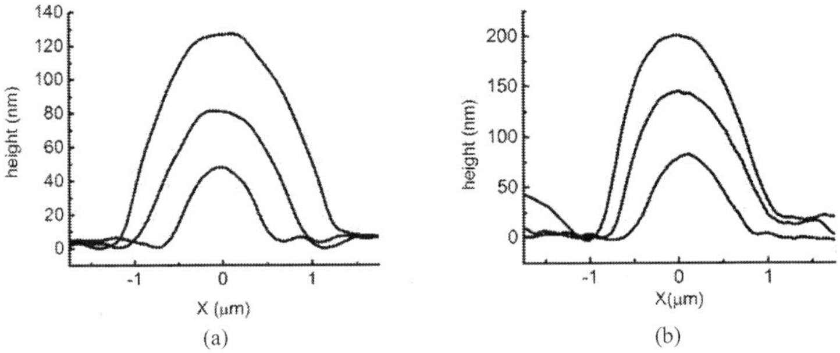

Figure 20. Line-cut profiles of droplets of glycerol on: a) mica b) stainless steel

Contact angle values corresponding to the droplets were plotted as a function of droplet height (Figure 21) [55]. A decrease of contact angle with droplet height is observed, which indicates that the surface potential *P*(*e*) is negative, i.e., the interaction forces between surfaces are attractive or hydrophobic.

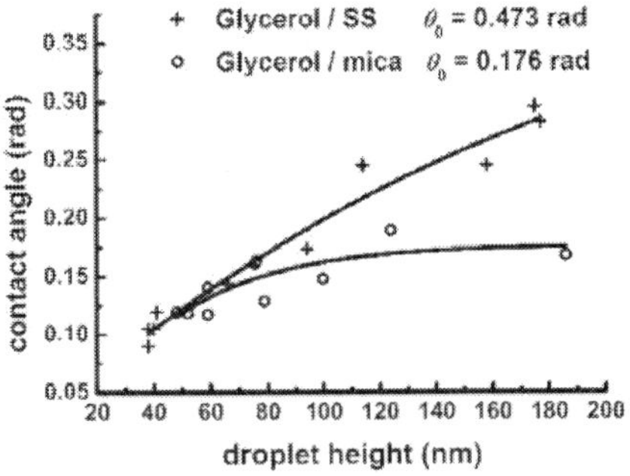

Figure 21. Dependence of contact angle on droplet height for glycerol on mica and stainless steel

Using relation (10) and the dependence of contact angle on droplet height (Figure 21), the height dependence of the term *P(e)–eP′(e)* for glycerol on mica and stainless steel was determined. The results are shown in Figure 22.

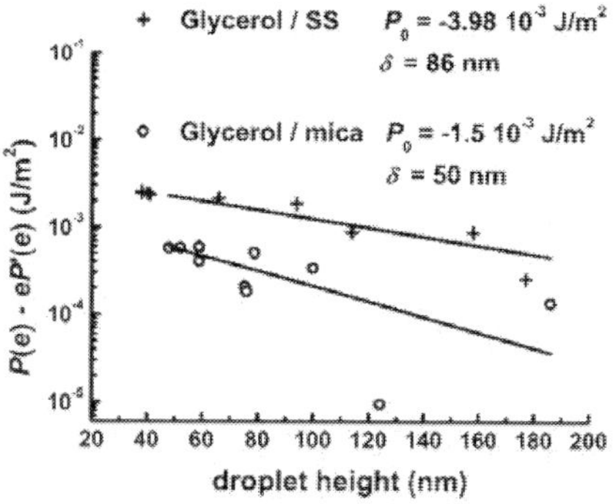

Figure 22. Semilog plot of $P(e)–eP'(e)$ vs. e for glycerol on stainless steel and mica

The decrease of contact angle with decreasing droplet height indicates that the surface potential *P(e)* is negative and an exponential dependence $P(e)=P_0\exp(-e/\delta)$ [52] with distance gives a good fit (Figure 22), where P_0 and δ are the fitting parameters. From the fitting parameters we determine the dependence of the potential energy on droplet height [55]:

$$P(e) = -1.4\cdot 10^{-3}\exp(-e/50\ \mathbf{nm})\,\mathbf{J/m^2}\ \textbf{for glycerol on mica}$$

$$P(e) = -3.98\cdot 10^{-3}\exp(-e/86\ \mathbf{nm})\,\mathbf{J/m^2}\ \textbf{for glycerol on stainless steel}$$

The exponential dependence of surface potential energy *P(e)* with distance indicates hydrophobic attractive forces between the glycerol–mica/stainless steel interfaces, having a decay length δ = 50 nm for glycerol on mica and δ = 86 nm for glycerol on stainless steel, values which dominate over the range of van der Waals forces. These forces may include double layer, solvation, and hydration forces.

The strength of the potential at e = 0 nm gives the spreading coefficient [55]:

$$S = P(0) = -1.4\cdot 10^{-3}\,\mathbf{J/m^2}\ \textbf{for glycerol on mica}$$

$$S = P(0) = -3.98\cdot 10^{-3}\,\mathbf{J/m^2}\ \textbf{for glycerol on stainless steel}.$$

In both cases, the values for spreading coefficient indicate a very weak hydrophobic interaction for these systems, compared to the surface tension of glycerol. These potential energies give a negative disjoining pressure Π of 0.28 atm for glycerol on mica and 0.46 atm for glycerol on stainless steel. As we see, in the case of the glycerol on stainless steel the strength of the disjoining pressure appears to be two times higher than that for glycerol on mica system.

5.3. Wetting properties of glycerol on silicon, native SiO_2, and bulk SiO_2

Glycerol was chosen for the formation of liquid droplets on silicon, native SiO_2, and bulk SiO_2 [56]. The substrates were prepared as follows: silicon (*p*-type Si(100)) and bulk SiO_2 were washed in ultrapure water and acetone, in an ultrasonic bath, for the removal of organic contamination. Silicon substrates were treated for one minute in HF solution (20% wt.), in order to remove the native oxide layer and to expose the bare silicon surface. The substrates were free of macroscopic defects and impurities over relatively large areas (several mm^2), as revealed by optical microscopy examination (images not shown).

Glycerol droplets were created on the substrates by condensation, as described in Section 4.a. Figure 23 shows optical microscopy images of the substrates after the same deposition time. As can be seen, the droplets are smaller and denser on silicon. We attributed this to the nanoscale roughness of the bare silicon surface resulting after the wet etching. The droplets might become pinned to the surface defects, which would prevent the small droplets coalescing into bigger ones [56].

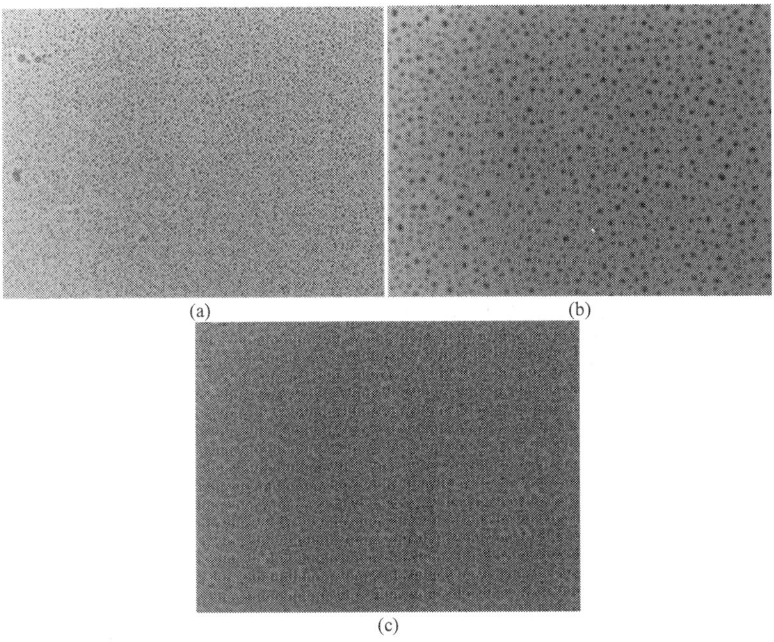

Figure 23. Optical microscopy images of the samples after the deposition of the nanodroplets: a) glycerol on Si; b) glycerol on native SiO_2; c) glycerol on bulk SiO_2. Field of view ~70μm × 55μm

Figure 24 shows typical SPFM-AC topography images of the samples after the deposition of the droplets on silicon, native SiO_2, and bulk SiO_2. The images show that the dispersion of the droplets is strongly influenced by the type of substrate.

http://dx.doi.org/10.5772/60886

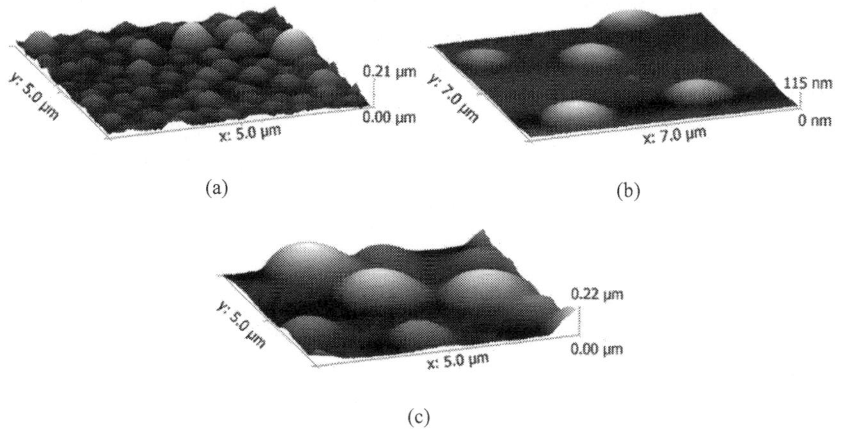

Figure 24. Typical SPFM-AC images of the samples after the deposition of the droplets: a) glycerol on Si, 5µm × 5µm; b) glycerol on native SiO_2, 7µm × 7µm; c) glycerol on bulk SiO_2, 5µm × 5µm

Some of the droplet profiles extracted from the SPFM topography images are plotted in Figure 25 [56].

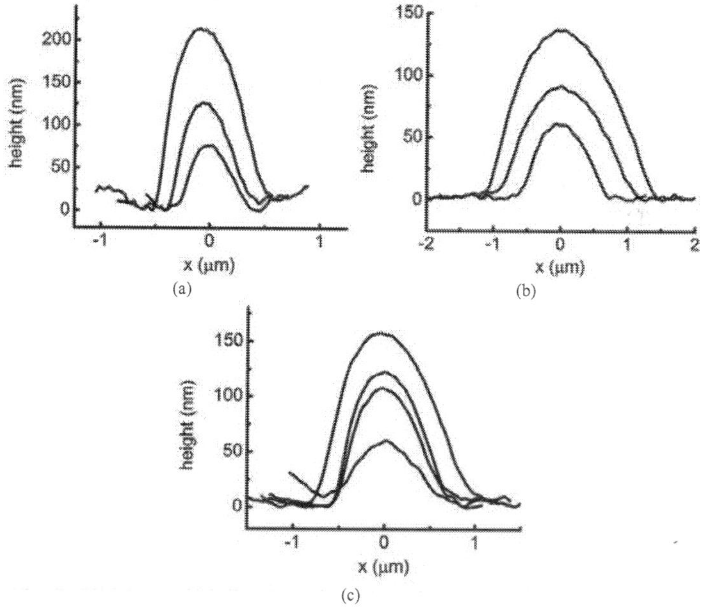

Figure 25. Line-cut profiles of droplets of glycerol on: a) silicon; b) native SiO_2; c) bulk SiO_2

Contact angle values corresponding to droplets such as those shown in Figure 24 (having a circularly symmetric shape) were plotted as a function of droplet height (Figure 26) [56].

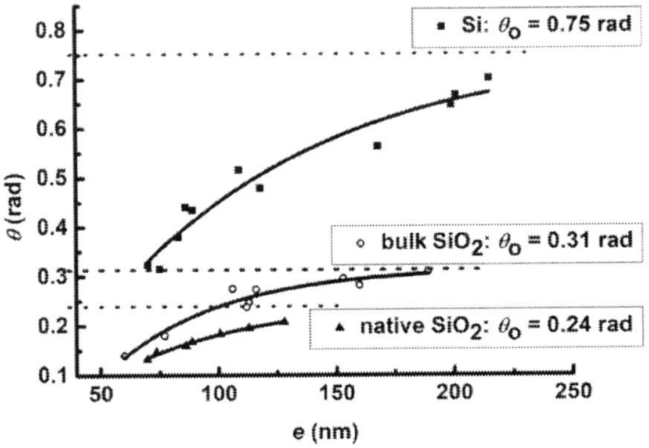

Figure 26. Dependence of contact angle on droplet height for glycerol on silicon, native SiO_2, and bulk SiO_2

The dependence of surface potential energy $P(e)$ on droplet height is calculated from relation $\theta^2 = \theta_0^2 + \frac{2}{\gamma}[P(e) + e\Pi(e)]$, taking into account the measured dependence of contact angle on droplet height. The term $P(e) - eP'(e)$ is plotted against e, as shown in Figure 27 [56].

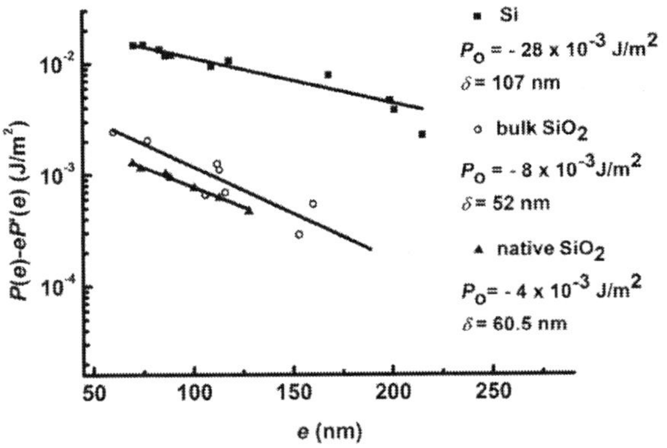

Figure 27. Semi-logarithmic scale plot of the term $P(e) - eP'(e)$ versus e

As the contact angle decreases with decreasing droplet height, an exponential distance dependence of the negative surface potential, $P(e) = P_0 \exp(-e/\delta)$, is expected to give a good fit of the data [52]. Here, P_0 and δ are the fitting parameters. From these parameters we determined the dependence of the potential energy on droplet height [56]:

$$P(e) = -28 \cdot 10^{-3} \exp(-e / 107 \text{ nm}) \text{J/m}^2 \textbf{ for glycerol on silicon}$$

$$P(e) = -4 \cdot 10^{-3} \exp(-e / 60.5 \text{ nm}) \text{J/m}^2 \textbf{ for glycerol on native SiO}_2$$

$$P(e) = -8 \cdot 10^{-3} \exp(-e / 52 \text{ nm}) \text{J/m}^2 \textbf{ for glycerol on bulk SiO}_2$$

The strength of the potential at $e = 0$ nm gives the spreading coefficient S for each case [56]:

$$S = P(0) = -28 \cdot 10^{-3} \text{J/m}^2 \textbf{ for glycerol on silicon}$$

$$S = P(0) = -4 \cdot 10^{-3} \text{J/m}^2 \textbf{ for glycerol on native SiO}_2$$

$$S = P(0) = -8 \cdot 10^{-3} \text{J/m}^2 \textbf{ for glycerol on bulk SiO}_2 .$$

The (negative) values of the disjoining pressure Π, resulting from these potential energies, are: 2.8 atm, 0.6 atm, and 1.5 atm, for glycerol on silicon, native SiO_2, and bulk SiO_2, respectively.

Table 1 summarizes the results obtained in the studies described in Sections 4 a–c [54-56], for the parameters P_0, δ, and Π. In order to make a direct comparison with other experiments, in Table 1 are also introduced the results on mica, stainless steel, HOPG, and aluminum substrates, described in Sections 4 a and 4 b. In all these cases, the values of spreading coefficient indicate a weak hydrophobic interaction for the systems, compared to the surface tension of glycerol and sulfuric acid [54-56].

liquid / substrate	P_0 (mJ/m²)	δ (nm)	Π (atm)
glycerol / silicon	–28	107	2.8
glycerol / native SiO_2	–4	60.5	0.6
glycerol / bulk SiO_2	–8	52	1.5
glycerol / mica	–1.4	50	0.28
glycerol / stainless steel	–3.98	86	0.46
glycerol / HOPG	–6	150	0.4
glycerol / aluminum	–15	316	0.47
H_2SO_4 / HOPG	–13	10	13

Table 1. Values of P_0, δ, and Π for glycerol on silicon, native and bulk SiO_2, mica, stainless steel, HOPG and aluminum, and for H_2SO_4 on HOPG.

5.4. Conclusions

SPFM, one of the most advanced noncontact techniques having nanometer spatial resolution, was successfully used, together with suitable theoretical models, to determine the potential energy, disjoining pressure and spreading coefficient from the dependence of the contact angle on droplet height, for micro- and nanodroplets. The contact angle was directly measured from the recorded topography images of the droplets. The testing liquids were chosen for their suitable properties: sufficiently high boiling point (290°C for glycerol, 337°C for 98% wt. sulfuric acid, at atmospheric pressure) and low vapor pressure (0.26 mbar for glycerol at 100°C; 1.3 mbar for sulfuric acid at 145.8°C). These ensure a low enough evaporation rate of the micro- and nanodroplets during measurements. The tested materials (HOPG, mica, stainless steel, aluminum, silicon, and silicon dioxide) were chosen based on specific considerations: *i*) HOPG and mica, due to the ease with which they can be cleaved, can be regarded as model surfaces as they can be readily prepared in the form of defect free, atomically clean and flat surfaces; *ii*) stainless steel and aluminum because of their importance and widespread use in industrial applications, from precision mechanics to heavy construction machinery and structures; *iii*) silicon and silicon dioxide as key materials in the continuously growing fields of MEMS/NEMS and understanding their wetting behavior at micro- and nanoscale is a key factor in the design and optimization of these technologies. A large range of technologies, all of which play continuously growing roles in the development of current and innovative applications, from the silicon-based electronics industry to MEMS/NEMS manufacturing, as well as lab-on-chip devices or dip pen nanolithography, are fully benefiting from the qualitative and quantitative information about wetting properties at the micro- and nanoscale on silicon-based materials, offered by the SPFM technique described above.

6. General conclusions

Wetting properties at the macroscopic scale benefited from extensive studies during the course of the last two centuries. A lot of scientific literature has been dedicated to theoretical advancements towards the understanding of the involved mechanisms and physical quantities, and to the development of suitable experimental techniques.

The basic and advanced techniques used to characterize the wettability of materials at the macro-, micro-, and nanoscale have been described in this chapter. Due to the high demand of understanding the wetting and adhesion at micro-scale, new and innovative techniques for measuring the wettability of surfaces continue to emerge.

The interest for studying wetting properties at the micro- and nanoscale started increasing with the rapid development of micro- and nanotechnologies in the last decades. At these scales, local inhomogeneity of the materials plays a significant role in the behavior of liquids coming into contact with solid surfaces. The SPM family of techniques, which actually made possible the evolution of nanotechnology, offers direct methods for determining nanoscale adhesion and wetting properties. SPFM, as a truly noncontact technique of this family, stands out in particular by enabling the investigation of the topography of diverse liquid surfaces, thus proving to be extremely useful for the study of wetting properties at micro- and nanoscale. It overcomes the difficulties inherent to "classical" AFM techniques (e.g., difficult to maintain stable feedback on liquid surfaces) and offers a direct means for the measurement of the microscopic contact angle. Nevertheless, the wetting behavior at the micro- and nanoscale is still an open field with many open questions related to wetting mechanisms and manipulation of micro- and nanodroplets.

Author details

Antoniu Moldovan[1,2] and Marius Enachescu[1*]

*Address all correspondence to: marius.enachescu@upb.ro

1 Center for Surface Science and Nanotechnology, University POLITEHNICA of Bucharest, Romania

2 National Institute for Laser, Plasma and Radiation Physics (INFLPR), Magurele – Bucharest, Romania

References

[1] Heo SY, Park BJ, Jeong JR, Yoon SG. Enhanced transmittance, mechanical durability, and anti-fingerprinting qualities of silver nanoparticles deposited onto glass substrates. Journal of Alloys and Compounds 2014; 602 255–260. DOI: 10.1016/j.jallcom.2014.03.019

[2] Kolattukudy PE. Natural waxes on fruits. Post Harvest Pomology Newsletter 1984; 2(2) 3–7

[3] Yuan Y, Randall Lee T. Contact angle and wetting properties. In: Bracco G, Holst B (ed.) Surface Science Techniques. Springer Series in Surface Science. Berlin, Heidelberg: Springer-Verlag 2013; 51 3–34. DOI: 10.1007/978-3-642-34243-1_1

[4] Méndez-Vilas A, Jódar-Reyes AB, González-Martín ML. Ultrasmall liquid droplets on solid surfaces: production, imaging, and relevance for current wetting research. Small 2009; 5(12) 1366–1390. DOI: 10.1002/smll.200800819

[5] Bush J. 18.357 Interfacial Phenomena, Fall 2010. (Massachusetts Institute of Technology: MIT OpenCourseWare), http://ocw.mit.edu (Accessed 17 Apr, 2015). License: Creative Commons BY-NC-SA

[6] Beebe DJ, Mensing GA, Walker GM. Physics and applications of microfluidics in biology. Annual Review of Biomedical Engineering 2002; 4 261–286. DOI: 10.1146/annurev.bioeng.4.112601.125916

[7] de Gans BJ, Duineveld PC, Schubert US. Inkjet printing of polymers: state of the art and future developments. Advanced Materials 2004; 16(3) 203–213. DOI: 10.1002/adma.200300385

[8] Bachofen H, Schürch S. Alveolar surface forces and lung architecture. Comparative Biochemistry and Physiology Part A: Molecular & Integrative Physiology 2001; 129(1) 183–193. DOI: 10.1016/S1095-6433(01)00315-4

[9] Hidalgo A, Cruz A, Pérez-Gil J. Barrier or carrier? Pulmonary surfactant and drug delivery. European Journal of Pharmaceutics and Biopharmaceutics 2015 (article in press). DOI: 10.1016/j.ejpb.2015.02.014

[10] Tabor D, Winterton RHS. The direct measurement of normal and retarded van der Waals forces. Proceedings of the Royal Society of London, A 1969; 312 435–450. DOI: 10.1098/rspa.1969.0169

[11] Israelachvili JN, Tabor D. The measurement of van der Waals dispersion forces in the range 1.5 to 130 nm. Proceedings of the Royal Society of London, A 1972; 331 19–38. DOI: 10.1098/rspa.1972.0162

[12] Israelachvili JN, Tabor D. Van der Waals forces: theory and experiment. Progress in Surface and Membrane Science 1973; 7 1–55. DOI:10.1016/B978-0-12-571807-3.50006-5

[13] Israelachvili JN. Intermolecular and Surface Forces. Burlington: Academic Press 1992

[14] Israelachvili JN, Min Y, Akbulut M, Alig A, Carver G, Greene W, Kristiansen K, Meyer E, Pesika N, Rosenberg K, Zeng H. Recent advances in the surface forces apparatus (SFA) technique. Reports on Progress in Physics 2010; 73 036601. DOI: 10.1088/0034-4885/73/3/036601

[15] Kristiansen K, McGuiggan P, Carver G, Meinhart C, Israelachvili JN. 3D force and displacement sensor for SFA and AFM measurements. Langmuir 2008; 24 1541–1549. DOI: 10.1021/la702380h

[16] Idziak SH, Safinya CR, Hill RS, Kraiser KE, Ruths M, Warriner HE, Steinberg S, Liang KS, Israelachvili JN. The x-ray surface forces apparatus—structure of a thin smectic liquid crystal film under confinement. Science 1994; 264 1915–1918. DOI: 10.1126/science.264.5167.1915

[17] Mukhopadhyay A, Zhao J, Chul Bae S, Granick S. An integrated platform for surface forces measurements and fluorescence correlation spectroscopy. Review of Scientific Instruments 2003; 74 3067–3072. DOI: 10.1063/1.1570947

[18] Joanny JF, de Gennes PG. A model for contact angle hysteresis. Journal of Chemical Physics. 1984; 81 552. DOI: 10.1063/1.447337

[19] Eral HB, 't Mannetje DJCM, Oh JM. Contact angle hysteresis: a review of fundamentals and applications. Colloid and Polymer Science 2013; 291 (2) 247. DOI: 10.1007/s00396-012-2796-6

[20] Gao L, McCarthy TJ. Contact angle hysteresis explained. Langmuir 2006; 22 (14) 6234–6237. DOI: 10.1021/la060254j

[21] Yazdanpanah MM, Hosseini M, Pabba S, Berry SM, Dobrokhotov VV, Safir A, Keynton RS, Cohn RW. Micro-Wilhelmy and related liquid property measurements using constant-diameter nanoneedle-tipped atomic force microscope probes. Langmuir 2008; 24 13753–13764. DOI: 10.1021/la802820u

[22] Bigelow WC, Pickett DL, Zisman WA. Oleophobic monolayers: I. Films adsorbed from solution in non-polar liquids. Journal of Colloid Science 1946; 1 513–538. DOI: 10.1016/0095-8522(46)90059-1

[23] Erbil HY, McHale G, Rowan SM, Newton MI. Determination of the receding contact angle of sessile drops on polymer surfaces by evaporation. Langmuir 1999; 15 7378–7385. DOI: 10.1021/la9900831

[24] Srinivasan S, McKinley GH, Cohen RE. Assessing the accuracy of contact angle measurements for sessile drops on liquid-repellent surfaces. Langmuir 2011; 27 13582–13589. DOI: 10.1021/la2031208

[25] Wan JM, Wilson JL. Colloid transport in unsaturated porous media. Water Resources Research 1994; 30 857–864. DOI: 10.1029/93WR03017

[26] Wu W. Baseline studies of The Clay Minerals Society source clays: colloid and surface phenomena. Clays and Clay Minerals 2001; 49 446–452

[27] Costanzo PM, Wu W, Giese RF, van Oss CJ. Comparison between direct contact angle measurements and thin layer wicking on synthetic monosized cuboid hematite particles. Langmuir 1995; 11 1827–1830. DOI: 10.1021/la00005a064

[28] Wan JM, Wilson JL, Kieft TL. Influence of the gas-water interface on transport of microorganisms through unsaturated porous media. Applied and Environmental Microbiology 1994; 60 509–516

[29] Abu-Lail NI, Camesano TA. The effect of solvent polarity on the molecular surface properties and adhesion of *Escherichia coli.* Colloids and Surfaces B: Biointerfaces 2006; 51 62–70. DOI: 10.1016/j.colsurfb.2006.05.009

[30] Kukavica-Ibrulj I, Darveau A, Jean J, Fliss I. Hepatitis A virus attachment to agri-food surfaces using immunological, virological and thermodynamic assays. Journal of Applied Microbiology 2004; 97 923–934. DOI: 10.1111/j.1365-2672.2004.02366.x

[31] Sundberg M, Månsson A, Tågerud S. Contact angle measurements by confocal microscopy for non-destructive microscale surface characterization. Journal of Colloid and Interface Science 2007; 313 454–460. DOI:10.1016/j.jcis.2007.04.067

[32] Fischer DG, Ovryn B. Interfacial shape and contact-angle measurement of transparent samples with confocal interference microscopy. Optics Letters 2000; 25 478–480. DOI: 10.1364/OL.25.000478

[33] Yu J, Wang H, Liu X. Direct measurement of macro contact angles through atomic force microscopy. International Journal of Heat and Mass Transfer 2013; 57 299–303. DOI: 10.1016/j.ijheatmasstransfer.2012.10.041

[34] Checco A, Guenoun P, Daillant J. Nonlinear dependence of the contact angle of nanodroplets on contact line curvature. Physical Review Letters 2003; 91 186101. DOI: 10.1103/PhysRevLett.91.186101

[35] Salmeron M. Nanoscale wetting and de-wetting of lubricants with scanning polarization force microscopy. In: Bhushan B (ed.) Fundamentals of Tribology and Bridging the Gap Between the Macro- and Micro/Nanoscales. NATO Science Series II. Dordrecht: Kluwer Academic Publishers 2001; 10 651–662

[36] Hu J, Xiao XD, Salmeron M. Scanning polarization force microscopy: a technique for imaging liquids and weakly adsorbed layers. Applied Physics Letters 1995; 67 476–478. DOI: 10.1063/1.114541

[37] Jung YC, Bhushan B. Technique to measure contact angle of micro/nanodroplets using atomic force microscopy. Journal of Vacuum Science and Technology A 2008; 26 777–782. DOI: 10.1116/1.2832409

[38] Hu J, Xiao XD, Ogletree DF, Salmeron M. Imaging the condensation and evaporation of molecularly thin films of water with nanometer resolution. Science 1995; 268 267–269. DOI: 10.1126/science.268.5208.267

[39] Xu L, Lio A, Hu J, Ogletree DF, Salmeron M. Wetting and capillary phenomena of water on mica. Journal of Physical Chemistry B 1998; 102 540–548. DOI: 10.1021/jp9722891

[40] Dai Q, Hu J, Salmeron M. Adsorption of water on NaCl (100) surfaces: role of atomic steps. Journal of Physical Chemistry B 1997; 101 1994–1998. DOI: 10.1021/jp9625772

[41] Luna M, Rieutord F, Melman NA, Dai Q, Salmeron M. Adsorption of water on alkali halide surfaces studied by scanning polarization force microscopy. Journal of Physical Chemistry A 1998; 102 6793–6800. DOI: 10.1021/jp9820875

[42] Dai Q, Hu J, Freedman A, Robinson GN, Salmeron M. Nanoscale imaging of a corrosion reaction: sulfuric acid droplets on aluminum surfaces. Journal of Physical Chemistry 1996; 100 9–11. DOI: 10.1021/jp952538a

[43] Rieutord F, Salmeron M. Wetting properties at the submicrometer scale: a scanning polarization force microscopy study. Journal of Physical Chemistry B 1998; 102 3941–3944. DOI: 10.1021/jp9801491

[44] Xu L, Salmeron M, Bardon S. Wetting and molecular orientation of 8CB on silicon substrates. Physical Review Letters 2000; 84 1519–1522. DOI: 10.1103/PhysRevLett.84.1519

[45] Xu L, Ogletree DF, Salmeron M, Tang H, Gui J, Marchon B. De-wetting of lubricants on hard disks. Journal of Chemical Physics 2000; 112 2952–2957. DOI: 10.1063/1.480868

[46] Dai Q, Vurens G, Luna M, Salmeron M. Lubricant distribution on hard disk surfaces: effect of humidity and terminal group reactivity. Langmuir 1997; 13 4401–4406. DOI: 10.1021/la970323h

[47] Reder-Christ K, Schmitz P, Bota M, Gerber U, Falkenstein-Paul H, Fuß C, Enachescu M, Bendas G. A dry membrane protection technique to allow surface acoustic wave biosensor measurements of biological model membrane approaches. Sensors 2013; 13 12392–12405. DOI: 10.3390/s130912392

[48] Calin M, Stan D, Schlesinger M, Simion V, Deleanu M, Ana Constantinescu C, Gan AM, Pirvulescu MM, Butoi E, Manduteanu I, Bota M, Enachescu M, Borsig L, Bendas G, Simionescu M. VCAM-1 directed target-sensitive liposomes carrying CCR2 antagonists bind to activated endothelium and reduce adhesion and transmigration of monocytes. European Journal of Pharmaceutics and Biopharmaceutics 2015; 89 18–29. DOI:10.1016/j.ejpb.2014.11.016

[49] Xu L, Salmeron M. Studies of wetting and capillary phenomena at nanometer scale with scanning polarization force microscopy. In: Rosoff M (ed.) Nano-Surface Chemistry. New York: Marcel Dekker 2001; ch. 6 243–287

[50] Martin Y, Abraham DW, Wickramasinghe HK. High-resolution capacitance measurement and potentiometry by force microscopy. Applied Physics Letters 1988; 52 1103–1105. DOI: 10.1063/1.99224

[51] de Gennes PG. Wetting: statics and dynamics. Reviews of Modern Physics 1985; 57 827–863. DOI: 10.1103/RevModPhys.57.827

[52] Xu L, Salmeron M. Scanning polarization force microscopy study of the condensation and wetting properties of glycerol on mica. Journal of Physical Chemistry B 1998; 102 7210–7215. DOI: 10.1021/jp981834v

[53] Derjaguin BNV, Churaev NV, Muller VM, Kitchener JA. Surface Forces. New York: Consultants Bureau 1987; ch. V

[54] Moldovan A, Bota M, Poteca TD, Boerasu I, Bojin D, Buzatu D, Enachescu M. Scanning polarization force microscopy investigation of contact angle and disjoining pressure of glycerol and sulfuric acid on highly oriented pyrolytic graphite and aluminum. European Physical Journal - Applied Physics 2013; 64 31302. DOI: 10.1051/epjap/2013130235

[55] Moldovan A, Bota M, Boerasu I, Dorobantu D, Bojin D, Buzatu D, Enachescu M. Wetting properties of glycerol on mica and stainless steel by scanning polarization force microscopy. Journal of Optoelectronics and Advanced Materials 2013; 15 1101–1105

[56] Moldovan A, Bota M, Dorobantu D, Boerasu I, Bojin D, Buzatu D, Enachescu M. Wetting properties of glycerol on silicon, native SiO_2, and bulk SiO_2 by scanning polarization force microscopy. Journal of Adhesion Science and Technology 2014; 28 1277–1287. DOI: 10.1080/01694243.2014.900907

Chapter 3

TiO_2 -Based Surfaces with Special Wettability – From Nature to Biomimetic Application

Jian-Ying Huang and Yue-Kun Lai

Additional information is available at the end of the chapter

http://dx.doi.org/10.5772/60826

Abstract

Super-wetting/antiwetting surfaces with extremely high contrast of surface energy and liquid adhesion have attracted a lot of interest in both fundamental research and industry. Various types of special wetting surfaces can be constructed by adjusting the topographical structure and chemical composition. In this chapter, recent advance of the super-wetting/antiwetting surfaces with special solid/liquid adhesion has been reviewed, with a focus on the biomimetic fabrication and applications of TiO_2-based surfaces. Special super-wettability examples include lotus-leaf-inspired surfaces with low adhesion, rose-petal-inspired surfaces with high adhesion, spider silk bio-inspired surfaces with directional adhesion, fish-scale-inspired underwater supero-leophobic surface, and artificial surfaces with controllable or stimuli-responsive liquid adhesion. In addition, we will review some potential applications related to artificial antiwetting surface with controllable adhesion, e.g., self-cleaning, antifogging/anti-icing, micro-droplet manipulation, fog/water collection, water/oil separation, anti-bioadhesion, micro-template for patterning, and friction reduction. Finally, the difficulty and prospects of this renascent and rapidly developing field are also briefly proposed and discussed.

Keywords: Wettability, Super-wetting/antiwetting, Adhesion, TiO_2-based surface, Application

1. Introduction

During the past decades, super-antiwetting surfaces, with a water contact angle (WCA) greater than 150 °, have attracted considerable interest due to their importance in both fundamental research and practical application [1-9]. Two types of extremely super-antiwetting cases exist in nature, which are the "sliding" super-antiwetting lotus leaves or water strider with ultralow water sliding resistance and the "sticky" super-antiwetting petal effect or gecko feet with extremely high liquid adhesion. Biomimetic studies demonstrate that the synergistic effect of topographical structures and chemical compositions plays a vital role on the special functional surfaces with various adhesions. These findings have inspired the creation of super-antiwetting functional surfaces with self-cleaning, anti-icing/antifogging, water/oil separation, micro-droplet manipulation, anti-bioadhesion, micro-template, and low-friction transportation [10-16].

Figure 1. Some natural cases in lotus leaf, mosquito eye, rose petal, spider silks, butterfly wings, gecko feet, desert beetle, and water strider and their potential applications of biomimetic super-antiwetting surfaces, such as self-cleaning, anti-icing/antifogging, micro-droplet manipulation, fog/water collection, water/oil separation, anti-bioadhesion, micro-template for patterning, and friction reduction

In the following two sections, we will give a summary of super-wetting surface and super-antiwetting surfaces with specific adhesions observed in nature, as demonstrated in Figure 1, such as lotus leaf, mosquito eye, rose petal, spider silks, butterfly wings, gecko feet, desert beetle, and water strider. The corresponding bio-inspired superhydrophobic surfaces with special wettability and solid/liquid adhesion are also reviewed and discussed (super-antiwetting surfaces with weak/strong adhesion, anisotropic adhesion, or controllable/switchable adhesion). The methods to construct or regulate the adhesion of artificial super-antiwetting surfaces can be considered from two strategies. One is to adjust the chemical component or change the topographical structure respectively or simultaneously. The other is to provide external stimulations to induce chemical component or surface structure transition on responsive material surface. We will also review some significant applications related to TiO_2-based surfaces with special wettability and adhesion. Finally, we briefly draw conclusion from the problems and challenges that existed on the synthesis processes and practical applications of super-antiwetting surface with special solid/liquid adhesion.

2. Surfaces with special wetting states

In general, compared to hydrophobic surfaces, artificial super-antiwetting surfaces obviously exhibit lower solid/liquid adhesion. For example, typical super-antiwetting surface usually has a low adhesive force and a sliding angle smaller than 10 °. However, some specific super-antiwetting surfaces, under comparable contact angle, have exhibited much stronger solid/liquid adhesion. Such special adhesion is highly related to the specific contact modes at the three-phase (solid/liquid/air) contact line (TCL) and the interaction at the solid/liquid interface. Recently, Wang et al. categorized five types of wetting states on super-antiwetting surfaces according to their liquid adhesion from high to low: Wenzel state and Gecko (petal) state, intermediate state, Cassie state, and lotus state [17]. A stronger adhesive force is demonstrated on the Wenzel state substrate because liquid penetrates to come in contact with the solid surface, resulting in the obvious increase of the solid/liquid real contact area. Simultaneously, the corresponding TCL becomes continuous and stable [18]. Thus with this state, the surface exhibits relatively high adhesion, and liquid droplet does not fall off easily even on the tilted substrate with a large angle. Droplets are not stable on such surface with weak solid/liquid interaction. In contrast, for the Cassie state, a large amount of air pocket is trapped between the solid and liquid interface, the real solid/liquid contact area is lower, and the TCL is highly discontinuous [19]. However, in most cases, an intermediate state between Cassie and Wenzel states is generally found with partial air trapped in the solid/liquid interface. Such an intermediate state is not stable and preferentially changed to the Wenzel contact state under external stimuli. This means that solid/liquid adhesion on such super-antiwetting substrate can be adjusted to some extent without the change of solid surface compositions and morphologies. Therefore, controllable solid/liquid adhesions can be realized base on the manipulation of surface outmost composition, topographical structure, and external stimuli.

Normally, the solid/liquid adhesive force can be qualitatively assessed by the sliding angle (contact angle hysteresis), which is usually defined as the contrast between advancing and receding contact angles. The sliding angle on superhydrophobic surfaces can be influenced by the topography, structure, chemical heterogeneity, TCL, etc. In the following sections, various types of super-wetting/antiwetting surfaces are taken into account according to their special adhesion items. Moreover, we mainly discuss the mechanism for TiO_2-based composites with special wetting states, and the corresponding wettability or adhesion adjustment and their practical or potential applications will be presented in other sections.

2.1. Super-wetting surfaces with high surface energy

In 1997, Fujishima et al. reported the UV-induced superamphiphilic ability of a single-crystal rutile TiO_2 surface [20]. Before UV illumination, the TiO_2 surface displayed a water contact angle of 72 °. However, the droplets completely spread on the film after UV illumination for certain duration. This is due to the creation of numerous high-energy domains with hydrophilic/oleophilic property on TiO_2 surface. Moreover, the wettability of either polycrystal or single-crystal TiO_2 surfaces could be reversible between hydrophobicity and superhydrophilicity under the alternation of long-term dark storage and UV light irradiation, independent of their photocatalytic properties. After the discovery of light-induced superamphiphilic ability on TiO_2 surfaces, great attention has been paid to explicate the mechanism of such unique wetting transition under UV light [21]. In addition to the pristine single-crystal TiO_2, the polycrystalline TiO_2-based thin film also exhibited photo-induced amphiphilicity. Although another inherent photo-induced phenomenon can take place simultaneously on the identical TiO_2-based surface, the photocatalysis-induced hydrophilicity demonstrated intrinsically different processes and mechanisms to that of abovementioned photo-induced hydrophilicity. It has been believed that the combination of structure/component changes and the photocatalytic degradation of low energy hydrocarbon groups at TiO_2-based surface is the main reason for the realization of photocatalysis-induced hydrophilic ability. Utilizing these two inherent abilities of TiO_2, a wide variety of super-wetting TiO_2 materials have been prepared, exhibiting environmental friendly applications in self-cleaning, antifogging, and other fields. In this section, we will focus on the recent developments in super-wetting TiO_2-based materials by various techniques with or without the assistance of UV light activation.

Fujishima et al. reported TiO_2/SiO_2 nanostructure composited films with super-wettability under UV irradiation [22]. The top TiO_2 layer with photocatalysis and photo-induced super-wetting and the bottom porous SiO_2 layer with low refractive index provided promising outdoor uses for self-cleaning and antifogging/reflection. Shang et al. reported N- and F-codoped TiO_2 nanotube arrays with palladium oxide nanoparticles decoration [23]. These codoped TiO_2 nanotube arrays exhibited a fast visible light-induced super-wetting due to its superior photocatalytic property and special vertical alignment, which has promising applications in self-cleaning.

Electro-wetting proved to be a versatile method to significantly decrease the contact angle of partially wetting droplets on solid substrates with an externally applied voltage [24]. Recently, Ma et al. reported that an electric polarization process would enhance surface hydrophilicity of microarc-oxidized TiO_2 coatings by taking advantage of the electro-wetting effect [25]. Although the electric polarization treatment could not alter the surface roughness, this approach resulted in surface electric fields and produced surface charges. Owing to the electric polarization, the electrostatic forces decreased the surface energy and the contact angle, resulting in the durable wettability.

Recent studies also demonstrated that the thermal treatment can increase the antiwetting ability of TiO_2 films. Meng et al. fabricated TiO_2 layers with thermo-induced hydrophilicity on silicon or quartz substrates by radio-frequency magnetron sputtering [26]. It was found that the crystal structure played a vital role in the surface wettability of TiO_2 layers. When the annealing temperature was about 900 °C, the mixture of anatase and rutile induced optimal hydrophilicity.

UV or visible light illumination is an effective approach for the creation of super-wetting TiO_2-based surfaces. However, it is promising to construct TiO_2 layers possessing persistent super-wetting without the need of external stimuli. Utilizing a facile electrophoretic process for the deposition of charged nanotube particles prepared by hydrothermal approach (Figure 2a, b), a super-wetting cross-stacked TiO_2-based thin films with good mechanical adhesion were successfully coated on a conductive glass substrate (Figure 2c, d) [27]. Without the use of UV illumination, the TiO_2 layers displayed superhydrophilic property with a rapid droplet spreading within 0.16 s (Figure 2e). This can be ascribed to the porosity effect and surface hydroxyl, which played an important role in controlling the interaction between the liquid and film.

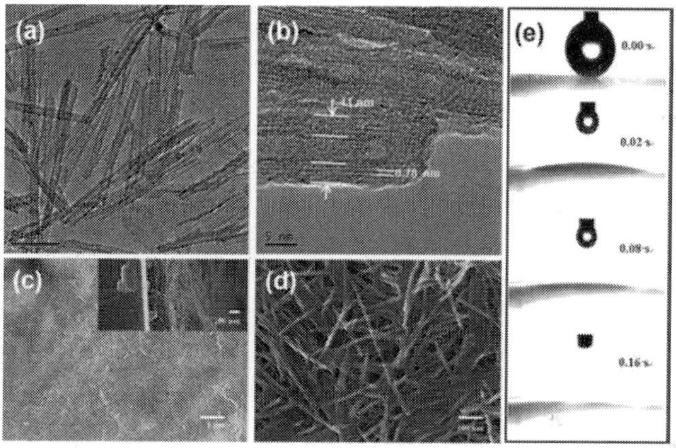

Figure 2. (a, b) TEM images of titanate nanotube prepared by hydrothermal approach; (c, d) the SEM images of titanate nanotube film by a facile electrophoretic deposition (EPD) process; and (e) the liquid wetting behavior on porous superhydrophilic TiO_2(B) surface. The inset figure is the corresponding cross-sectional image of deposition film

Fu et al. fabricated superhydrophilic surfaces with the micro-nano dual-scale hierarchical structures using SDBS-modified TiO_2 nanoparticles [28]. These films exhibited superhydrophilicity without the requirement of UV irradiation, arising from the high surface roughness. This work opened an avenue to construct functional TiO_2 for promising applications in self-cleaning and antifogging without UV irradiation. Funakoshi et al. fabricated superhydrophilic TiO_2 thin films on glass substrate surfaces through the titanium alkoxide hydrolysis [29]. Zeng et al. prepared transparent nanosized crack-free TiO_2 thin films with good adhesion to glass substrates through a self-assembly approach. This method consisted of the preparation of TiO_2 colloidal particles and adsorptive self-assembly in the colloidal stock and followed by heat treatment [30]. The resultant films exhibited superhydrophilicity with a water contact angle of about 0 ° and still retained at about 15 ° even after the dark storage for a week.

Moreover, these self-assembled TiO_2 films presented high transmittance and excellent photocatalytic activity.

2.2. Super-antiwetting surfaces with low adhesion

In nature, many biological creatures possess the combination of low-energy surfaces and hierarchical structures that endow the super-antiwetting properties with weak adhesion. Among them, the water stride and lotus leaf are two of the most classic examples [31, 32]. The phenomenon of collecting the dirty debris by the sliding of water droplets on lotus leaf surface is commonly known as the "lotus effect." The lotus leaf therefore always keeps itself clean by this self-cleaning behavior. Bio-inspired by the self-cleaning effect of lotus leaves, different artificial superhydrophobic TiO_2-based surfaces with low adhesion have been created by the construction of surface-roughened structures with suitable chemical compositions [33-40].

Bhushan et al. created different self-cleaning superhydrophobic structures by the replication of a micropatterned silicon substrate using an epoxy resin and hydrophobic alkane coating [41]. They investigated the effect of lotus-like structures on wetting and self-cleaning efficiency. The main factor of hierarchical structure for superhydrophobicity with low adhesion was also revealed. Zhou et al. fabricated superhydrophobic metallic oxide layer on titanium or aluminum substrates with a combination of electrochemical anodizing and self-assembly technique [42]. In comparison with the smooth surface, the porous metallic oxide exhibited excellent self-cleaning ability and can be stabilized over a wide range of pH value. Jiang's group successfully constructed artificial water strider leg with a ribbed conical nano-needle structure containing oriented nano-grooves sculptured on the lateral nano-needle surface [43]. Such artificial leg exhibited good dynamic stability under loading and during load relaxation. In addition to the implication for the design of stable superhydrophobic structure surfaces, this work provided inspiration for the applications in drag-reduction fields and miniaturized aquatic devices.

While many techniques have been developed to fabricate superhydrophobic surfaces by creating roughened surfaces and/or changing the surface energy, these techniques are limited by the types of materials to be treated. The direct coating of highly hydrophobic powder by spinning or spraying on various substrates was verified to be a facile method to construct surface with superhydrophobic property. For example, to demonstrate the versatility of the current approach, we have prepared the superhydrophobic on a silicon substrate, lab rubber glove, and A4-size printing paper by spraying, spin coating, or dip coating [44]. It is also interesting to note that the cross-stacked microscale/nanoscale fluorinated titanium nanobelt structures are formed on substrate (Figure 3a, b). Water droplet is easy to roll off due to the ultralow solid/liquid adhesion and bounce off the decorated silicon surface when it is dropped from a certain height (Figure 3c).

2.3. Super-antiwetting surfaces with high adhesion

In contrast to conventionally super-antiwetting surfaces (lotus leaves and water stride's legs) with low adhesion, there are some other biological surfaces (rose petals and gecko toes) on which water droplets are firmly pinned meanwhile kept with a contact angle larger than 150 °

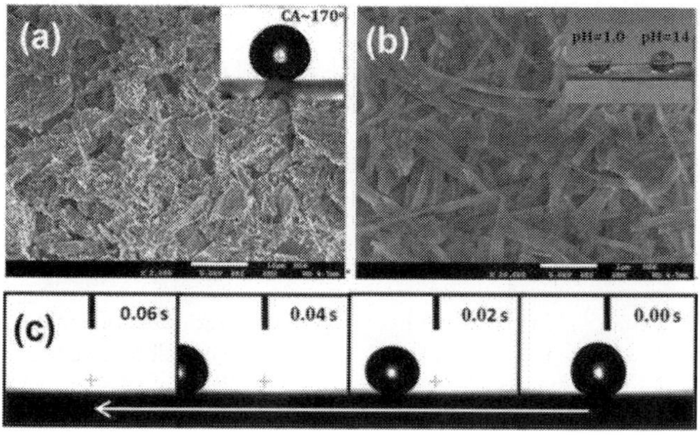

Figure 3. SEM image of cross-stacked microscale/nanoscale fluorinated titanium nanobelt particles formed on silicon substrate and the corresponding water droplet behavior on particle-decorated surface

[45-47]. To discover the mechanism of such special high adhesion, Feng et al. investigated the microstructures of rose petal and found that topographical surfaces of rose petal are made up with a periodic array of micro-papillae with compactly arranged structures and numerous nanoscale cuticular folds on micro-papillae [47]. Water droplets are expected to penetrate into the microscale papillae, but a large amount of air trapped inside the nanoscale folds resulted in the formation of specific petal-like wetting state with high adhesion. Inspired by the adhesion mechanism of rose petals and geckos, several efforts have been paid to prepare sticky super-antiwetting surfaces, on which a droplet does not slide even with a tilting angle of 90 degrees [48-53]. For example, Jiang et al. successfully fabricated well-aligned super-antiwetting polystyrene nanotube array films with high hysteresis by using anodized alumina membrane templates [48]. They believed that the main reason responsible for the great adhesion is the enhancement of van der Waals forces between the densely packed nonpolar polystyrene nanotubes in close contact with droplet. Guo et al. reported a facile method to construct sticky superhydrophobic surface via etching of an aluminum alloy surface and eliminating its loose layer [49]. They subsequently demonstrated that a much stronger adhesion by the joint action of capillary forces and van der Waals forces from the micro-orifices and hydrophilic nanoparticle composite structures.

Recently, Lai et al. reported a facile electrochemical anodizing process to construct TiO_2 nanopore structure surface on Ti substrate (Figure 4a, b) [54]. Compared to the hydrophobic of smooth Ti surface, the nanopore structure surface exhibited a superhydrophobic property (Figure 4c, d). Moreover, such specific superhydrophobic structure displayed a very strong adhesive force of 75.6 μN (Figure 4e), which resulted from the van der Waals attraction between the water molecules and fluoroalkylsilane molecules and from the negative pressure caused by the volume change of air sealed in the nanopore units. As a result, the bead was firmly pinned on such sticky superhydrophobic surface without any movement when it was tilted to a vertical position (Figure 4f) or even when turned upside down.

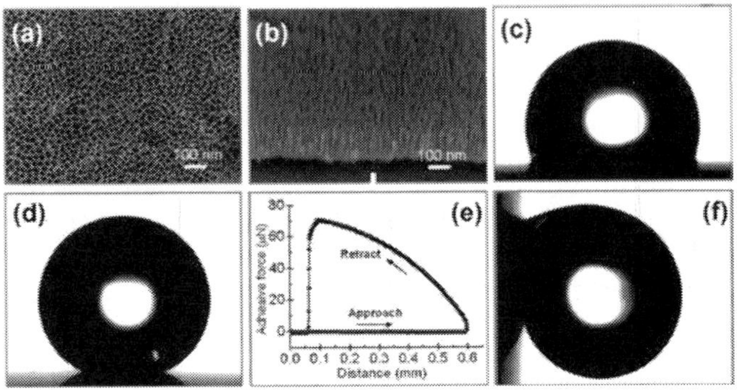

Figure 4. (a, b) Top-view and side-view SEM images of TiO_2 nanopore structure. (c) Photo of a water droplet on the TiO_2 flat surface with PTES modification. (d) Droplet on the superhydrophobic TiO_2 nanopore film. (e) The force-distance curves recorded before and after the superhydrophobic surface made contact with a droplet. (f) A water droplet on a highly adhesive superhydrophobic surface vertically tilted

2.4. Super-antiwetting surfaces with controllable adhesion

2.4.1. Effect of morphology and structure

Based on the theoretical background of wetting, it is easy to understand that the great effect of morphological and structural parameters on the determination of wetting models and contact TCL situation and resultant to influence the droplet adhesion on super-antiwetting surfaces. For the identical material with certain surface energy, the solid/liquid adhesive force can be controlled by tailoring the morphology or the scale of the structures on the surface [55-59]. Accordingly, we designed and constructed three kinds of unique super-antiwetting TiO_2 nanostructure models, e.g., a nanotube array (NTA), a nanopore array (NPA), and a nanovesuvianite structure (NVS) by a facile electrochemical anodizing process (Figure 5) [54]. These different structures with various wetting states and controllable porosities could effectively change the air-pocket ratio in sealed and open systems, so that the solid/liquid adhesive force could be achieved in a wide range. Capillary adhesive force plays a dominant role in imparting adhesive behavior on NTA and NPA nanostructures due to the formation of sealed air pockets, while the NVS nanostructures exhibit extremely low adhesion due to the absence of sealed air pockets. In addition, adhesive forces could be tuned by changing the nanotube diameter, especially the tube length, due to the negative pressure caused by the volume change of air sealed in the nanotubes. The cooperation effect between the negative pressures induced by the volume change of sealed air pockets and the van der Waals attraction at solid/liquid interface contributes to the water adhesion. These findings are valuable to deepen insight into the roles of nanostructures in tailoring water repellency and adhesion for exploring new applications. Lee et al. further verified that the liquid adhesion could be controlled by tuning the formation ratio of dead-end nanopores on superhydrophobic anodized aluminum oxide [60, 61].

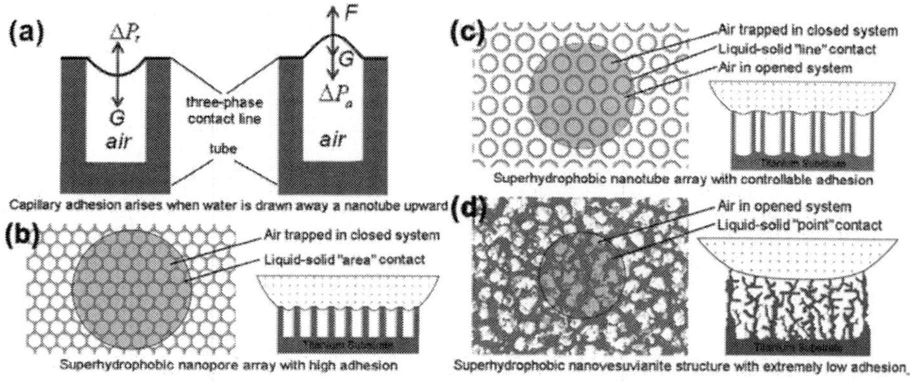

Figure 5. Schematic illustration of three types of super-antiwetting nanostructure models with different adhesions. (a) Schematic illustration of capillary adhesive force caused by negative pressure when a water droplet is drawn away an NPA or NTA close system; (b) super-antiwetting NPA with high adhesion; (c) super-antiwetting NTA with controllable adhesion; and (d) super-antiwetting NVS with extremely low adhesion

2.4.2. Effect of chemical composition

The chemical component with certain surface energy is another key factor in determining the wettability and solid/liquid adhesion. For a super-antiwetting surface with identical morphology, the solid/liquid adhesive force could be greatly controlled by utilizing the difference in the alkyl chain length or by adjusting the ratio of high-energy hydrophilic areas [62-66]. Recently, we reported a successful case of the super-antiwetting spongy-like TiO_2 film exhibiting adhesion ranging from 5.0 μN (ultralow) to 76.6 μN (extremely high) by adjusting the dosage concentration of high-surface-energy materials of nitrocellulose (Figure 6) [62]. The super-antiwetting films without the introduction of NC displayed an exceptionally low adhesion with a sliding angle of only approximately 0.8 °. However, the nitrocellulose modified super-antiwetting film can firmly catch liquid droplets even when it is turned upside down, indicating high adhesion to droplets. Such significant adhesion enhancement was ascribed to the fact that the introduction of hydrophilic nitro groups not only leads to the formation of hydrogen bonding with the hydroxyl groups at the solid/liquid interfaces but also result in the disruption of the dense alignment of hydrophobic silane molecule.

Other techniques for constructing sticky super-antiwetting layer are based on terminating of high-surface-energy groups on rough structures or site-selective discomposing of low-surface-energy chains to obtain patterned surface with micrometer-scale chemically heterogeneous composition [67-69]. For example, Wang et al. realized switchable wettability and reversible adhesion between sticky superoleophobicity and sliding superoleophobicity by alternating annealing treatment and UV illumination on TiO_2 nanostructure surfaces [67]. However, most of the abovementioned techniques cannot rapidly and in situ change the wettability and adhesion.

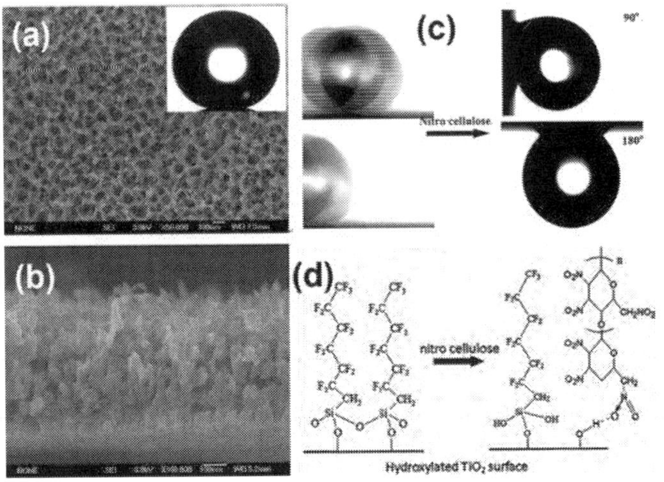

Figure 6. Top-view (a) and side-view (b) SEM image of a sponge-like TiO_2 nanostructure film on Ti substrate. The inset of (a) shows the droplet image on corresponding superhydrophobic TiO_2 surface. (c) Behavior of a water droplet on sponge-like TiO_2 surface before and after NC introduction. (d) Model of molecular self-assembly before and after NC introduction

In contrast to the top-down photo-cleavage of chemical components to control the liquid adhesion on superhydrophobic surface, herein, we realized that both water wettability and adhesion change with a large degree of contrast on superhydrophobic TiO_2 nanotube array film via a facile, in situ bottom-up ink combination to simultaneously regulate topographical structure or chemical components at targeted places (Figure 7) [70]. The site-specific ink covering and removal allows the wettability and reversible adhesion transitions to be realized. These techniques are helpful to effectively construct novel functional nanomaterials with customer-tailored surface wettability and adhesion [71].

2.5. Super-antiwetting surfaces with responsive adhesion

Recently, intelligent surfaces with responsive wettability and adhesion that can reversibly change between highly hydrophilic (strong adhesion) and hydrophobic (weak adhesion) have attracted much attention due to their practical applications. As we have known, the wetting state has a strong influence of droplet adhesion on super-antiwetting surface. The change of contact angle hysteresis (decreasing of the receding angle) is a key factor to control droplet mobility on surfaces. In general, Wenzel-state super-antiwetting surfaces exhibit higher droplet adhesion, while the adhesion will be dramatically decreased on Cassie wetting-state surfaces. There exist certain energy barriers to prevent the spontaneous transition between these two wetting states. If a molecule can reversibly alter its surface energy on a hierarchical surface to induce the change of contact model, such smart surface whose wettability can be modulated between highly hydrophilic and hydrophobic, or even super-wetting and super-antiwetting, can be realized (Figure 8a). In contrast, if a specific molecule can alter its polarity

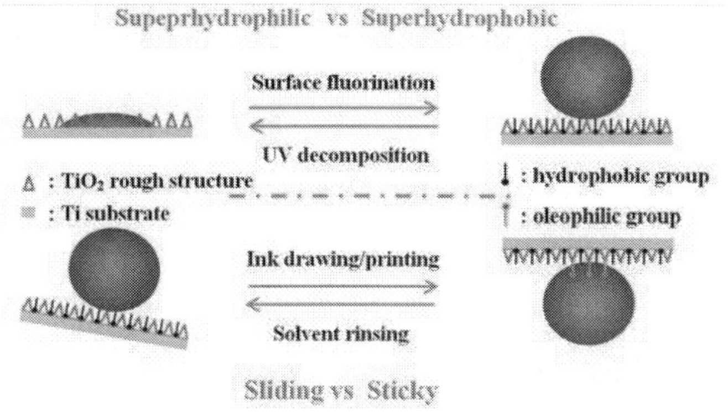

Figure 7. Schematic representation of the mechanism of wettability and adhesion on robust superhydrophobic TiO_2 rough structure prepared via a facile electrochemical anodizing strategy for TiO_2 nanotube construction, surface fluorination, and reversible site-specific ink printing/removing

but without greatly changing the surface energy of a hierarchical structured surface, a super-antiwetting surface with controllable adhesion can be achieved. The mobility of droplet behavior on such surfaces can be controllably switched from sticky to a sliding state (Figure 8b). Based on this strategy, various types of smart-responsive materials, such as polymers, liquid crystal molecules, and semiconducting inorganic oxides can be utilized to achieve wettability switches that are controlled by external stimuli. Therefore, if an external stimulus can be applied to induce the transition, the wettability and droplet adhesion will be reversibly switched.

Recently, stimuli-responsive surface wettability has been extensively studied under external stimulus such as light irradiation [72-74], electric field [75-79], thermal treatment [80, 81], magnetic field [82, 83], transformation [84, 85], pH [86], and multi-responsive factors [87, 88]. Among them, light irradiation and photoelectric cooperative wetting can be facilely used to achieve liquid patterns for printing, which is also promising to gear up the scientific investigation in the fields of biological and health sciences and broadened their practical applications. Inorganic TiO_2 semiconductor is an important photo-responsive multifunctional material. For example, responsive water-droplet adhesion between sliding and pinned states on a superhydrophobic polythiophene film constructed by a versatile electrodeposition method is realized by adjusting electric potential [89]. In recent years, smart wettability-responsive surfaces with reversibly switching between superhydrophilicity and superhydrophobicity have attracted increasing interests. Recently, Zhou et al. demonstrated a facile method to achieve reversible change of adhesion to water droplets without loss of superhydrophobicity [90]. The super-antiwetting TiO_2 surface was illuminated with UV through a mask to create alternatively arranged hydrophilic/superhydrophobic domains. Photo-induced hydrophilic regions increased water adhesion, and surrounding super-antiwetting regions prevented the spread of water droplets and preserved super-antiwetting. The recovery was very fast upon heating at a relatively high temperature or the use of microwaves. The TiO_2 nanotube array surface presented a UV-thermal-induced reversible adhesion switching attributed to the displacement of C-F groups and introduced -OH groups upon site-specific UV irradiation, which can increase droplet interaction with superhydrophobic surface. The heat treatment resulted in the decrease of high-energy Ti-O-H bonds and the restoration of hydrophobic C-F bonds with low surface energy, showing the primary Cassie superhydrophobic state.

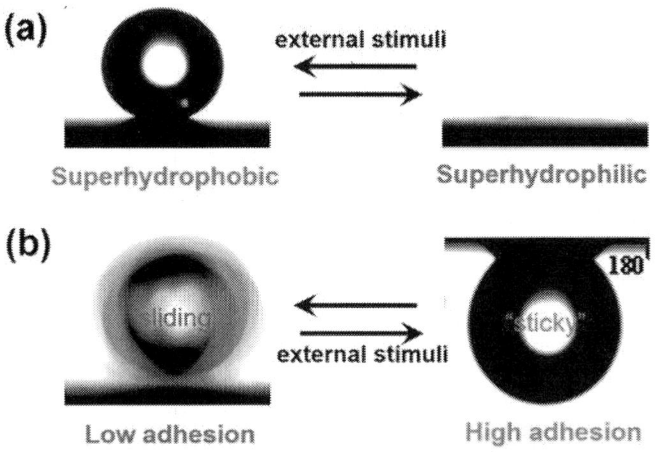

Figure 8. (a) Reversible switching wettability on smart surfaces between superhydrophobicity (left) and superhydrophilicity (right) by external stimuli. (b) Reversible switching of solid/liquid adhesion with extremely high contrast on a superhydrophobic smart surface

2.6. Super-antiwetting surfaces with anisotropic adhesion

If a surface displays different contact angles or sliding angles under various directions, such surface is stated to be anisotropic in wettability. Directional droplet manipulation as a vital process for biological system is ubiquitous in nature. For example, anisotropic super-antiwetting surfaces allow spider silk to collect water, butterflies to shed off droplets, and plants to trap insects and pollen. Butterfly wings have directional adhesion [91]. Droplets can easily roll off the wings along the radial outward direction of the central axis of the butterfly body while it is firmly pinned in the opposite direction. Recent investigations verified that anisotropic wettability and adhesion were highly related to the anisotropic three-phase contact line [92, 93]. Recently, anisotropic wetting surfaces have attracted a great amount of attention in various fields due to their specific advantage, such as microfluidic devices, and self-assembled pattern synthesis.

Micropatterned wettability surfaces displaying anisotropic sliding and adhesion in a high contrast to droplets have the ability to directionally manipulate micro-droplets and many promising applications in lab-on-a-chip and high-throughput fluidic devices [94]. By taking advantage of adhesion anisotropy, the on-demand manipulation of droplets is feasible. Figure

9 displays the effects of ink unit (dot or line) size and liquid volume on mobility of micro-droplet on a superamphiphobic film. In general, the sliding angle increased upon decreasing the volume of the droplet, regardless of tilting direction and pattern shape (Figure 9a). For dot patterning surface, the mean value of sliding angle along various directions is found very closely, indicating similar isotropic wetting property. However, the line patterning surface additionally exhibited a directionally dependent droplet shape and sliding angle (Figure 9b). On the tilted substrate, small droplets located on the thicker line against the orthogonal direction are stable than that on the thinner line along with the parallel direction. Such strong anisotropy of sliding angle and droplet distortion on line patterning surfaces was ascribed to the high contrast of energy barrier, solid/liquid/air phase contact line, and wetting behavior along an orthogonal or parallel direction.

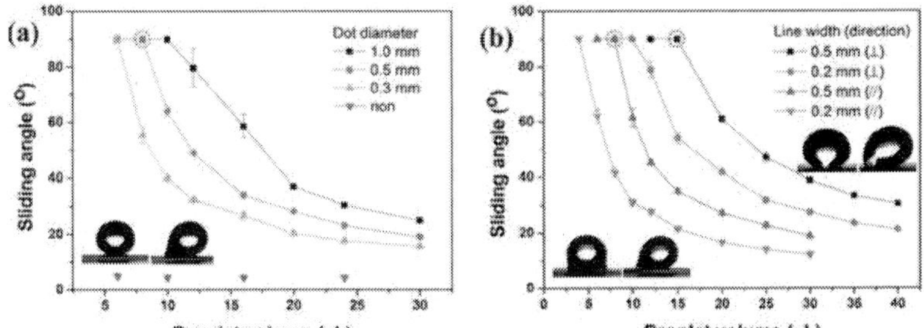

Figure 9. Droplet sliding angles on a) dot and b) line patterned superamphiphobic surface with respect to patterning size, droplet volume, and tilting direction. Insets in (a) are side-view optical images of an 8 μL droplet on a 0.5 mm ink dot before (left) or after (right) tilting with an angle of 90 °. Left insets of b) are the side-view images of an 8 μL droplet on a line with 0.5 mm in width before or after a tilting angle of 90 ° from an orthogonal direction. Right insets of b) are the side-view images of a 15 μL droplet on a line with 0.5 mm in width before or after a tilting angle of 90 ° from parallel direction (right)

3. Applications of surfaces with special wettability

Wettability is a main property of surface governed not only by geometrical structure but also by chemical composition. Recently, super-wetting and antiwetting-derived material interfaces, such as superhydrophilicity, patterned superhydrophobicity, or superhydrophobicity with special adhesion (ultralow/ultrahigh adhesion or anisotropic adhesion), have attracted much attention. Superhydrophilicity and superhydrophobicity are defined based on the conventional liquid contact angle measurement. A superhydrophilic surface generally displays a water contact angle lower than 5 °. The opposite superhydrophobic case refers to surface with contact angle greater than 150 °. Such two extreme cases and their corresponding wettability patterning surfaces have attracted great interest due to their importance in both theoretical research and practical applications.

TiO_2-based nanostructures as one of the hottest semiconductor nanomaterials have been applied in many fields including environmental/energy research and biomedical implants, owing to their unique merits, e.g., good chemical and mechanical stability, as well as excellent biocompatibility, anticorrosion, and photocatalytic properties. In this section, recent progress of the bio-inspired surfaces with special wettability has been discussed, with a focus on the emerging potential applications of TiO_2-based layers. Examples include lotus-leaf-inspired low adhesion, rose-petal-inspired high adhesion, spider silk bio-inspired super-antiwetting layers with directional liquid adhesion, and biomimetic super-antiwetting layers with adjustable and smart stimuli-responsive liquid adhesion.

While the artificial super-antiwetting surfaces with special adhesion have been introduced and constructed by many research groups as discussed above, attention has also been shifted to the potential or practical applications of multifunctional surface with special wettability and adhesion in the recent few years. Some important applications, such as self-cleaning [95-100], antifogging/icing [101-110], micro-droplet transportation [111-115], water/fog collection [116-122], oil/water separation [123-129], anti-bioadhesion [130-139], micro-template for patterning [140-149], and friction reduction [150-159], could be realized through the improved understanding of super-antiwetting with various degrees of adhesion. Selected examples are discussed below. Finally, the challenges and prospects of this renascent and rapidly developing field are also briefly addressed and discussed.

3.1. Super-wetting surfaces for self-cleaning and antifogging

Since the report of TiO_2 with super-wetting under UV light illumination in 1997, a great number of applications have been focused on this multifunctional material. Water droplets can uniformly spread or be absorbed quickly into superhydrophilic substrates. As discussed in the above section, super-wetting TiO_2 surfaces can be easily constructed through UV illumination, even displaying superamphiphilic for both water and oil. Utilizing this strategy, various super-wetting TiO_2-based nanostructure materials have been fabricated, exhibiting environmental-friendly applications such as self-cleaning, antifogging, underwater superoleophobicity, and other items.

Zhai et al. recently demonstrated a multifunctional porous TiO_2 nanotube membrane with superhydrophilic-induced underwater superoleophobicity for oil-water separation, flow-through photocatalysis, and self-cleaning by taking advantage of its excellent photocatalytic oxidation activity for the degradation of pollutant or contaminated organic molecules [160]. Lai et al. prepared super-wetting cross-stacked TiO_2-based thin films with good mechanical adhesion on a conductive glass substrate [161]. Without the use of UV irradiation, the porous TiO_2 films showed superhydrophilicity with an ultrafast droplet spreading (Figure 10a). This can be ascribed to the porosity effect and high-energy-surface hydroxyl groups, which played an important role in controlling the interaction between the liquid and film. As shown in Figure 10b, the untreated quartz slide (upper) fogged immediately, and the words below are blurred by strong light scattering. In contrast, the treated slide with the coating of TiO_2(B) films (bottom) can significantly prevent fogging formation by almost instantaneously spreading and penetrating the condensed water droplets to form a thin water membrane (Figure 10c, d).

Therefore, the slide with the TiO$_2$(B)-coated glass remains highly transparent, and the words below are clearly seen.

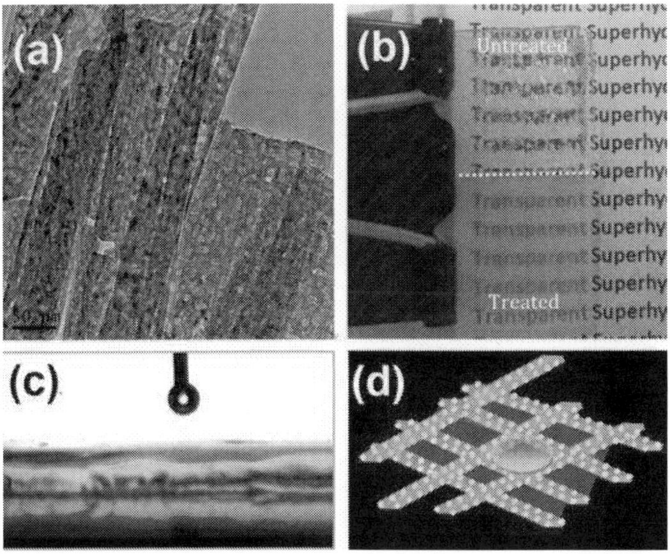

Figure 10. (a) The TEM image of porous superhydrophilic TiO$_2$(B) surface after calcination; (b) the liquid wetting behavior on porous superhydrophilic TiO$_2$(B) surface; and (c) photograph of an ITO substrate deposited with superhydrophilic coatings (bottom) and a control ITO substrate without any coating deposition (upper) taken from a refrigerator (-4 °C) to the humid laboratory air (ca. 50 % RH). (d) Schematic illustration of the ultrafast water spreading on the microscale/nanoscale pore-like superhydrophilic TiO$_2$(B) film. The 3D cross-aligned TiO$_2$(B) network and microscale and nanoscale pores result in great capillary effect to quick water invasion

3.2. Applications of Super-antiwetting surfaces

3.2.1. Self-cleaning

In nature, there exist many kinds of self-cleaning surfaces from the land to sea after hundreds of millions of years of the evolution process. As we all have known, wettability is highly related to functional surfaces with self-cleaning ability, which suggests many important hints for antifouling. Artificial technologies for self-cleaning have developed since the late twentieth century, and some achievements have led to practical applications. The applications of self-cleaning technology are very broad and vary from window glasses to solar cell panels. Three types of surfaces, including superhydrophobic, superoleophobic, and even superhydrophilic, offer solution to keep surface cleaning. Surface self-cleaning is a significant application of super-antiwetting surface. In the case of super-antiwetting rough surfaces, the solid/water interface is minimized. Water forms a spherical droplet and easily rolls off, and the droplet collects the particles from the surface. Therefore, the super-antiwetting surfaces with low adhesion always exhibit a very low degree of contamination, which is known as self-cleaning [162, 163]. For a normal hydrophobic surface, because of the nonslip boundary condition, the water drop falls across the dirt particles and the dirt particles are mainly displaced to the sides of the droplet and redeposited behind the droplet. Ding et al. reported a long-term superhydrophobic self-cleaning coating prepared by simply blending ambient-cured fluorinated polysiloxane binder with TiO_2 nanoparticles. The obtained coating exhibited excellent wetting stability in various environments and photocatalytic self-cleaning property for practical applications [164]. Yamashita et al. fabricated super-antiwetting surfaces with photocatalytic self-cleaning abilities through coating a nanocomposite TiO_2 photocatalyst and hydrophobic layer onto a structure substrate with a co-deposition strategy [165]. This coating realizes adequate photocatalytic activity for self-cleaning and inducing unique wettability changes. The nanocomposite can contain multiple functions of energy-saving and maintenance-free characteristics. On the super-wetting surface, the water droplets spread rapidly and go beneath the contaminants to bring them away.

In addition to super-antiwetting and super-wetting, super-antiwetting of oil is also a promising characteristic for self-cleaning surfaces, which can be easily achieved in hydrophilic surfaces in water or constructed with the combination of suitable structure and chemical components. Moreover, low drag, slippery surfaces, anti-fouling, and photocatalytic activities play an important role in the self-cleaning effect. It is necessary to further develop the characterization techniques for self-cleaning surfaces. Besides pure TiO_2, some materials derived from TiO_2, such as titanate particles by hydrothermal treatment of TiO_2 particles, came into use generally for more wide applications. For example, a stable titanate nanobelt (TNB) particle suspension was prepared by a hydrogen-bond-driven assembly of pre-hydrolyzed fluoroalkylsilane (FAS) on its surface. Using this kind of TNB/FAS nanobelt, a one-step electrophoretic deposition process was applied to fabricate a cross-aligned superhydrophobic film with high transparency (up to 80 %) on glass substrate, as shown in Figure 11. Furthermore, the as-prepared coating displayed good self-cleaning ability and high chemical stability in a wide range of pH solutions [161].

3.2.2. Antifogging, anti-icing, and anticorrosion

Super-antiwetting compound eyes of mosquitoes are comprised of the smart design of elaborate microstructures and nanostructures: nanoscale of hexagonally non-close-packed nipples and microscale of hexagonally close-packed ommatidia. Such composite structures can prevent fog drops from condensing and being trapped on the super-antiwetting surface to provide an effective protective mechanism for maintaining clear vision under humid condition. Gao et al. successfully fabricated artificial super-antiwetting compound eyes by employing soft lithography methods to transfer polydimethylsiloxane micro-hemispheres and silica nano-spheres [166]. The easy rolling of water or fog droplets may take the dust particles away from the super-antiwetting compound eye-like surface with a low adhesive force, thereby achieving self-cleaning and antifogging under humid environment.

Besides the good antifogging performance, super-antiwetting surfaces can also retard the formation of frost and ice due to the metastable state of the three-phase line, compared with

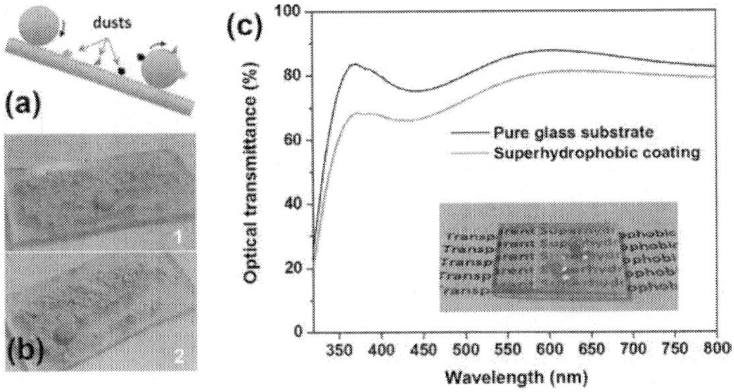

Figure 11. (a) Schematic illustration of robust superhydrophobic surface for self-cleaning. (b) Time sequence of the self-cleaning process on the superhydrophobic coating with low water adhesion. (c) Transmittance of glass substrate with or without TNB/silane hybrid films. The inset photograph of the transparent electrodeposited films with highly super-hydrophobicity.

that on a smooth hydrophobic surface [104-108]. Very recently, He et al. reported the uniform ZnO nanorod array surfaces kept superhydrophobic not only to sessile macro-droplets at room temperature but also to condensed micro-droplets at temperatures below the freezing point of 5 °C or -10 °C [108]. The time of condensed droplets maintaining the liquid state (retardation period) increases with the increasing of contact angle on the superhydrophobic ZnO surfaces, indicating an obvious retardation and prevention of ice/frost formation. In general, the study of the anti-icing and antifogging of superhydrophobic surfaces is just at its beginning stage [109, 110]. Further research is needed to understand the controlling factors to achieve optimized performance. Shen et al. investigate the icephobic property of fluoroalkylsilane (F17)-modified hierarchical Ti6Al4V alloy substrates prepared by the combination of sand blasting and hydrothermal treatment [167]. They concluded that the icing-delay duration of droplets on the superhydrophobic rough Ti6Al4V substrate was many times longer than that of droplets on the smooth hydrophilic substrate, and the ice adhesion strength on superhydrophobic substrate was significantly decreased due to the Cassie state of droplets. Additionally, water droplets always bounced off the superhydrophobic surfaces before freezing under subzero conditions give rise to the promising anti-icing application.

Super-antiwetting layers have also been found to have increased resistance to the microbial-induced corrosion and fouling in sea water due to its capability to reduce bacterial adhesion [168]. Mahalakshmi et al. measured super-antiwetting titanium surface in seawater and found out that microbes did not attach to the surface and the corrosion resistance of titanium in seawater was significantly increased [169].

Metals as important engineering materials are widely used in industrial fields. However, the metal corrosion phenomenon is ubiquitous. Many techniques have been developed to prevent the corrosive process. For example, the use of chromate-containing pigments and paints as protective layers on metal surface for anticorrosion. Concerning the processing and environmental effects, more active techniques should be invented to solve this global corrosion issue. Recently, superhydrophobic surfaces were found to serve as an effective barrier to restrain water, chloride ions, and oxygen from attaching and reacting with the metal surface. For example, Shen et al. fabricated highly hydrophobic TiO_2 nanoparticle films on 316L stainless steel by using a combination of sol-gel and dip-coating process. Even after immersion in an oxygen-saturated Ringer's solution for a duration of 42 days, no metallic oxides are observed in the layers, demonstrating the superior anticorrosion property of such highly hydrophobic TiO_2 coatings [170, 171].

While many techniques have been developed to fabricate superhydrophobic surfaces by creating roughened surfaces and/or changing the surface energy, these techniques are limited by the types of materials to be treated. Our approach can be simply applied to any substrate after the as-prepared highly hydrophobic TNB/FAS powder has dried as a result of the solvent evaporation. To demonstrate the versatility of the current approach, we have also prepared the superhydrophobic porous cross-stacked TNB/FAS structures on a silicon substrate, lab rubber glove, and A4-size printing paper by spraying, spin coating, or dip coating [44]. Figure 12 shows the stable wettability tested through water contact angles under different pH values. This diagram indicated that the as-prepared highly hydrophobic TNB/FAS powder-coated samples remained superhydrophobic with water contact angle larger than 150 ° in a wide pH range from 0 to 14, which suggested that the stable wettability resulted from the high chemical stability of TNB/FAS film over a wide pH range.

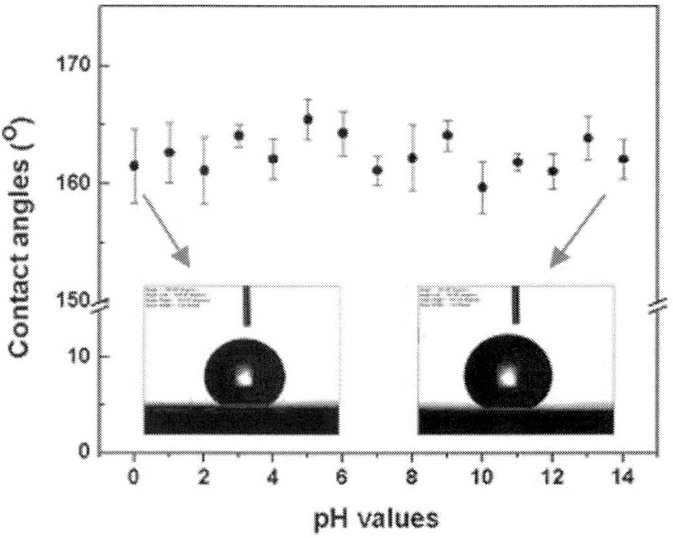

Figure 12. Relationship between pH values and water contact angles of the TNB/FAS films on silicon plate. The contact angle stability indicated that highly hydrophobic TNB/FAS powder coating endowed an excellent anticorrosion property for the coated substrate

3.2.3. Micro-droplet manipulation

The efficient manipulation of micro-droplets offers many potential applications in relation to biomedical and chemical tests and protocols. Super-antiwetting surfaces with in situ controllable/responsive adhesion in a high contrast demonstrate promising applications in micro-droplet manipulation [111-115]. Hence, efforts have also been paid to enhance the transferring process. For example, Jiang et al. prepared high adhesive super-antiwetting polystyrene nanotube layers and used them to catch micro-droplet from super-antiwetting surface with low adhesion. The droplet was then released to a hydrophobic/hydrophilic target [48]. Following this work, they further developed a more effective technique to transfer droplets by the deformation-induced reversible adhesion transition on super-antiwetting polydimethylsiloxane surface [84]. They changed the surface curvature to adjust water droplet adhesion on polydimethylsiloxane pillar array surface from a low adhesive "roll-down" state to a highly adhesive "pinned" state. This provides the possibility for precise, no-loss manipulation of droplets.

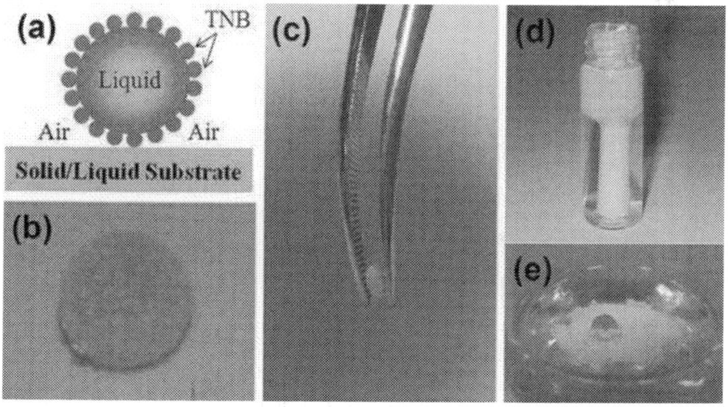

Figure 13. Model (a) and digital image (b) of a liquid marble composed of an RhB colored water droplet coated with TNB/FAS micropowder/nanopowder; (c) images showing a liquid marble being picked up with a pair of tweezers; (d) TNB layer at the air/water interface response to intrusion by a hydrophilic glass stick; and (e) a liquid droplet sitting on the air/liquid interface covered by a layer of highly hydrophobic TNB powder

More interestingly, micro-droplets capsulated by hydrophobic particles to form liquid marbles can be manipulated/transported on various substrates without being contaminated, as shown in Figure 13a, b [44]. The spontaneous encapsulation of TNB/FAS particles around on the water droplet interface is due to the minimization of the free energy of the surface. Another remarkable feature of the marble by such highly hydrophobic titanate nanobelt shell is its strong mechanical strength and high deformability. For example, these marbles can be easily held by a pair of tweezers with large deformation without rupture (Figure 13c). Furthermore, a tightly packed layer self-assembled on liquid surface created a highly robust and flexible superhydrophobic barrier to prevent direct contact between an inserted hydrophilic glass stick and water solution (Figure 13d). Water droplets larger than 3.0 mm in diameter can also stand on such tightly packed TNB layer (Figure 13e).

3.2.4. Fog, water, and oil collection

In many countries, the spring mornings are graced with spectacular phenomena of dew drops hanging on leaves and on spiders' webs [116-118]. Some plant leaves, in particular, spot particularly large droplets that last well into the morning. Recently, Jiang et al. fully discussed the detailed insights into the directional water collection on spider silk [116]. They discovered that the silk structure made a "wet-rebuilt" change to periodic spindle-knots with random nanofibrils and was separated by joints with aligned nanofibrils after wetting. Guided by the spider silk, they successfully prepared an artificial fiber with a similar structure with the spider silk to realize the directional water collection.

Another example lies with some beetles in the water-limited Namib Desert who collect drinking water from fog-laden wind on their backs [119-122]. Large droplets form by virtue of the insect's bumpy surface which consists of alternating wax-coated hydrophobic and non-waxy hydrophilic regions. Inspired by the water-collecting structure of the desert beetle's back, Cohen et al. fabricated hydrophilic or superhydrophobic patterns on superhydrophobic surfaces with similar fog-collecting characteristics with the desert beetle [120]. Hydrophilic patterns on superhydrophobic surfaces were created with water/2-propanol solutions of a polyelectrolyte to produce surfaces with extreme hydrophobic contrast. Selective deposition of multilayered films onto the hydrophilic patterns introduces different properties to the area.

Recently, inspired by living creatures with special wettability and their precise arrangement of structure and chemical component, we described a facile one-step approach to large-scale construct superamphiphobic TiO_2 film [172]. The as-prepared TiO_2 film displays excellent superamphiphilic property in air, changes to superamphiphobicity with good dynamical stability after silane modification (Figure 14a). Moreover, such superamphiphobic surface reversibly switched from superoleophobic to superoleophilic in air and underwater, respectively (Figure 14b, c). The 3D functional surface would be a versatile platform in a wide range of applications, e.g., self-cleaning, friction reduction, and repeatable water/oil separation. For example, we applied the fluoroalkylsilane-modified pinecone-like TiO_2 particle plates (superoleophobic in air and superoleophilic underwater) as "oil capture hands" to gather oil droplets in water (Figure 14d, process 1-3). The collected oil was then extracted from the water easily to achieve effective oil/water separation (process 4-6). More importantly, it is worth noting that the plate kept self-cleaning without any oil residue due to the excellent superamphiphobic property in air. This is different from the conventional superhydrophobic surface that is completely oil contaminated after oil absorption in water, which makes it difficult for repeated use. This study opens up a new strategy for the treatment of oily wastewater, which has a significant potential for future industrial applications.

3.2.5. Oil/water separation

Oil/water separation technology is very important for a wide range of environmental, agricultural, industrial, and biomedical applications [123-129]. Oil/water separation is a

http://dx.doi.org/10.5772/60826

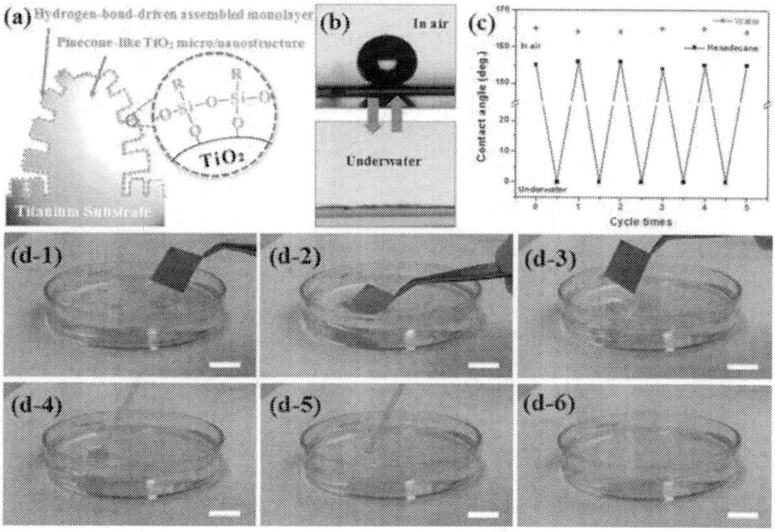

Figure 14. (a) The self-assembled process on the hydroxylated TiO$_2$ particles surface. (b) The oil droplet image on the modified TiO$_2$ surface in air or underwater environment. (c) The reversible wetting change of oil droplet on the TiO$_2$ surface by changing the environment. (d) The oil capture and collection process with a superoleophobic plate. Process 1-3: a superoleophobic plate touches, captures, and collects the sprayed oil drops. Process 4-6: the collected oil droplets are separated from the water. The oil was dyed pink for clear observation of the manipulation process

worldwide challenge due to the increasing industrial oily wastewater and the frequent oil spill accidents. The conventional technique for oil/water separation is dependent on the special porous membranes with both superhydrophobicity and superoleophilicity for the selective separation of oil from the oil/water mixture [123-125]. Ben et al. reported the combination of porous polyurethane and polystyrene microspheres to prepare composite film with highly hydrophobic/oleophilic property for oil/water separation [127]. However, among these approaches and materials used, the fabrication process usually was time consuming, and wettability contrast may decrease under harsh environment. Inspired by the surface geometry and composition of the lotus leaf with its self-cleaning behavior, Huang et al. prepared a TiO$_2$@fabric composite via a facile strategy for preparing marigold flower-like hierarchical TiO$_2$ particles through a one-pot hydrothermal reaction [173]. In comparison with hydrophilic cotton fabric, the fluoroalkylsilane (F17)-modified TiO$_2$@fabric exhibited a high superhydrophobic activity with a contact angle of highly about 160 ° and a sliding angle lower than 10 °. The robust superhydrophobic fabric demonstrated good self-cleaning ability and high efficient oil/water separation due to its extreme wettability contrast of water and oil (Figure 15). Another important water/oil separation process is the preparation of high porous membranes with superoleophilic that can fast absorb oil with high absorption capacity. Stellacci et al. developed a self-assembly process to construct nanowire membranes with high-contrast water/oil wettability [128]. Such superhydrophobic membranes can selectively absorb oils up to 20 times the material's weight by taking advantage of its high porosity and capillary action. Moreover, the paper-like nanowire membranes demonstrated excellent recyclable separation performance.

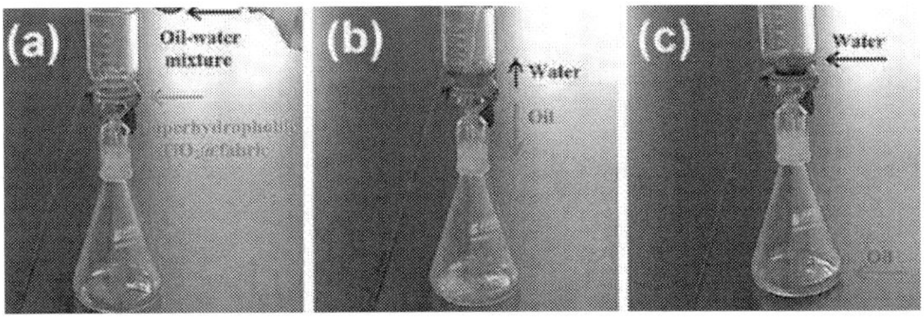

Figure 15. Time sequence of the oil-water separation process with the superhydrophobic F17/TiO_2@fabric

3.2.6. Anti-bioadhesion and biomedical application

Functional TiO_2-based substrates have positive effects in biomedical fields such as biomedical implants, bioactive scaffolds, biosensors, vascular stents, and drug delivery systems [175, 176]. Many factors, such as charge, wettability, composition, and topography, are vital properties for biological applications [176]. Surface with unique wettability is one of the most important items influencing the adsorption of protein and adhesion of platelet, bacteria, and cell [130-139]. Antiwetting surfaces, both smooth and having suitable surface roughness of varying length scales to create super-antiwetting, were incubated in protein solution. The samples were then exposed to flow shear in a device designed to simulate a microfluidic environment. It is reported that a similar amount of protein adsorbed onto smooth and nanometer-scale rough surfaces with static incubation, although a greater amount was found to adsorb onto super-antiwetting surfaces with micrometer-scale roughness, resulted from the increasing binding strength of hydrophobic interactions between bovine serum albumin and the super-antiwetting surface. However, incubated in a flow cell similar to that used in microfluidics, flow shear removed a considerably larger proportion of adsorbed protein from the super-antiwetting surfaces than from the smooth ones, with almost all of the protein being removed from some nanoscaled surfaces. This type of surface may therefore be useful in microfluidics, where protein sticking is a problem. Moreover, they reported that super-antiwetting surfaces with larger roughness dimensions (~4 μm or 800 nm particle size) caused increased adsorption of protein than the copper oxide nano-pillars about 60 nm wide and 10 nm thick.

Platelet attachment and activation on biomedical implant surfaces may lead to blood coagulation and thrombosis. Therefore, regulating the surface wettability to increase the blood compatibility is an effective way for the decrease of platelet adhesion. Recently, Lin et al. [132] studied the effect of surface energy on TiO_2 on antiplatelet activity. The in vitro blood compatibility experiment indicated that the super-antiwetting nanotube TiO_2 layers exhibit a remarkable ability in reducing platelet adhesion (Figure 16c). Compared to the relatively high amount of platelets on superhydrophilic TiO_2 nanotube surface (Figure 16b, 22 ± 1.5 per 5,

000μm²) and flat Ti surface (Figure 16a, 77 ± 7.4 per 5, 000 μm²), only very few platelets (1 ± 0.8 per 5, 000 μm²) were found to adhere on the super-antiwetting TiO_2 nanotube layers. Moreover, even though some platelets were occasionally attached to the super-antiwetting surface, they looked smooth without any growth of pseudopods, implying that the platelets adhered on the super-antiwetting TiO_2 nanotube surface remain inactive and hardly grow and spread out. From the in vitro evaluation, the super-antiwetting TiO_2 nanotube layers exhibited excellent blood compatibility and remarkable performance in preventing platelets from adhering to the implant surface. Tang et al. also observed the similar bacterial adhesion on TiO_2 structures with various surface wettabilities [177]. Compared to ordinary hydrophilic or hydrophobic film, the as-prepared superhydrophobic structured film can greatly decrease the adhesion and activation of *Staphylococcus aureus*. This indicates that the special topography and wettability play critical role in blood compatibility.

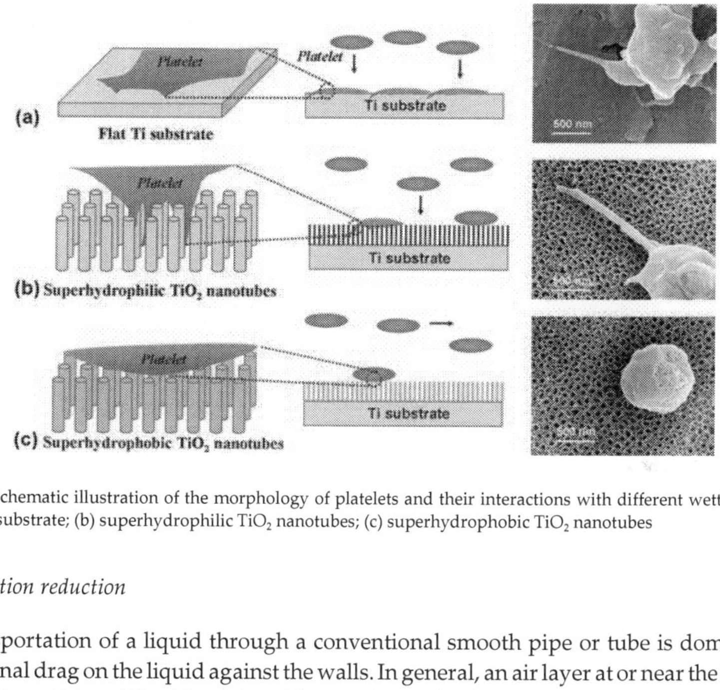

Figure 16. Schematic illustration of the morphology of platelets and their interactions with different wetting surfaces. (a) Plain Ti substrate; (b) superhydrophilic TiO_2 nanotubes; (c) superhydrophobic TiO_2 nanotubes

3.2.7. Friction reduction

The transportation of a liquid through a conventional smooth pipe or tube is dominated by the frictional drag on the liquid against the walls. In general, an air layer at or near the boundary between the solid and liquid, achieved by producing bubbles at the interface, by the vaporization of liquid or by a cushion of air (e.g., below a hovercraft), can reduce the resistance to flow against a solid. However, these methods require a continuous energy consumption. Recently, nonstick super-antiwetting surfaces demonstrating high static contact angles, low contact area, and lateral friction to liquid droplets on natural surfaces have attracted much attention of researchers [156-165]. Watanabe and Zou et al. described a more reduced flow adhesion and friction phenomena when water is passed through a super-antiwetting surface in comparison to lower antiwetting surfaces [155, 156]. The flow resistance reduction is attributed to the reduced molecular attraction and the resultant solid/liquid contact area by the air layer trapped in the rough microstructured/nanostructured super-antiwetting surfaces. To the best of our knowledge, most reports used small sections of lithographically patterned substrates and rarely considered the pressure contrasts or varying flow rates. McHale et al. found that the super-antiwetting nanoribbon coating allowed greater flow at low pressure differences, but the effect on friction reduction disappeared as the pressure increased [158]. Large pressure and friction forces are known to be detrimental in microelectromechanical system applications due to large surface-to-volume ratio and minute spacing [160, 161].

3.3. Super-wetting/antiwetting TiO_2-based surfaces

Titanium dioxide (TiO_2), as another active ingredient for self-cleaning and forming superhydrophobic surfaces, has been widely employed in recent years. By constructing a TiO_2 film and modifying the resultant film with low-surface-energy fluoroalkylsilane molecules, a super-antiwetting TiO_2 film would be easily fabricated with additional photo-patternable properties. Such uniform film was able to be conveniently patterned with ultraviolet light to one-step create super-wetting/antiwetting surface micropatterns. Precise adjustment over the chemistry and geometry of functional materials based on patterned template with highly wetting contrast is of significant importance for a wide range of applications. For example, the wetting pattern could be utilized to site specifically immobilized biomolecules, and metal or semiconductor particles, for multifunctional applications.

3.3.1. Template for micropatterning/printing

It is still a primary challenge for uniform assembling functional inorganic nanomaterials. In this project, nature smartly adopts a perfect approach in biomineralization, where "matrix" macromolecules induce nucleation and growth of inorganic crystals at specific positions with controlled shape, size, morphology, and even orientation. Recently, patterned thin films had received considerable attention due to their interesting properties for widely potential applications [178, 179]. Compared to the traditional patterning techniques of physical/chemical vapor deposition or sputtering, solution-based deposition approaches are becoming popular for the patterning films creation due to the facile synthesis parameters and conditions, less energy, and time requirement [180, 181]. Although the common photolithographic patterning technique is excellent for preparing submicron-sized template in solution, it is a complex multistep process that has to remove great part of the formed film and the supplemental photoresist [182]. Super-wetting/antiwetting pattern with an extreme wettability contrast by photolithography, particularly the rewritable pattern by using high photocatalytic activity of anatase TiO_2 [140-144], is a potentially powerful and economical approach to precisely construct various multifunctional nanostructures in aqueous solution [145-149]. However, such photolithography technique require high-energy light sources of X-ray or laser and ultrahigh vacuum to directly destroy the low-energy materials to some extent or only suitable for semiconductor substrate with inherent photocatalytic activity. For example, Li et al. applied femtosecond laser to one-step construct a three-dimensional (3D) on-demand wettability pattern on a superhydrophobic surface (Figure 17) [175]. We also demonstrated that 3D binary pattern with extremely high contrast would be applied as microfluidic manipulators and biomedical scaffolds to guide droplet transportation and cell site-selective growth, respectively. Moreover, the wetting patterns prepared by UV photocatalytic decomposed surface low-energy groups or monolayer is invisible under ambient condition [94]. Such highly contrast wetting patterns would be developed or exposed under a certain condition (Figure 18), e.g., humid air atmosphere, underwater, fluororesin labeling, and solvent site-selective assembling.

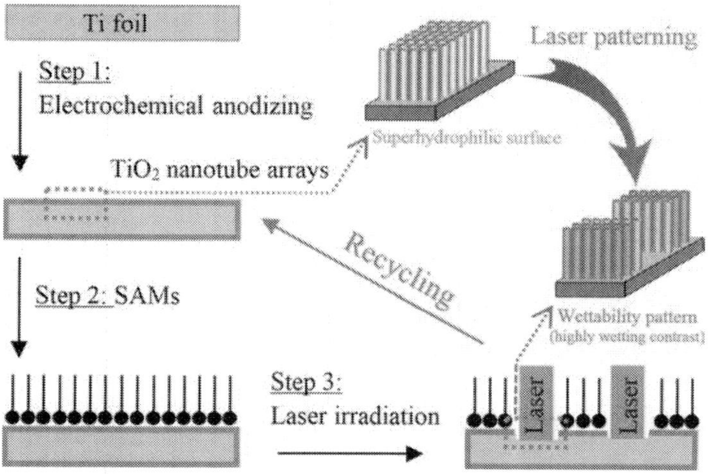

Figure 17. Schematic outline of the procedures to construct superhydrophilic-superhydrophobic micropattern by an ultrafast laser processing technique

Based on ink patterning template with an extreme wettability, adhesion, and conduction contrast (superamphiphobic/amphiphilic), we develop versatile patterning strategies without the requirement of seed/catalyst to construct shape-controlled (positive/negative) ZnO nanomaterial patterns [94]. Specifically, ZnO nanorods are site-selective deposited by a facile liquid-phase deposition process on the ink covering regions to prepare a positive micropattern with clear boundaries (Figure 19a). The obvious suppression phenomenon in superamphiphobic regions is ascribed to the air pocket trapped between the liquid solution and superamphiphobic substrate to highly inhibit the direct solid/liquid interaction. By taking advantage of ink coating with nonconducting character, we realize the rapid construction of an unconventional negative ZnO micropattern on the superamphiphobic regions instead of the superamphilphic regions under the assistance of electric field (Figure 19b). Such wetting micropattern with a high contrast by ink-printing technique is particularly useful and promising in constructing dual-scale or even complicate nanostructures optoelectronic device on a large scale, e.g., photodetectors and photo-sensors.

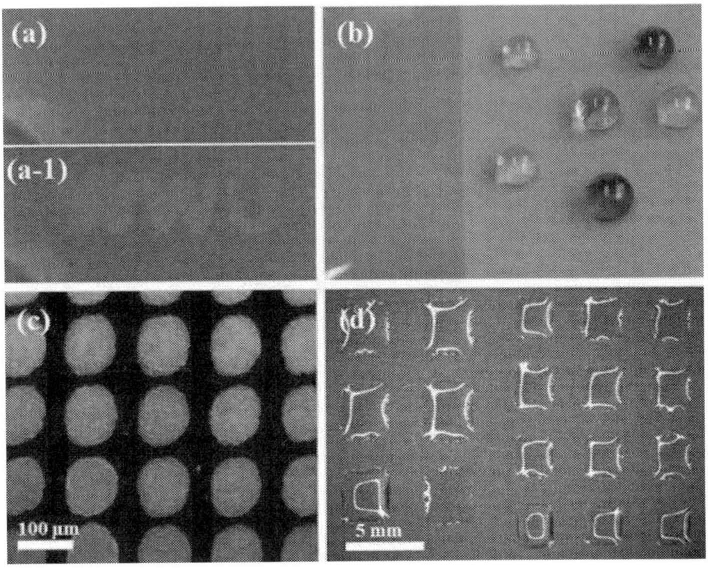

Figure 18. (a) Optical image of an invisible pattern on a superamphiphobic surface constructed by a photo-cleavage strategy and (a-1) the exposed image of "WWU" under humid air atmosphere. Typical wetting patterns exposed with (b) water wetting, (c) fluororesin labeling, and (d) site-selective solvent assembling

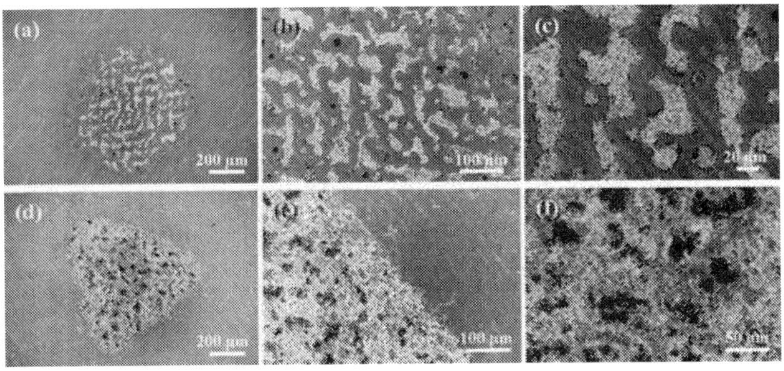

Figure 19. FESEM images of ZnO nanoparticles deposited on ink-patterned superamphiphobic TiO_2 surface. (a-c) Site-selective deposition of ZnO nanoparticles on ink-covered areas by a wet chemical deposition technique to form a positive pattern; (d-f) site-selective deposition on ink-free regions by an electrochemical technique to create a contrary negative pattern

On the basis of photolithography patterning technique on TiO_2 structures, we applied the as-prepared two-dimensional (2D) superhydrophilic-superhydrophobic pattern as scaffold to direct and guide the selective attachment of cell for high-throughput cell behavior evaluation, bioassay, and other related applications (Figure 20) [183]. Following this principle, arrays of

TiO_2 [149], ZnO [150], crystalline CdS materials [151], molecular probes [152], and bioactive CaP layer [153] were nucleated and assembled directly from solution onto Ti substrates at the desired locations precisely. Such patterned structures can then be fabricated into matrix devices for large-scale microelectronic applications or bio-compatible coatings where drugs could be encapsulated in specific regions of the coating [154]. This strategy for micropatterned composites will be helpful to develop various micropatterned functional nanostructured materials. However, a big challenge for widespread practical applications remains to be the development of high-throughput, low-cost, and easily controlled techniques to achieve desired orientation of the nanoscale units.

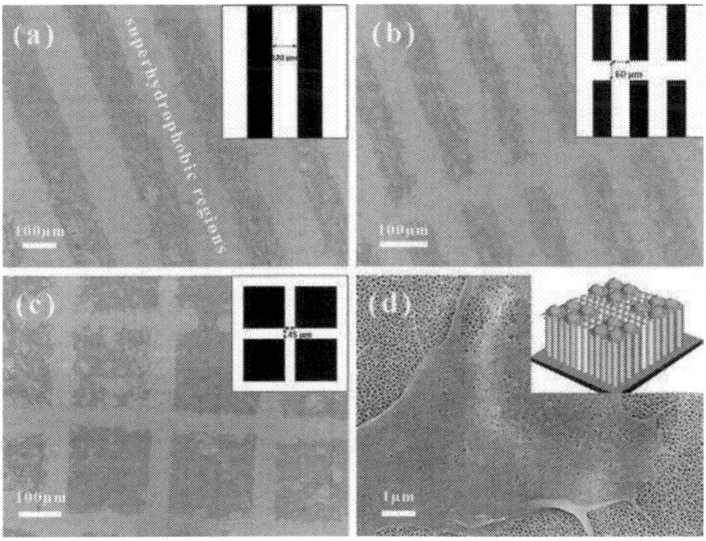

Figure 20. (a-c) SEM images of 3T3 cell adhesion on the super-antiwetting micropatterns with various shapes and sizes. (d) Magnified image of cell on super-wetting region and the schematic diagram of the site-selective cell attachment on scaffold

3.3.2. Molecular sensing device

A biosensor is a device incorporating specific molecular recognition on the basis of the affinity between sensing units and targeted enzyme, antibody or molecular. The novel dual micro-nanoscale patterning surfaces described in previous sections have a wide range of potential applications. As an example, Lai et al. investigated the surface-enhanced Raman scattering (SERS) sensing property of the silver nanoparticles (AgNP) decoration on TiO_2 patterns with a electrodeposition process (Figure 21a, b) [183]. Using rhodamine 6G probing molecule, the patterned AgNP@TiO_2 sample displays an excellent SERS effect with a high SERS contrast between superhydrophilic and superhydrophobic areas (Figure 21c). The AgNPs within the superhydrophilic areas exhibited an obvious and uniform SERS enhancement. This is ascribed to the novel 3D rough structure TiO_2 substrate with a high surface area for high-density loading of AgNPs and absorption of probing molecules. On the contrary, the SERS intensity was negligible at the superhydrophobic areas. This is attributed to the fact that an air pocket at the solid/liquid interface has a strong resistance for AgNPs coating and rhodamine 6G loading in the superhydrophobic areas. Furthermore, based on the high efficient photocatalytic ability of anatase TiO_2 on neat superhydrophilic SERS substrate, the AgNP@TiO_2 can be obtained by rapidly decomposing the absorbed organic molecules into inorganic molecules, such as H_2O and CO_2 (Figure 21d). Such recyclable micropatterned AgNP@TiO_2 devices, by taking advantage of high contrast wettability of TiO_2 patterns, can be a highly sensitive and reproducible SERS sensor candidate for high-throughput molecules detection and biorecognition.

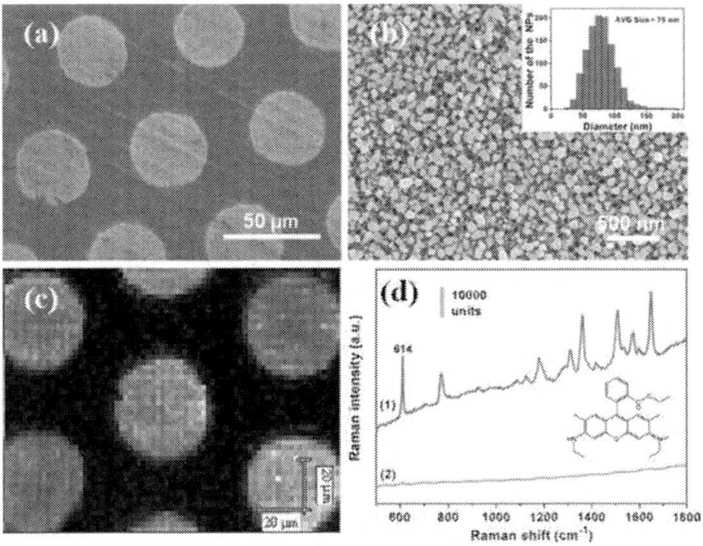

Figure 21. SEM image of AgNP@TiO_2 micropattern (a) and the corresponding magnified view of Ag particles assembled on the superhydrophilic areas (b). SERS mapping of AgNP@TiO_2 pattern with a rhodamine 6G probe using the reference peak of 614 cm^{-1} (c). (d) Raman spectra of the rhodamine 6G molecules loaded on AgNP@TiO_2 with (2) or without UV irradiation (1). The inset of (b) shows the corresponding AgNP size distribution

3.3.3. *Microfluidic manipulation*

As discussed above, most of wetting patterns are invisible and require to develop under certain conditions, e.g., fluororesin/solvent labeling, underwater, or humid environment. Recently, we reported a facile and in situ bottom-up surface modification method to create high-surface-energy ink patterning on superhydrophobic TiO_2 substrates by pen drawing or using commercially available desktop printing technology [70]. Such one-step bottom-up process can realize the simultaneous adjustment of surface topography and chemical component to obtain visible wetting patterns with extremely high adhesion contract. The visible tunability of the superhydrophobic surface wetting and adhesion are suitable to implement four basic operations for the manipulation of liquid drops from storage, transfer, transport, and mixing (Figure 22). In their two-dimensional lab-on-paper prototype, liquid droplets adhere to the porous substrate, rather than absorbing into the substrate; as a result, liquid droplets remain accessible for further quantitative testing and analysis after performing simple qualitative on-chip testing.

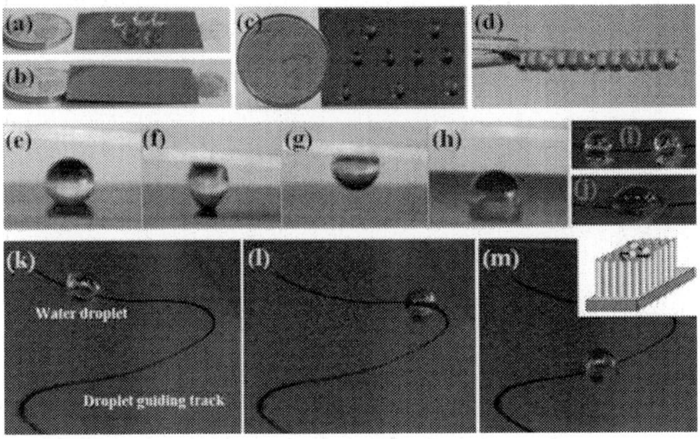

Figure 22. The uncontrollable and unstable liquid droplets on uniform superhydrophobic TNA surface (a, b). Implementation of adhesion contrast on some typical micro-droplet manipulation processes based on the ink-patterned superhydrophobic TNA surfaces: storage (c, d), transfer (e-h), mixing (i, j), and collection (k-m)

4. Summary and outlook

Despite the insightful studies and many promising applications of nanostructures with special wettability and adhesion, great challenges still remain in applying them in practical applications [184]. For example, TiO_2-based surfaces tend to lose their super-wetting ability on a cloudy day or in the dark because the photo-induced superhydrophilic ability of TiO_2 is mainly active with UV light. The photo-induced super-wetting TiO_2 surfaces will recover to a more hydrophobic level within minutes to hours, which restricts their practical applications in our daily life. To construct super-wetting TiO_2-based surfaces without the requirement of UV illumination, various synthesis processes or modification techniques should been developed to prepare visible light active TiO_2-based surfaces in the future. As above stated, a micro-/nano-surface roughness and a low surface energy is prerequisite for super-antiwetting; thus, they generally suffer from weak durability due to the mechanically fragile hierarchical structure, and the rapid degradation of surface functional chemistry resulted from the high photocatalytic activity of anatase TiO_2. Any damage of the micro-/nano-topographical structures or the loss of the low-surface-energy layer may result in a reduction of its super-antiwetting ability. Although extensive studies have been carried out, developing processes to create superhydrophobic surfaces, research on the durability aspect of superhydrophobic surface only started a few years ago. Several concepts to improve the mechanical durability of superhydrophobic surfaces are suggested, for example, by using hierarchical roughness in which robust microscale bumps can provide protection to the more fragile nanoscale roughness, using 3D uniform buck materials and incorporating self-healing materials.

The chapter reviews the recent advance on the natural and bio-inspired super-antiwetting surfaces with different kinds of special adhesion properties, such as low adhesion, high adhesion, anisotropic adhesion, and stimuli-responsive adhesion. Studies on the biological structure surfaces and the biomimetic fabrication of artificial surfaces reveal that the combination of topography structure and chemical component results in these special adhesion states. For example, the tailoring of topography and the scale of microstructures and nanostructures to achieve certain contact models can effectively change the solid/liquid adhesion from low to high levels. On the other hand, the specific arrangement of the microstructures and the chemical component may result in anisotropic solid/liquid adhesion. Furthermore, reversible switching of adhesion between the low-adhesive rolling state and high-adhesive pinning state for water droplets on a superhydrophobic surface could be achieved via cooperation of the stimuli-responsive materials and surface roughness. In addition, potential applications of super-antiwetting surfaces with special adhesion were also discussed, including self-cleaning, water/oil separation, microfluidic manipulation, and micro-templates.

Super-antiwetting surface with responsive liquid adhesion has attracted much attention; great progress has been achieved recently. However, up to now, most adhesion changes reported are achieved ex situ, which require a different droplet to study the adhesion on smart responsive surfaces. Many problems, such as stability and cost of materials with multifunctionalities, need to be addressed before industrial applications can be realized. Therefore, stable, renewable, and smart responsive super-antiwetting surface with solid/liquid adhesion adjustment in a high contrast under single or multiple stimuli responses is a key issue for future investigation. Moreover, the in situ and fast adhesive force measurement on the super-antiwetting substrates is still needed to be improved.

Acknowledgements

The authors thank the Natural Science Foundation of Jiangsu Province of China (BK20130313; BK20140400), and Priority Academic Program Development of Jiangsu Higher Education Institutions (PAPD) for financial support of this work.

Author details

Jian-Ying Huang and Yue-Kun Lai*

*Address all correspondence to: yklai@suda.edu.cn

National Engineering Laboratory for Modern Silk, College of Textile and Clothing Engineering, Soochow University, Suzhou, China

References

[1] Liu KS, Yao X, Jiang L. Chem. Rev. 2015; 114: 10044.

[2] Feng L, Li SH, Li YS, Li HJ, Zhang LJ, Zhai J, Song YL, Liu BQ, Jiang L, Zhu DB. Adv. Mater. 2002; 14: 1857.

[3] Lafuma A, Quéré D. Nat. Mater. 2003; 2: 457.

[4] Sun TL, Feng L, Gao XF, Jiang L. Acc. Chem. Res. 2005; 38: 644.

[5] Feng XJ, Jiang L. Adv. Mater. 2006; 18: 3063.

[6] Roach P, Shirtcliffe NJ. Newton MI. Soft Matter 2008; 4: 224.

[7] Xia F, Jiang L. Adv. Mater. 2009; 20: 2842.

[8] Zhu H, Guo ZG, Liu WM. Chem. Commun. 2014; 50: 3900.

[9] Xin BW, Hao JC, Chem. Soc. Rev. 2010; 39: 769.

[10] Teisala H, Tuominen M, Kuusipalo J. Adv. Mater. Interface 2014; 1: 1300026.

[11] Guo ZG, Liu WM, Su BL. J. Colloid Interface Sci. 2011; 353: 335.

[12] del Campo A, Arzt E. Chem. Rev. 2008; 108: 911.

[13] Darmanin T, de Givenchy ET, Amigoni S, Guittard F. Adv. Mater. 2013; 25: 1378.

[14] Wang XF, Ding B, Yu JY, Wang MR. Nano Today 2011; 6: 510.

[15] Liu MJ, Zheng YM, Zhai J, Jiang L. Acc. Chem. Res. 2010; 43: 368.

[16] Bhushan B, Jung YC. Prog. Mater. Sci. 2011; 56: 1.

[17] Wang S, Jiang L. Adv. Mater. 2007; 19: 3423.

[18] Wenzel RN. Ind. Eng. Chem. 1936; 28: 988.

[19] Cassie ABD, Baxter S. Trans. Faraday Soc. 1944; 40: 546.

[20] Wang R, Hashimoto K, Fujishima A, Chikuni M, Kojima E, Kitamura A, Shimohigoshi M, Watanabe T. Nature 1997; 388: 431.

[21] Fujishima A, Zhang XT, Tryk DA. Surf. Sci. Rep. 2008; 63: 515.

[22] Zhang XT, Fujishima A, Jin M, Emeline AV, Murakami T. J. Phys. Chem. B 2006; 110: 25142.

[23] Li Q, Shang JK. Environ. Sci. Technol. 2010; 44: 3493.

[24] Mugele F, Baret JC. J. Phys. Condens. Matter. 2005; 17: R705.

[25] Ma CF, Nagai A, Yamazaki Y, Toyama T, Tsutsumi Y, Hanawa T, Wang W, Yamashita K. Acta Biomater. 2012; 8: 860.

[26] Meng FM, Xiao L, Sun ZQ. J. Alloys Compd. 2009; 485: 848.

[27] Lai YK, Chen YC, Tang YX, Gong GG, Chen Z, Lin CJ. Electrochem. Commun. 2009; 11: 2268.

[28] Song S, Jing LQ, Li SD, Fu HG, Luan YB. Mater. Lett. 2008; 62: 3503.

[29] Funakoshi K, Nonami T. J. Am. Ceram. Soc. 2006; 89: 2782.

[30] Xi BJ, Verma LK, Li J, Bhatia CS, Danner AJ, Yang H, Zeng HC. ACS Appl. Mater. Interfaces 2012; 4: 1093.

[31] Liu KS, Zhang ML, Zhai J, Wang J, Jiang L. Appl. Phys. Lett. 2008; 92: 183103.

[32] Gao XF, Jiang L. Nature 2004; 432: 36.

[33] Guo ZG, Zhou F, Hao JC, Liu WM. J. Am. Chem. Soc. 2005; 127: 15670.

[34] Xi JM, Feng L, Jiang L. Appl. Phys. Lett. 2008; 92: 053102.

[35] Nosonovsky M, Bhushan B. Curr. Opin. Colloid Interface Sci. 2009; 14: 270.

[36] Tuteja A, Choi W, Ma ML, Mabry JM, Mazzella SA, Rutledge GC, McKinley GH, Cohen RE. Science 2007; 318: 1618.

[37] Wang DA, Wang XL, Liu XJ, Zhou F. J. Phys. Chem. C 2010; 114: 9938.

[38] Bhushan B, Jung YC, Niemietz A, Koch K. Langmuir 2009; 25: 1659.

[39] Xie QD, Xu J, Feng L, Jiang L, Tang WH, Luo XD, Han CC. Adv. Mater. 2004; 16: 302.

[40] Meng HF, Wang ST, Xi JM, Tang ZY, Jiang L. J. Phys. Chem. C 2008; 112: 11454.

[41] Koch K. Bhushan B. Jung YC, Barthlott W. Soft Matter 2009; 5: 1386.

[42] Wu WC, Wang XL, Wang DA, Chen M, Zhou F, Liu WM, Xue QJ. Chem. Commun. 2009; 9: 1043.

[43] Yao X, Chen QW, Xu L, Li QK, Song YL, Gao XF, Quéré D, Jiang L. Adv. Funct. Mater. 2010; 20: 656.

[44] Lai YK, Tang YX, Huang JY, Wang H, Li HQ, Gong DG, Ji XB, Gong JJ, Lin CJ, Sun L, Chen Z. Soft Matter 2011; 7: 6313.

[45] Autumn K, Liang YA, Hsieh ST, Zesch W, Chan WP, Kenny TW, Fearing R, Full RJ. Nature 2000; 405: 681.

[46] Bhushan B, Her EK. Langmuir 2010; 26: 8207.

[47] Feng L, Zhang YA., Xi JM, Zhu Y, Wang N, Xia F, Jiang L. Langmuir 2008; 24: 4114.

[48] Jin MH, Feng XJ, Xi JM, Zhai J, Cho KW, Feng L, Jiang L. Macromol. Rapid Commun. 2005; 26: 1805.

[49] Guo ZG, Liu WM. Appl. Phys. Lett. 2007; 90: 223111.

[50] Boscher ND, Carmalt CJ, Parkin IP. J. Mater. Chem. 2006; 16: 122.

[51] Zhao WJ, Wang LP, Xue QJ. J. Phys. Chem. C 2010; 114: 11509.

[52] Park SG, Moon HH, Lee SK, Shim J, Yang SM. Langmuir 2010; 26: 1468.

[53] Zhu SJ, Li YF, Zhang JH, Lu CL, Dai X, Jia F, Gao HN, Yang BJ. Colloid Interface Sci. 2010; 344: 541.

[54] Lai YK, Gao XF, Zhuang HF, Huang JY, Lin CJ, Jiang L. Adv. Mater. 2009; 21: 3799.

[55] Farsinezhad S, Waghmare PR, Wiltshire BD, Sharma H, Amiri S, Mitra SK, Shankar K. RSC Adv. 2014; 4: 33587.

[56] Cheng ZJ, Gao J, Jiang L. Langmuir 2010; 26: 8233.

[57] Gao LC, McCarthy TJ, Zhang X. Langmuir 2009; 25: 14100.

[58] Balu B, Kim JS, Breedveld V, Hess DW. J. Adhesion Sci. Technol. 2009; 23: 361.

[59] Ting WH, Chen CC, Dai SA, Suen SY, Yang IK, Liu L, Chen FMC, Jeng RJ. J. Mater. Chem. 2009; 19: 4819.

[60] Park BG, Lee W, Kim JS, Lee KB. Colloids Surf. A 2010; 370: 15.

[61] Lee W, Park BG, Kim DH, Ahn DJ, Park Y, Lee SH, Lee KB. Langmuir 2010; 26: 1412.

[62] Lai YK, Lin CJ, Huang JY, Zhuang HF, Sun L, Nguyen T. Langmuir 2008; 24: 3867.

[63] Di Mundo R, Palumbo F, d'Agostino R. Langmuir 2010; 26: 5196.

[64] Sun W, Shen LY, Wang LM, Fu K, Ji J. Langmuir 2010; 26: 14236.

[65] Huang JY, Lai YK, Wang LN, Li SH, Ge MZ, Zhang KQ, Fuchs H, Chi LF. J. Mater. Chem. A 2014; 2: 18531.

[66] Yang J, Zhang ZZ, Men XH, Xu XH, Zhu XT. J. Colloid Interface Sci. 2010; 346: 241.

[67] Wang DA, Liu Y, Liu XJ, Zhou F, Liu WM, Xue QJ. Chem. Commun. 2009; 45: 7018.

[68] Liu XJ, Ye Q, Yu B, Liang YM, Liu WM, Zhou F. Langmuir 2010; 26: 12377.

[69] Liu XJ, Wu WC, Wang XL, Luo ZZ, Liang YM, Zhou F. Soft Matter 2009; 5: 3097.

[70] Lai YK, Pan F, Xu C, Fuchs H, Chi LF. Adv. Mater. 2013; 25: 1682.

[71] Zhang LB, Wu JB, Hedhili MN, Yang XL, Wang P. J. Mater. Chem. A 2015; 3: 2844.

[72] Nishimoto S, Becchaku M, Kameshima Y, Shirosaki Y, Hayakawa S, Osaka A, Miyake M. Thin Solid Films 2014; 558: 221.

[73] Xu QF, Liu Y, Lin FJ, Mondal B, Lyons AM. ACS Appl. Mater. Interface 2013; 5: 8915.

[74] Uchida K, Nishikawa N, Izumi N, Yamazoe S, Mayama H, Kojima Y, Yokojima S, Nakamura S, Tsujii K, Irie M. Angew. Chem. Int. Ed. 2010; 49: 5942.

[75] Xu L, Ye Q, Lu X. Lu Q. ACS Appl. Mater. Interfaces 2014; 6: 14736.

[76] Zhao XD, Fan M, Luo J, Ding J, Liu XY, Zou BS, Feng YP. Adv. Funct. Mater. 2011; 21: 184.

[77] Liu MJ, Nie FQ, Wei ZX, Song YL, Jiang L. Langmuir 2010; 26: 3993.

[78] Verplanck N, Galopin E, Camart JC, Thomy V, Coffinier Y, Boukherroub R. Nano Lett. 2007; 7: 813.

[79] Lapierre F, Thomy V, Coffinier Y, Blossey R, Boukherroub R. Langmuir 2009; 25: 6551.

[80] Li C, Guo RW, Jiang X, Hu SX, Li L, Cao XY, Yang H, Song YL, Ma YM, Jiang L. Adv. Mater. 2009; 21: 4254.

[81] Chen L, Liu MJ, Lin L, Zhang T, Ma J, Song YL, Jiang L. Soft Matter 2010; 6: 2708.

[82] Hong X, Gao X, Jiang L. J. Am. Chem. Soc. 2007; 129: 1478.

[83] Cheng ZJ, Feng L, Jiang L. Adv. Funct. Mater. 2008; 18: 3219.

[84] Wu D, Wu SZ, Chen QD, Zhang YL, Yao J, Yao X, Niu LG, Wang JN, Jiang L, Sun HB. Adv. Mater. 2011; 23: 545.

[85] Zhang JL, Lu XY, Huang WH, Han YC. Macromol. Rapid Commun. 2005; 26: 477.

[86] Zhang QL, Xia F, Sun TL, Song WL, Zhao TY, Liu MC, Jiang L. Chem. Commun. 2008; 10: 1199.

[87] Tian D, Chen Q, Nie FQ, Xu JJ, Song YL, Jiang L. Adv. Mater. 2009; 21: 3744.

[88] Xia F, Ge H, Hou Y, Sun TL, Chen L, Zhang GZ, Jiang L. Adv. Mater. 2007; 19: 2520.

[89] Xu LY, Lu XM, Li M, Lu QH. Adv. Mater. Interfaces 2014; 1: 1400011.

[90] Wang DA, Liu Y, Liu XJ, Zhou F, Liu WM, Xue QJ. Chem. Commun. 2009; 45: 7018.

[91] Zheng YM, Gao XF, Jiang L. Soft Matter 2007; 3: 178.

[92] Chen Y, He B, Lee JH, Patankar NA. J. Colloid Interface Sci. 2005; 281: 458.

[93] Gao XF, Yao X, Jiang L. Langmuir 2007; 23: 4886.

[94] Huang JY, Lai YK, Pan F, Yang L, Wang H, Zhang KQ, Fuchs H, Chi LF. Small 2014; 10: 4865.

[95] Furstner R, Barthlott W, Neinhuis C, Walzel P. Langmuir 2005; 21: 956.

[96] Nakajima A, Hashimoto K, Watanabe T, Takai K, Yamauchi G, Fujishima A. Langmuir 2000; 16: 7044.

[97] Bhushan B, Jung YC, Koch K. Langmuir 2009; 25: 3240.

[98] Qi DP, Lu N, Xu HB, Yang BJ, Huang CY, Xu MJ, Gao LG, Wang ZX, Chi LF. Langmuir 2009; 25: 7769.

[99] Nystrom D, Lindqvist J, Ostmark E, Antoni P, Carlmark A, Hult A, Malmstrom E. ACS Appl. Mater. Interface 2009; 1: 816.

[100] Xiu YH, Hess DW, Wong CP, J. Adhes. Sci. Technol. 2008; 22: 1907.

[101] Mishchenko L, Hatton B, Bahadur V, Taylor JA, Krupenkin T, Aizenberg J. ACS Nano 2010; 4: 7699.

[102] Kulinich SA, Farhadi S, Nose K, Du XW. Langmuir 2011; 27: 25.

[103] Cao LL, Jones AK, Sikka VK, Wu JZ, Gao D. Langmuir 2009; 25: 12444.

[104] Z. L. Liu, Y. J. Gou, Wang JT, Cheng SY. Int. J. Heat Mass. Tran. 2008; 51: 5975.

[105] Gao XF, Yan X, Yao X, Xu L, Zhang K, Zhang JH, Yang B, Jiang L. Adv. Mater. 2007; 19: 2213.

[106] Zhang JH, Yang B. Adv. Funct. Mater. 2010; 20: 3411.

[107] He M, Wang JX, Li HL, Jin XL, Wang JJ, Liu BQ, Song YL. Soft Matter 2010; 6: 2396.

[108] He M, Wang JX, Li HL, Song YL. Soft Matter 2011; 7: 3993.

[109] Jung S, Dorrestijn M, Raps D, Das A, Megaridis CM, Poulikakos D. Langmuir 2011; 27: 3059.

[110] Meuler AJ, Smith JD, Varanasi KK, Mabry JM, McKinley GH, Cohen RE. ACS Appl. Mater. Interface 2010; 2: 3100.

[111] Balu B, Berry AD, Hess DW, Breedveld V. Lab Chip 2009; 9: 3066.

[112] Shi LT, Jiang CG, Ma MJ, Wu CW. Biomicrofluidics 2010; 4: 041101.

[113] Chunder A, Etcheverry K, Longde G, Cho HJ, Zhai L. Colloids Surf. A 2009; 333: 187.

[114] Bormashenko E, Pogreb R, Bormashenko Y, Musin A, Stein T. Langmuir 2008; 24: 12119.

[115] ZhaoY, Fang J, Wang HX, Wang XG, Lin T. Adv. Mater. 2010; 22: 707.

[116] Zheng YM, Bai H, Huang ZB, Tian XL, Nie FQ, Zhao Y, Zhai J, Jiang L. Nature 2010; 463: 640.

[117] Bai H, Tian XL, Zheng YM, Ju J, Zhao Y, Jiang L. Adv. Mater. 2010; 22: 5521.

[118] Shirtcliffe NJ, McHale G, Newton MI. Langmuir 2009; 25: 14121.

[119] Parker AR, Lawrence CR. Nature 2001; 414: 33.

[120] Zhai L, Berg MC, Cebeci FC, Kim Y, Milwid JM, Rubner MF, Cohen RE. Nano Lett. 2006; 6: 1213.

[121] Dorrer C, Ruhe J. Langmuir 2008; 24: 6154.

[122] Garrod RP, Harris LG, Schofield WCE, McGettrick J, Ward LJ, Teare DOH, Badyal JPS. Langmuir 2007; 23: 689.

[123] Lee CH, Johnson N, Drelich J, Yap YK. Carbon 2011; 49: 669.

[124] Li HL, Wang JX, Yang LM, Song YL. Adv. Funct. Mater. 2008; 18: 3258.

[125] Zhu Q, Tao F, Pan QM. ACS Appl. Mater. Interfaces 2010; 1: 3141.

[126] Feng L, Zhang ZY, Mai ZH, Ma YM, Liu BQ, Jiang L, Zhu DB. Angew. Chem. Int. Ed. 2004; 43: 2012.

[127] Ben W, Liang WX, Guo ZG, Liu WM. Chem. Soc. Rev. 2015; 44: 336.

[128] Yuan JK, Liu XG, Akbulut O, Hu JQ, Suib SL, Kong J, Stellacci F. Nat. Nanotechnol. 2008; 3: 332.

[129] Gui CX, Wei JW, Wang KL, Cao AY, Zhu HW, Jia Y, Shu Q, Wu DH. Adv. Mater. 2010; 22: 617.

[130] Koc K, de Mello AJ, McHale G, Newton MI, Roach P, Shirtcliffe NJ. Lab Chip 2008; 8: 582.

[131] Zhang XX, Wang L, Levanen E. RSC Adv. 2013; 3: 12003.

[132] Yang Y, Lai YK, Zhang QQ, Wu K, Zhang LH, Lin CJ, Tang PF. Colloids Surf. B 2010; 79: 309.

[133] Huang YX, Lai YK, Lin LX, Sun L, Lin CJ. Acta Phys.-Chim. Sin. 2010; 26: 2057.

[134] Ishizaki T, Saito N, Takai O. Langmuir 2010; 26: 8147.

[135] Chen L, Liu MJ, Bai H, Chen PP, Xia F, Han D, Jiang L. J. Am. Chem. Soc. 2009; 131: 10467.

[136] Genzer J, Efimenko K. Biofouling 2006; 22: 339.

[137] Zhou M, Yang JH, Ye X, Zheng AR, Li G, Yang, Zhu Y, Cai L. J. Nano Res. 2008; 2: 129.

[138] Mao C, Liang CX, Luo WP, Bao JC, Shen J, Hou XM, Zhao WB. J. Mater. Chem. 2009; 19: 9025.

[139] Hou XM, Wng XB, Zhu QS, Bao JC, Mao C, Jiang LC, Shen JA. Colloids Surf. B 2010; 80: 247.

[140] Lai YK, Sun L, Chen YC, Zhuang HF, Lin CJ, Chin JW. J. Electrochem. Soc. 2006; 153: D123.

[141] Zhuang HF, Lin CJ, Lai YK, Sun L, Li J. Environ. Sci. Technol. 2007; 41: 4735.

[142] Lai YK, Huang JY, Zhang HF, Subramaniam VP, Tang YX, Gong DG, Sundar L, Sun L, Chen Z, Lin CJ. J. Hazard. Mater. 2010; 184: 855.

[143] Soliveri G, Annunziata R, Ardizzone S, Cappelletti G, Meroni D. J. Phys. Chem. C 2012; 11: 26405.

[144] Pittrof A, Bauer S, Schmuki P. Acta Biomater. 2011; 7: 424.

[145] Lai YK, Lin CJ, Wang H, Huang JY, Zhuang HF, Sun L. Electrochem. Commun. 2008; 10: 387.

[146] Nakata K, Nishimoto S, Yuda Y, Ochiai T, Murakami T, Fujishima A. Langmuir 2010; 26: 11628.

[147] Nishimoto S, Sekine H, Zhang XT, Liu ZY, Nakata K, Murakami T, Koide Y, Fujishima A. Langmuir 2009; 25: 7226.

[148] Kinoshita H, Ogasahara A, Fukuda Y, Ohmae N. Carbon 2010; 48: 4403.

[149] Lai YK, Huang JY, Gong JJ, Huang YX, Wang CL, Chen Z, Lin CJ. J. Electrochem. Soc. 2009; 156: D480.

[150] Lai YK, Lin ZQ, Huang JY, Sun L, Chen Z, Lin CJ. New J. Chem. 2010; 34: 44.

[151] Lai YK, Lin ZQ, Chen Z, Huang JY, Lin CJ. Mater. Lett. 2010; 64: 1309.

[152] Huang YX, Sun L, Xie KP, Lai YK, Biu BJ, Ren B, Lin CJ. J. Raman Spectrosc. 2011; 42: 986.

[153] Lai YK, Huang YX, Wang H, Huang JY, Chen Z, Lin CJ. Colloids Surf. B 2010; 76: 117.

[154] Bekou S, Mattia D. Curr. Opin. Colloid Interface Sci. 2011; 16: 259.

[155] Watanabe, Yanuar, Udagawa H. J. Fluid Mech. 1999; 381: 225.

[156] Song Y, Nair RP, Zou M, Wang YA. Thin Solid Films 2010; 518: 3801.

[157] Cottin-Bizonne C, Barrat JL, Bocquet L, Charlaix E. Nat. Mater. 2003; 2: 237.

[158] Shirtcliffe NJ, MaHale G, Newton MI, Zhang Y. ACS Appl. Mater. Interface 2009; 1: 1316.

[159] Maboudian R, Howe RT. J. Vac. Sci. Technol. B 1997; 15: 1.

[160] Li L, Liu ZY, Zhang QQ, Meng CH, Zhang TR, Zhai J. J. Mater. Chem. A 2015; 3: 1279.

[161] Lai YK, Tang YX, Gong JJ, Gong DG, Chi LF, Lin CJ, Chen Z. J. Mater. Chem. 2012; 22: 7420.

[162] Wang F, Zhang G, Zhao Z, Tan H, Yu W, Zhang X, Sun Z. RSC Adv. 2015; 5: 9861.

[163] Kartini I, Santosa SJ, Febriyanti E, Nugroho OR, Yu H, Wang LZ. J. Nanoparticle Res. 2014; 16: 2514.

[164] Ding XF, Zhou SX, Gu GX, Wu LM. J. Mater. Chem. 2011; 21: 6161.

[165] Kamegawa T, Shimizu Y, Yamashita H. Adv. Mater. 2012; 24: 3697.

[166] Gao XF, Yan X, Yao X, Xu L, Zhang K, Zhang JH, Yang B, Jiang L. Adv. Mater. 2007; 19: 2213.

[167] Shen YZ, Tao HJ, Chen SL, Zhu LM, Wang T, Tao J. RSC Adv. 2015; 5: 1666.

[168] Hu YW, Huang SY, Liu S, Pan W. Appl. Surf. Sci. 2012; 258: 7460.

[169] Mahalakshmi PV, Vanithakumari SC, Gopal J, Mudali UK, Raj B. Curr Sci. 2011; 101: 1328.

[170] Shen GX, Chen YC, Lin L, Lin CJ, Scantlebury D. Electrochim. Acta 2015; 50: 5083.

[171] Shen GX, Chen YC, Lin CJ. Thin Solid Films 2005; 489: 130.

[172] Lai YK, Tang YX, Huang JY, Pan F, Chen Z, Zhang KQ, Fuchs H, Chi LF. Sci. Rep. 2013; 3: 3009.

[173] Huang JY, Li SH, Ge MZ, Wang LN, Xing TL, Chen GQ, Liu XF, Al-Deyab SS, Zhang KQ, Chen T, Lai YK. J. Mater. Chem. A 2015; 3: 2825.

[174] Wu SL, Weng ZY, Liu XM, Yeung KWK, Chu PK. Adv. Funct. Mater. 2014; 24: 5464.

[175] Li HQ, Lai YK, Huang JY, Tang YX, Yang L, Chen Z, Zhang KQ, Wang XC, Tan LP. J. Mater. Chem. B 2015; 3: 342.

[176] Huo KF, Gao B, Fu JJ, Zhao LZ, Chu PK. RSC Adv. 2014; 4: 17300.

[177] Tang PF, Zhang W, Wang Y, Zhang BX, Wang H, Lin CJ, Zhang LH. J. Nanomater. 2011; 2011: 178921.

[178] Amos FF, Morin SA, Streifer JA, Hamers RJ, Jin S. J. Am. Chem. Soc. 2007; 129: 14296.

[179] Segalman RA. Mater. Sci. Eng. R 2005; 48: 191.

[180] Liu SH, Wang WCM, Mannsfeld SCB, Locklin J, Erk P, Gomez M, Richter F, Bao ZN. Langmuir 2007; 23: 7428.

[181] Yoshimura M, Gallage R. J. Solid State Electrochem. 2008; 12: 775.

[182] Bessonov A, Kim JG, Seo JW, Lee JW, Lee S. Macromol. Chem. Phys. 2010; 211: 2636.

[183] Lai YK, Lin LX, Pan F, Huang JY, Song R, Huang YX, Lin CJ, Fuchs H, Chi LF. Small 2013; 9: 2945.

[184] Zhou, SX, Ding, XF, Wu, LM. Prog. Org. Coat. 2013; 76: 563.

Chapter 4

Increased Wettability and Surface Free Energy of Polyurethane by Ultraviolet Ozone Treatment

Ping Kuang and Kristen Constant

Additional information is available at the end of the chapter

http://dx.doi.org/10.5772/60798

Abstract

The wettability of polyurethane (PU) was altered using ultraviolet ozone (UVO) treatment. The effect of UVO treatment on PU surface chemistry was investigated with various experiments. The direct measurement of sessile drops was employed to quantify the static contact angle of different wetting liquids on homogeneous PU films with various UV ozone treatment times. The contact angle of DI water droplets was decreased to 17.2º from 70.04º after 5min UVO treatment. The surface free energy of PU films was 51.46mN/m prior to treatment and was increased to 71.5mN/m after being fully treated. X-ray Photoelectron Spectroscopy (XP) analysis shows a significant amount of polar functional species (C-O and C=O bonding) were formed on the PU surface by UVO treatment. Atomic Force Microscopy (AFM) characterization shows the PU surface morphology was different before and after UVO treatment. The effect of water washing on UVO treated surface was also investigated. An aging effect study indicates the UV ozone modification can sustain the improved wettability with limited hydrophobic recovery, where the DI water contact angle remains constant at around 22º after the UVO treatment.

Keywords: UV ozone treatment, Surface energy, Contact angles, Surface wettability

1. Introduction

Polyurethane (PU) is a versatile polymeric material which, due to its wide range of molecular weights, can exist in different solid forms. In industry, PU is commonly used as rigid and flexible foams, adhesives, coatings, and molds [1-2]. PU coatings and molds are formed by the reaction of organic isocyanates, high molecular weight polyols, and low molecular weight chain extenders, and they are usually elastomers [3]. Because elastomeric PU is frequently used in contact with various materials, optimizing the PU surface properties is critical for enhancing their performance. Furthermore, complete understanding and characterization of PU surface properties, such as its wettability, is helpful for various practical applications. Moreover, a wide range of novel applications can be realized if the PU surface properties can be modified and tailored [4].

One novel application of PU is for fabricating patterned microstructures using microtransfer molding, which is one of various soft lithography techniques [2]. It was shown that, when using PU microstructures as molds in soft lithography, modification of the surface is important to achieve improved wetting for the infiltration of slurries or sol-gel materials. The purpose of this work was to utilize a simple and economical way to improve the surface wettability of PU molds for the infiltration of aqueous solutions including slurries of oxide nanoparticles.

Surface treatments are frequently used to modify the surface chemistry and improve wetting characteristics of polymers prior to use in a range of technological applications. Such treatments are necessary because the intrinsic activities of the polymer surfaces are frequently too low to allow satisfactory adhesion of surface coatings or laminates. Oxidation processes in the gaseous phase introduce a range of functionalities at the polymer surface therefore increasing the polar component of free energy. Gas phase treatments require the impingement of reactive species, such as oxygen radicals, on the polymer surface. In order to alter PU membranes or coatings to obtain a PU surface with better wettability, different surface modification techniques such as multicomponent poly-addition reactions, oxygen plasma, and ultraviolet (UV) irradiation have been used [3-5].

Another simple yet effective method for polyurethane surface modification is UV ozone (UVO) treatment. UVO treatment has been used to modify the surface wettability of various polymers [6-13]. A summary of selected previous work appears in Table 1. However, no experimental work had been done on polyurethane before our work.

Here, we show that UVO can have a much more significant impact on polyurethane surface. In UVO treatment for polymer surface modification, there are two different wavelengths of UV radiation present, 184.9 nm (λ_1) and 253.7 nm (λ_2) [14]. When the molecules on the surface of, as an example, a thin PU film are exposed to UV light, the shorter-wavelength UV radiation (λ_1) will induce excitation and dissociation of the polymeric molecules. This is known as a photo-sensitized oxidation process [11]. Wettability of the film surface may be changed if the excited and dissociated PU molecules acquire different surface energies during this process. In addition, atomic oxygen is simultaneously generated when the oxygen molecules in air are dissociated by the 184.9 nm UV light and the ozone molecules by the 253.7 nm UV light. Upon dissociation of the oxygen molecules in air by the 184.9 nm UV radiation, the atomic oxygen will readily react with molecular oxygen to form ozone. Most hydrocarbons in PU and the ozone molecules can absorb the 253.7 nm UV radiation. Photolysis of ozone occurs and more highly reactive oxygen atoms are produced. Therefore, in UVO treatment when both wave-

http://dx.doi.org/10.5772/60798

Year	Paper	Materials
1996	"Improved adhesion to polymers by UV/ozone surface oxidation", Mathieson and Bradley, Int. J. Adhes. Adhes. [6]	Polyethylene (PE), polyetheretherketone (PEEK)
1998	"Surface modification of polyolefine by UV light/ozone treatment", Hu et al., J. Appl. Polym. Sci. [7]	Polyethylene (PE), polypropylene (PP)
1999	"Surface characterisation of ultraviolet-ozone treated PET using atomic force microscopy and X-ray photoelectron spectroscopy", Ton-That, C. et al., Surf. Sci. [8]	Polyethylene terephthalate (PET)
1999	"Atomic force microscopy study of polypropylene surfaces treated by UV and ozone exposure: modification of morphology and adhesion force", Nie et al., Appl. Surf. Sci. [9]	Polypropylene (PP)
2000	"Surface characterization and ageing of ultraviolet–ozone-treated polymers using atomic force microscopy and x-ray photoelectron spectroscopy", Bradley et al., Surf. Interface Anal. [10]	Poly(ethylene terephthalate) (PET), polystyrene (PS)
2002	"Surface Modification of Sylgard-184 Poly(dimethyl siloxane) Networks by Ultraviolet and Ultraviolet Ozone Treatment", Genzer et al., J. Coll. & Int. Sci. [11]	Poly(dimethyl siloxane) (PDMS)
2005	"Addition of ozone in the UV radiation treatment of a synthetic styrene-butadiene-styrene (SBS) rubber", Martin-Martinez et al., J. Adh. & Adh. [12]	Styrene-butadiene-styrene (SBS) rubber
2008	"Surface modification of poly (lactic acid) by UV/Ozone irradiation", Koo and Jang, Fiber Polym. [13]	Poly(lactic acid) (PLA)

Table 1. List of previous work of UVO treatment on various polymers

lengths are present, oxygen atoms are continuously created and ozone is continuously created and destroyed. Most importantly, the highly reactive gaseous species, the atomic oxygen and ozone molecules, are oxidizing agents that may react with polymer surfaces to form peroxy and hydroxyl radicals, hydro-peroxide, carbonyl, and carboxyl functionalities, which are responsible for the increased wettability of treated polymer surfaces [5, 12]. The number of these radical functionality groups produced during the treatment is dependent on the ozone concentration and exposure time, the presence of water vapor, and the distance between the PU surface and the UV radiation source, among other factors [12, 14]. While the surface may be greatly modified, the bulk of the PU films remain unchanged. The purpose of this study is to characterize the change of wettability of PU films under different UVO treatment times. It has been well established that contact angle goniometry can be used to examine the wettability and measure the change in surface energy of a polymeric surface [15]. Static contact angle measurement was used to investigate the changes of the wettability in this study. The surface chemistry and morphology were studied using X-ray photoelectron spectroscopy (XPS) and atomic force microscopy (AFM). The results of these investigations show that the UVO treatment significantly changes the surface properties of the PU films.

1.1. Wettability study of PU films by contact angle measurements

For static sessile droplet contact angle measurement, PU films were placed on the sample stage of the contact angle measuring system, shown in Figure 1. Distilled water and diiodomethane (CH_2I_2) (99%, Sigma-Aldrich) liquid were used for the contact angle measurements.

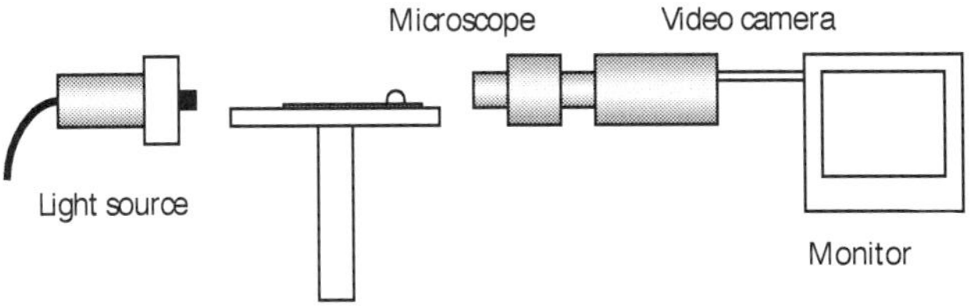

Figure 1. Contact angle measuring system

Contact angles of the two liquids on PU films with no UVO treatment and with maximum UVO treatment at 600 seconds are shown in Figure 2. It is clear that after UVO treatment, the water contact angle is dramatically reduced from ~70° to ~18°. For the non-polar CH_2I_2, with UVO treatment, the contact angle is moderately increased from ~16° to ~39°.

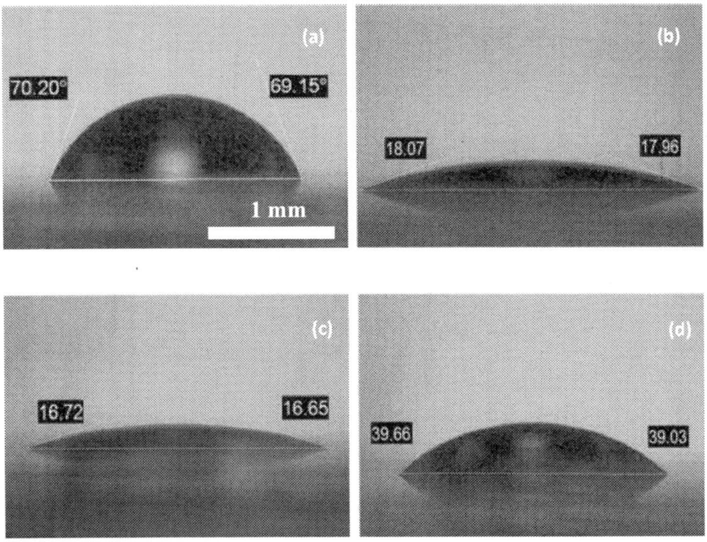

Figure 2. Contact angle images of (a) DI water droplet on non-treated PU film, (b) DI water droplet on 600sec UV-ozone treated PU film, (c) CH_2I_2 droplet on non-treated PU film, and (d) CH_2I_2 droplet on 600sec UV-ozone treated PU film

http://dx.doi.org/10.5772/60798

UVO time	0sec	10sec	20sec	40sec	60sec	80sec	100sec	300sec	600sec
Liquid									
Distilled water	*71.04*	*49.21*	*48.61*	*43.40*	*33.76*	*25.23*	*16.84*	*17.17*	*17.64*
Diiodomethane	*17.20*	*46.80*	*48.95*	*48.34*	*46.64*	*43.30*	*41.61*	*40.06*	*39.95*

Table 2. Average values of contact angles (degrees) on PU films with different UVO treatment times

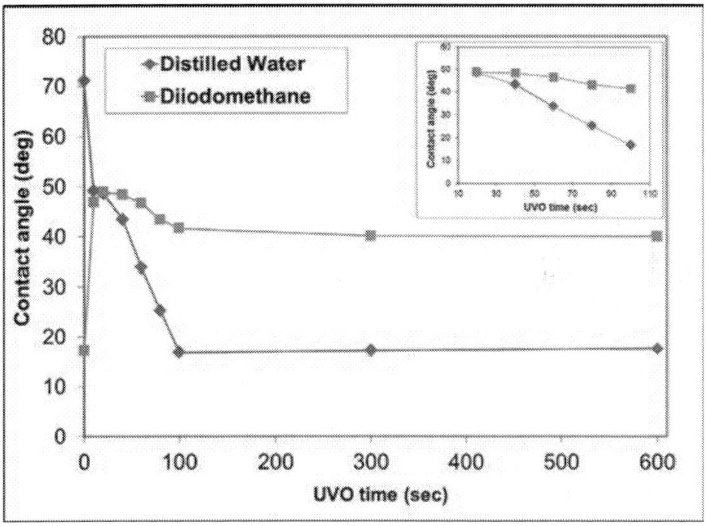

Figure 3. Contact angles of distilled water and diiodomethane on PU films (Insert shows the middle regime when UVO time is between 20 and 100 seconds)

For quantitative study, values of the average contact angles for two liquids, distilled water, and diiodomethane, on PU films with various UVO treatment times, were measured and tabulated in Table 2. The UVO treatment times were chosen as intervals from 0 to 600 seconds for different samples to determine the saturation time. Each sample was only treated once for the specific UVO treatment time. At least three data sets (three droplets) were taken for each sample, which consisted of both left and right angles. Therefore, at least six contact angle values for each UVO treatment time were recorded and the average value was used for data analysis. For distilled water droplets on the PU surface, the contact angle is 71.04º before any UVO treatment. Therefore, without any surface modification, the PU surface is close to hydrophobic and the water droplets only partially wet the surface. Values measured here are within the range of contact angles reported in the literature, which vary between 65º and 75º for this system [16, 17]. When the PU surface is UVO treated, the water contact angle decreases. It was found that as the UVO treatment time was increased, the contact angle of distilled water decreases continuously, up to a point. Therefore, the PU surface becomes less hydrophobic and more hydrophilic. In contrast, before UVO treatment, the contact angle is 17.2º for the diiodomethane liquid. Hence, diiodomethane wets the PU surface well without any surface modification. However, after UVO treatment, the contact angle for diiodomethane increases. The standard deviation of each data set was less than 2º for distilled water and 1º for diiodomethane.

The contact angles for both DI water and diiodomethane on PU are plotted in Figure 3 for different UVO treatment time. Even at a very short UVO treatment time (10 seconds), the contact angles were changed significantly for both liquids (an increase of ~30º for diiodomethane and an increase of ~22º for DI water). This implies that in the very beginning of the UVO exposure, when the PU film was placed in the UVO chamber, the ozone quickly reacts with the PU surface and changes its surface chemical composition immediately. For DI water liquid, when 20-second UVO treatment was carried out on a different PU film, the contact angle was further reduced but at a slower rate. This is because the ozone reaction on the PU surface is very quick and reaches its maximum effect within a very short period of time (< 20 sec), and the decrease in contact angle is not proportional to the ozone exposure time. However, for longer UVO exposure times, for example, between 20 and 100 seconds, it can be seen that there is a linear relationship between the DI water contact angle and the UVO treatment time (Fig. 3 insert). This is mainly due to the effect of UV light radiation on the PU surface. Because the UV light dose is linearly proportional to exposure time, the PU surface chemistry is altered in the same fashion. In contrast, when the UVO exposure is less than 20 seconds, the ozone reaction is dominating so the linear correlation between the time of UV light radiation and the DI water contact angle cannot be clearly seen. However, for the treatment times between 20 and 100 seconds, UV light radiation becomes the dominant source for UV surface chemistry modification, and the decrease in contact angle is linearly proportional to the treatment time, with the water contact angle reaching the minimum of ~ 17°. For UVO treatment times greater than 100 seconds, the PU surface is completely changed by both ozone molecules and UV light, and the contact angle remains at ~ 17°. On the other hand, for the diiodomethane liquid, the contact angle at first increases from 17.2° to a maximum value of 48.9° with 20-second treatment time. This indicates that the ozone gas also has the maximum effect on the PU surface and the effect is saturated within 20 seconds. Also, the PU film surface becomes less wettable for diiodomethane. However, for UVO treatment times between 20 and 100 seconds, the contact angle of diiodomethane droplets decreased as the treatment time increased, and the decrease is also linearly proportional to the treatment time. This indicates that the PU surface also becomes more wetting for diiodomethane after longer UV light exposure. However, the rate of decrease is lower than that of DI water, and the contact angle approaches the constant value of ~ 40° for UVO treatment greater than 100 seconds, when the UVO treatment effect is saturated.

Therefore, we can conclude that there are three regimes in UVO treatment on PU surfaces. The first stage (< 20 seconds) is when the ozone effect is dominant. For the second stage, the ozone effect diminishes and a linear UV effect takes over (20–100 seconds). Finally, in the third stage, the PU surface is fully treated (> 100 seconds).

1.2. PU surface free energy calculation

For a liquid droplet on a solid surface in equilibrium, the contact angle can be calculated by the well-known Young's equation:

$$\cos\theta = \frac{\gamma_{sv} - \gamma_{sl}}{\gamma_{lv}} \tag{1}$$

where θ is the contact angle, γ_{sv}, γ_{lv}, and γ_{sl} are the surface energy of solid against vapor, the surface energy of liquid against vapor, and the surface tension at the solid–liquid interface, respectively. Furthermore, Owens and Wendt proposed a general equation for calculating the surface free energy of solids based on Young's equation, which can also be applied for low surface energy materials such as polymers [18]. It can be seen that the surface tension at the solid–liquid interface, γ_{sl}, can be eliminated for the calculation from the equation:

$$1 + \cos\theta = 2\sqrt{\gamma_{sv}^{d}}\left(\frac{\sqrt{\gamma_{lv}^{d}}}{\gamma_{lv}}\right) + \sqrt{\gamma_{sv}^{h}}\left(\frac{\sqrt{\gamma_{lv}^{h}}}{\gamma_{lv}}\right) \tag{2}$$

where γ_{lv}^{d}, and γ_{lv}^{h} are the dispersive component (non-polar) and hydrogen bonding (polar) component of the surface energy of a given liquid against vapor, respectively $\left(\gamma_{lv} = \gamma_{lv}^{d} + \gamma_{lv}^{h}\right)$, and γ_{sv}^{d}, γ_{sv}^{h} are the dispersive component and hydrogen component of the surface free energy of the solid against vapor. The dispersive component is contributed to by the dispersive van der Waals forces between the liquid and solid and the hydrogen component includes nondispersive forces such as polar forces and hydrogen bonding forces. Additionally, $\left(\gamma_{sv} = \gamma_{sv}^{d} + \gamma_{sv}^{h}\right)$. The surface free energy of the PU film with different UVO treatment times can be calculated, since the component and the total surface free energy values are known for two liquids (Table 3), and the contact angle for each liquid droplet on PU surface at different UVO treatment time was measured. The total surface energies for PU films at different UVO treatment times, as well as their dispersive and hydrogen component, are plotted in Figure 4. For PU film without any UVO treatment, the total surface free energy is 51.5 mN/m, and the dispersive and hydrogen component of the surface energy is 46.1 mN/m and 5.4 mN/m, respectively. Without any treatment, the PU surface consists of long-chain molecules and minimum amount of high energy hydrogen bonds and other radical groups. Therefore, the hydrogen component of the surface free energy is very small, and polar liquids, such as water, will not wet the surface well when they are placed in contact with PU. On the contrary, the contact angle of diiodomethane on a non-treated PU surface is very small (~17.2°). This is because diiodomethane is a non-polar liquid, so it easily wets the untreated, non-polar PU surface, which has a high dispersive (non-polar) component value. As soon as the PU film is placed into the UVO chamber (UVO time < 20 seconds), the ozone molecules immediately reacts with the PU molecules on the film surface, leading to a significant increase of the radical groups with hydrogen bonds. The direct result is an increased hydrogen component and a decreased non-polar component of the surface free energy. Chain scission at the PU surface happens quickly and the long chains of polymer molecules are broken by the highly reactive ozone gas. Therefore, a large number of high-energy polar bonds of the broken chains are generated and exposed on the surface. The direct result is a much improved wettability for DI water, which is evident by the much smaller contact angle. In contrast, when non-polar liquids such as diiodomethane are placed on the treated PU surface, the polymer surface is less wetting. In addition, the number of long chain molecules decreases as more polar ends and hydrogen radical groups are generated on the PU surface. Therefore, the dispersive component of surface free energy decreases. For UVO treatment time 20–100 seconds, when the PU surface is modified mainly by the UV radiation, the hydrogen component (23.2–39.1 mN/m) also increases linearly, corresponding to the linear decrease of the DI water contact angles. Similarly, the small increase in the dispersive component (30.1–32.5 mN/m) corresponds to the decrease of diiodomethane contact angles (Fig. 4). Increases in both components indicates that the UV ozone improves both van der Waals interactions and hydrogen bond interactions between the solid and liquid. At UVO treatment time longer than 100 seconds, the PU surface is completely modified and the surface free energy reaches the maximum value of 71.5 mN/m, which is an 38.8% increase compared to an untreated PU surface.

Liquid	γ_{lv}	γ_{lv}^{d}	γ_{lv}^{h}
Water	72.8	21.8	51.0
Diiodomethane	50.8	50.4	0.4

Table 3. Surface tensions (in mN/m) of the two testing liquids [19]

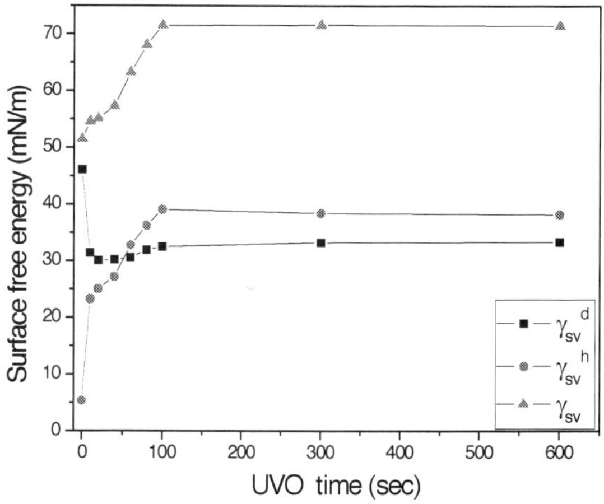

Figure 4. Surface energies of PU solid films at different UVO times

1.3. Characterization of surface chemistry change

In order to better understand the change in the PU surface by UVO treatment, an XPS study was done for untreated and 5 min UVO treated PU films. Low resolution scans of both films have the characteristic peaks corresponding to carbon (C 1s), oxygen (O 1s), and nitrogen (N1s) (scans not shown) [20]. Furthermore, the scans show additional peaks corresponding to sulfur (S 1s), which is a common element in typical polyurethane. It also shows that, after 5 minutes of UVO treatment, the oxygen level on the PU surface was considerably increased.

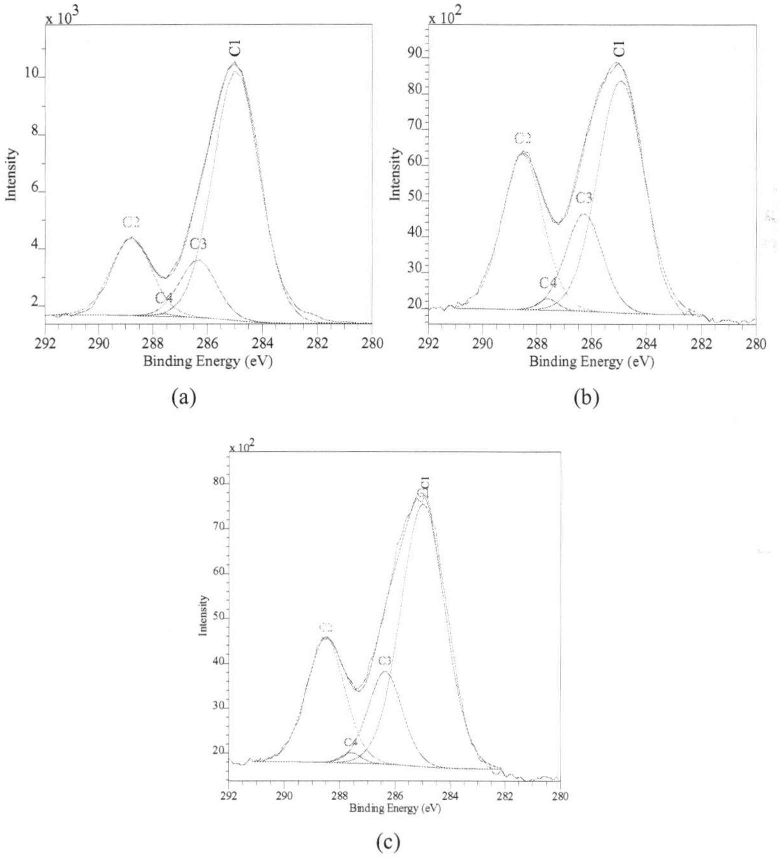

Figure 5. High resolution XPS spectra showing the deconvoluted C1s envelopes for (a) untreated polyurethane film and (b) 5 min UVO treated PU film, and (c) 5 min UVO treated PU film after washing

For detailed chemical analysis, Figure 5 (a) and (b) show the high-resolution spectra of the O 1s peaks for the untreated PU film and the 5min UVO treated PU film. The C 1s spectrum of the untreated PU film can be deconvoluted into four sub-peaks (Fig. 5(a)) [20]. The high-resolution spectra of the C 1s peaks for the samples were plotted in Figure 5 for further surface chemical characterizations. The C 1s spectrum of the untreated PU film can be deconvoluted into four sub-peaks (Fig. 5(a)) [20]. The peak at the lowest binding energy (285.0 eV) corresponds to (-C-C-) and (-C-H-) bonding (denoted **C1** in Fig. 1). The (-N-CO-O-) group corresponds to the second peak (**C2**) and is located at around 288.5 eV [4]. The third peak at 286.3 eV (**C3**) corresponds to the (-C-O-C-) group, where carbon atoms are single-bonded to oxygen atoms. A very small peak (**C4**) can be located at 287.6 eV, which corresponds to urea groups (-N-CO-N-). The C 1s spectrum of PU film shows significant chemistry change after 5 min UVO treatment (Fig. 5(b)). For the 5 min UVO treated PU surface, the oxygen-carbon double bonding (**C2, C4**) and single bonding (**C3**) peaks are much higher than those of the untreated PU surface. The significant increase of the oxygen content in PU surface is the direct result of UV ozone treatment, which breaks the long chain molecules, and atomic oxygen or ozone gas readily react and form oxygen-carbon single and double bonding groups. For the 5 min UVO treated PU, the number of oxygen-carbon bonding groups is saturated. Therefore, the XPS spectrum shows the maximum intensity of the peaks for various oxygen-carbon bonding species. The quantitative studies of the different species are shown in Table 4.

Peak area (%)	C1	C2	C3	C4	C1/C2	C1/C3
Untreated	*68.35*	*18.69*	*12.68*	*0.28*	**3.66**	**5.39**
UVO treated	*51.46*	*30.15*	*17.29*	*1.10*	**1.71**	**2.98**
UVO treated DI water washed	*58.92*	*23.92*	*16.11*	*1.05*	**2.46**	**3.66**

Table 4. Surface composition (area %) of untreated and 5 min UVO treated polyurethane thin films

Table 4 shows the surface chemical compositions for both untreated and treated PU films, represented by the integrated XPS intensity areas under each peak. The area of **C1** is reduced from 68% to 51%, while the areas of **C2**, **C3**, and **C4** increase for the 5 min treated sample. For comparison, the ratio of the areas of (C-O) and (C=O) bonds to the area of the (-C-C-) bonding (**C1/C2** and **C1/C3**) was taken for both samples. The ratio of the integrated areas **C1/C2** and **C1/C3** were 3.66 and 5.39, respectively, for the untreated sample. This is because carbon–carbon bonding is the dominant species on the pristine polyurethane surface. For the 5 min UVO treated PU surface, the ratios of **C1/C2** and **C1/C3** were reduced to 1.71 and 2.98, respectively. Therefore, the C-C bonds were reduced by about 50% after 5 min UVO treatment. The decrease in the area ratio after UVO treatment also implies the amount of C-O and C=O bonding species is increased. The increase of hydrophilic carbon-oxygen bonds is the main reason for the increased wettability of the PU films.

1.4. Surface morphology change by UVO

The 5μm x 5μm 3D AFM images in Figure 6(a) and (b) show the surface morphology of as-prepared, untreated and 5 min UVO treated PU films, respectively. The AFM image shows a very smooth surface morphology for the pristine PU surface. The area roughness study showed the root mean square roughness, R_{rms}, was 0.392 nm. The 5 min UVO treated sample showed a different surface morphology. The z-axis of the 3D image is 20 nm, compared to the 6 nm for the untreated sample. It shows many straight pillars with a height of around 3–8 nm. The R_{rms} was also increased significantly to 2.073 nm, more than 5 times rougher than the untreated sample. The pillars are low molecular weight oxidized species and possibly the tips of broken chains of PU molecules created after UVO exposure. The AFM images (Figure 6) were taken within 1 hour of the 5 min UV ozone treatment in a dry environment. Therefore, the low molecular weight oxidized species could still be observed. It is intuitive to state that, from previous studies of different short UVO treatment times (Table 2), as the UVO treatment time increases (10–100 sec), more and more low molecular weight oxidized species were created on the surface, thus the contact angle of water on treated PU surface gradually decreases (from 49° to 17°). At treatment times longer than 100 sec, the low molecular weight oxidized species are saturated on the PU film surface, and the water contact angle stabilizes around 17°. Furthermore, the low molecular weight oxidized species can be identified in the XPS plot, which shows the fraction of oxygen-containing species is significantly increased for the 5 min UVO treated samples. Therefore, the oxygen groups (**C2**, **C3**, and **C4**) are created by the UV ozone where chain scission takes place and oxygen-containing free radicals are formed, and they are observed as straight pillars in the AFM image.

The advantage of UVO treatment is that it does not physically remove material from the surface. This is preferred to oxygen plasma etching, which can remove a significant amount of material [20]. UVO treatment only affects the surface of the PU film by breaking the long chain (-C-C-) bonds and inserting atomic oxygen and ozone molecules at the chain ends to create C-O and C=O bonding, resulting in a much more hydrophilic surface. This non-destructive surface modification method provides another simple and economical approach for polymer surface wettability functionalization.

1.5. Effect of ultrasonic washing for the UVO treated sample

The effect of water washing on the UVO treated PU film was also characterized. Similar studies have been done for other polymer materials with UVO treatment on the surface [10]. In this study, for a polyurethane film surface with 5 min UVO treatment, the sample was immediately placed in an ultrasonic DI water bath for 5 minutes right after UVO treatment. The sample was then dried by blowing air, and characterization was carried out within an hour. As shown in Figure 5 (c), the XPS spectrum of the washed sample has the same carbon (C 1s) peaks as the untreated sample and the 5 min UVO treated sample without washing. Compared to the treated sample without washing (Fig. 5(b)), the intensity of the peaks for oxygen-containing species (**C2, C3**) is considerably reduced for the washed sample. When compared to the pristine PU surface (Fig. 5(a)), the intensity of those peaks is still noticeably higher for the washed sample. Table 4 shows the quantitative surface composition study and indicates the **C1** area is increased to ~59% for washed sample, compared to 51% for the treated sample without washing. **C1/C2** and **C1/C3** ratios of the integrated areas were 2.46 and 3.66, respectively. This indicates that some of the low molecular weight oxygen-containing species created on the PU

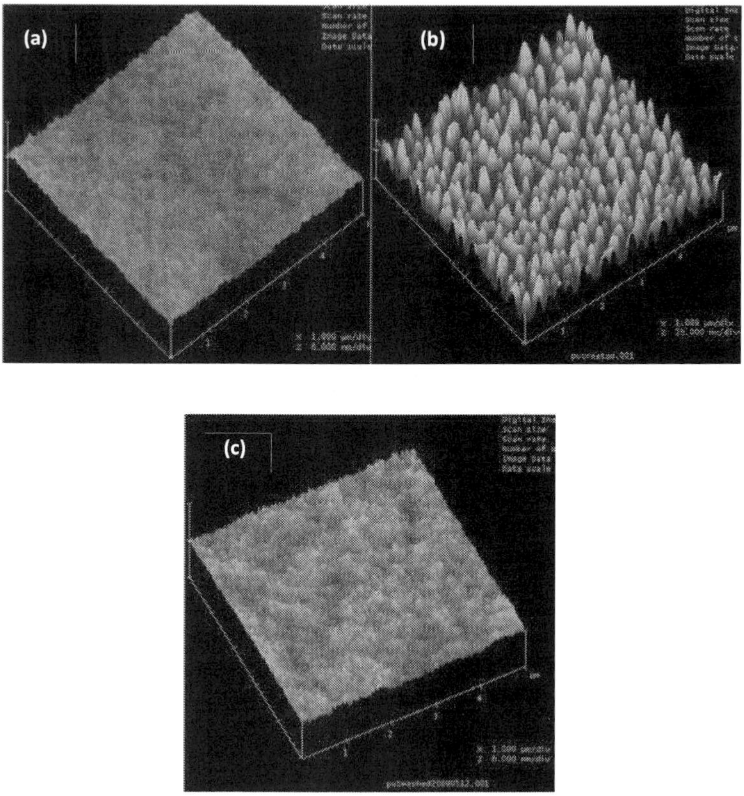

Figure 6. The 5μm×5μm 3D AFM images of (a) untreated polyurethane thin film with rms roughness w= 0.392 nm, (b) polyurethane thin film treated by 5 min UV ozone with w=2.073 nm, and (c) 5 min UV ozone treated PU thin film after washing with w=0.321 nm.

surface by UV ozone treatment were removed by water washing. Nevertheless, the washed sample still shows more oxygen content in the XPS spectrum than the untreated sample, and the oxygen content level is sustained days after washing. Furthermore, the AFM image of the UVO treated sample after washing is shown in Figure 6(c). Indeed, the surface roughness of the washed sample is quite similar to that of the pristine PU surface, and R_{rms} is 0.321 nm, which is slightly lower than that of the untreated sample. This is because the low molecular weight oxidized species have been dissolved and removed by washing in the ultrasonic DI water bath, which resulted in a flat, smooth PU surface, as observed by the AFM studies. The XPS results confirmed that the oxygen content levels on the PU surface for the washed sample were lower than for the UVO treated sample. The DI water contact angle for the washed sample was ~48°, which is higher than the UVO fully treated sample (17°, Table 2), but it is still lower than the PU films without UVO treatment (70°).

http://dx.doi.org/10.5772/60798

1.6. Hydrophobic recovery

Aging of the UVO treatment on PU surfaces was also investigated to study the hydrophobic recovery. The DI water contact angles on 5 min UVO treated PU films were measured at different aging times under ambient conditions (Fig. 7). The DI water contact angle on a 24 hour old PU film was 17.2º, which was nearly the same as the contact angle measured immediately after the UVO treatment (dotted line, Fig. 7). The contact angle increased to 19.8º after 2 days of aging and reached a constant value of about 21–22º after 3 days. Nevertheless, the increase in the water contact angle is negligible, since the contact angle for the untreated PU films is significantly higher at ~70° (Fig. 7 dashed line). This implies that the effect of UVO treatment is sustained with much improved wettability, and the hydrophobic recovery is insignificant. This is because in polar and hydrophilic species, the single and double carbon-oxygen bonds stay intact on the PU surface permanently after UVO treatment. It was also noted that the transparent PU film yellowed after the 5 min UVO treatment, but the yellow color disappeared after a few days. The yellowing effect is possibly caused by the high UV light exposure on the PU surface during the UVO treatment.

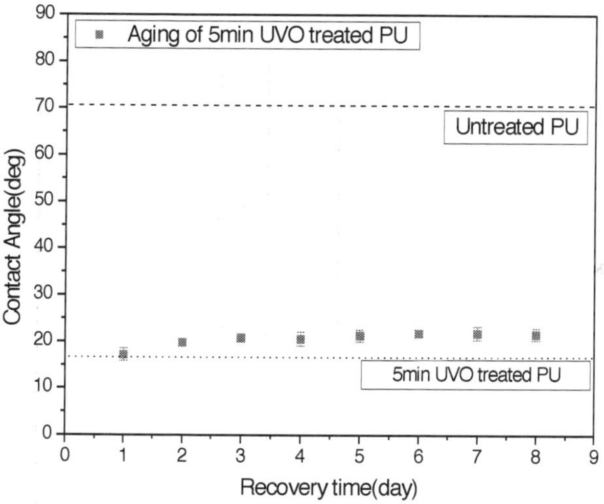

Figure 7. Change of the water static contact angle values versus aging time for polyurethane thin films after 5 minutes of UV ozone exposure

1.7. Wetting on PU grating structure

Microstructures with increased surface area and modulations possess more complicated surface wetting properties than flat surfaces. In general, due to the topography, such structures tend to be more hydrophobic with much larger apparent contact angles for different liquids. It is of particular interest to understand and manipulate the surface wettability of complex structures for specific surface wetting needs. Surface treatment on polymeric structures can increase the surface free energy and, thus, improve the wettability. For this investigation, a one-layer (1L) PU grating structure with 2.5 μm periodicity and 1:1 aspect ratio were fabricated by microtransfer molding (Fig. 8). The water contact angle on the 1L PU grating is increased significantly to ~138° (Fig. 9 (a)) compared to ~70° of the PU flat surface. This indicates the PU grating has become much more hydrophobic due to the surface corrugation, approaching a superhydrophobic state. Such a hydrophobic surface poses as a serious problem for complete conformal wetting of the surface and for removal of air gaps and liquid infiltration of gratings. After UVO treatment (720 sec) was carried out on the 1L PU grating structure, the water contact angle was reduced to ~67°, and the surface has changed from hydrophobic to hydrophilic state. This is mainly caused by increased surface free radicals and polar bonds on the polymer chains on the PU grating surface. Hence, the water droplet spread out and easily penetrated into the trenches of PU gratings. It has been shown that this surface treatment can benefit the infiltration of liquids into 3D microscale polymeric molding structure [21]. Therefore, UVO treatment can be a simple and economical surface modification method for various surfaces and structures.

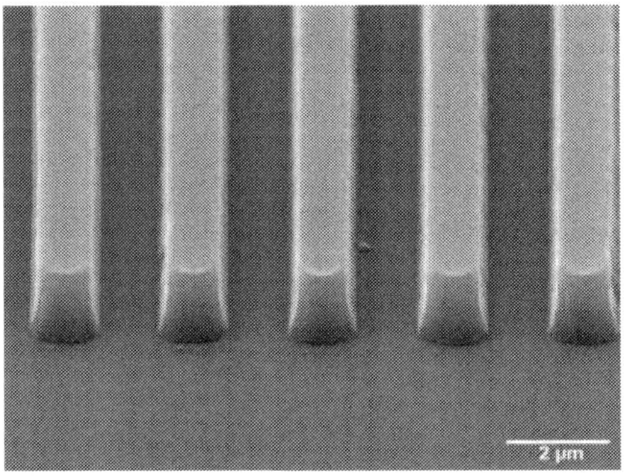

Figure 8. SEM image of a 1L PU grating structure with 2.5 μm periodicity and 1:1 aspect ratio

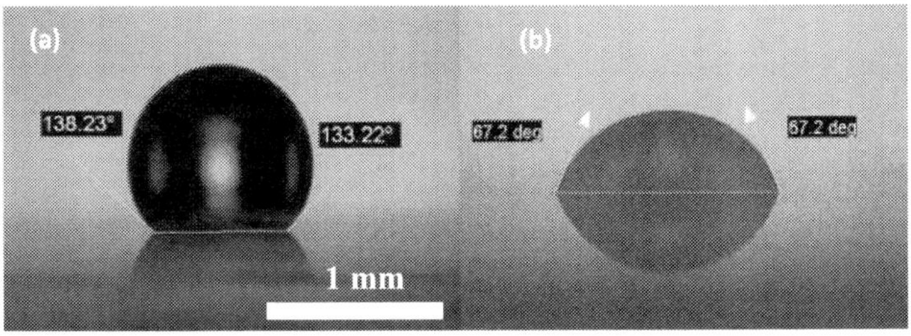

Figure 9. Water contact angles of 1L PU grating (a) without UVO treatment and (b) with 720 sec UVO treatment

1.8. Polyurethane fibrillar tip array as bio-inspired adhesives

Recently, a microstructured PU surface has been investigated for improved wettability and adhesion for novel applications using UVO treatment [22, 23]. Animal perching systems are being mimicked to realize similar locomotion capabilities for landing on rough or smooth surfaces with various inclinations. Bio-inspired adhesives comprising fibrillar arrays are similar to those used by geckos, spiders, and flies. Fibrillar bio-inspired adhesives provide high adhesion against a great variety of surfaces with different moisture conditions. Such microscale fibrillar structures can be easily fabricated using standard soft lithography techniques with polymeric elastomers such as polyurethane (Fig. 10). Subsequently, the fibrillar tips are enlarged by an "inking process" [23].

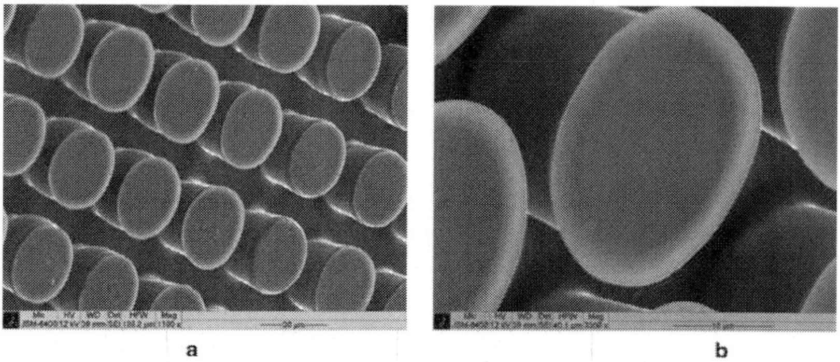

Figure 10. (a) SEM image of the polyurethane fibrillar array integrated into the perching/bio-inspired adhesion mechanism, (b) higher magnification of fibril enlarged tip [23] (Permission from reference 23)

The fibrillar arrays are hydrophobic in nature due to the low surface energy of PU surface and its surface corrugation. Surface treatment is necessary to improve its wettability and increase its surface free energy for better adhesion. In this reference, PU flat films exhibit similar wettability for water as illustrated in section 1 of this work, where the water angle is ~70° before UVO treatment and <15-17° after UVO treatment. For the fibrillar array patterns, the water contact angle is ~120° before UVO treatment. With UVO treatment, the water contact angle is significantly reduced to ~15-17° as well (Fig. 11 (a)). Unlike the flat PU film, the shear and tensile adhesion capacities of PU fibrillar arrays after UVO treatment were also enhanced significantly acting as bio-inspired adhesives. For demonstration, a perching system was made with the UVO treated fibrillar arrays placed on top of four flat elastomer pads, which were attached to the toes of four shape memory alloy (SMA) wires. The perching system was placed at a flat and smooth polyethylene sheet surface and then inclined at 75° with respect to the horizontal direction. The system can be sustained on the surface without falling off even with a 20 gram mass.

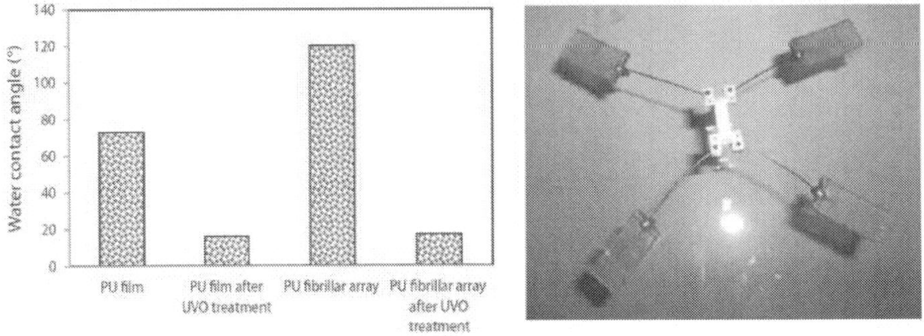

Figure 11. (a) Water contact angles of PU plain film and fibrillar array prior to and after UVO treatment. (b) Perched system with PU fibrillar array on four elastomer pads adhered to a smooth polyethylene sheet inclined at 75° with respect to the horizontal direction [23]. (Permission from reference 23)

2. Conclusion and summary

Studies have shown that, by using UVO treatment as a surface modification method, the wettability of polyurethane film surface can be significantly improved. The two liquids tested in this study showed different wetting phenomena. It was observed that, in the very beginning of the UVO treatment (< 20 seconds), the ozone gas molecules had an immediate effect on the PU surface and the contact angle for DI water was noticeably reduced. In the second stage of the UVO treatment (20–100 seconds), the UVO effect is linearly related to the treatment time, and the contact angles for both testing liquids decrease in a linear fashion until the PU surface is fully modified. Both polar and non-polar interactions between the liquids are improved by the UVO treatment in the second stage, and the total surface free energy of the PU surface is increased linearly until it is fully treated. The total surface free energy of a fully treated PU surface is 71.5 mN/m, which is a ~38.8% increase over the untreated PU surface. XPS analysis shows significant hydrophilic C-O and C=O bonding species were created on the PU surface by UVO treatment, which is the cause for higher surface free energy and improved surface wettability. Surface morphology study by AFM shows the PU surface roughness is increased by UVO treatment with increased amount of low molecular weight oxidized species and broken ends of long chain molecules, which can be removed by strong agitation such as water washing in ultrasonic bath. But the oxygen content on the washed sample is still higher than a pristine PU surface. Additional experiments with distilled water were also conducted for one-layer PU grating molds (2.5μm pitch) made by the micro-transfer molding (μTM) technique. It was shown that on the structured non-treated one-layer PU mold, hydrophobic wetting occurred and the contact angle of a water droplet was about 138 degrees. When sufficient UVO treatment was done on the mold, the water droplet had a much smaller contact angle and was pulled into the channels because of the improvement on the wetting characteristics of the PU mold. This opened up a door for future studies on the infiltration of aqueous

http://dx.doi.org/10.5772/60798

slurry into 3-D polymeric structures at micro-size, and for improved adhesion for novel applications [21, 23].

Acknowledgements

The authors would like to thank J. Anderegg for the assistance on XPS analysis and K.S. Nalwa on AFM analysis, and for meaningful discussions with Professor Zhiqun Lin. This work at the Ames Laboratory was supported by the U.S. Department of Energy—Basic Energy Sciences under contract No. DE-AC02-07CH11358.

Author details

Ping Kuang[1*] and Kristen Constant[2]

*Address all correspondence to: kuangp2@rpi.edu

1 Department of Physics, Applied Physics, and Astronomy, Rensselaer Polytechnic Institute, Troy, New York, USA

2 Department of Materials Science & Engineering, Iowa State University of Science and Technology, Ames, Iowa, USA

References

[1] Randall D., Lee S. The polyurethanes book, John Wiley & Sons, New York, 2002.

[2] Lee J.-H., Kim C.-H., Ho K.-M., Constant K. Two-polymer microtransfer molding for highly layered microstructures. Advanced Materials 2005; 17(20): 2481–2485.

[3] Ozdemir Y., Hasirci N., Serbetci K. Oxygen plasma modification of polyurethane membranes. Journal of Materials Science: Materials in Medicine 2002; 13(12): 1147–1152.

[4] Muller J. P., Mitzner E., Goering H., Gahde J. Surface modification of polyurethanes by multicomponent polyaddition reaction. Journal of Materials Science Letters 1998; 17(2): 115–118.

[5] Singh R. P., Tome N. S., Bhadraiah S. V. Photo-oxidation studies on polyurethane coating: effect of additives on yellowing of polyurethane. Polymer Degradation and Stability 2001; 73(3): 443–446.

[6] Mathieson I., Bradley R. H. Improved adhesion to polymers by UV/ozone surface oxidation. International Journal of Adhesion and Adhesives 1996; 16(1): 29–31.

[7] Gongjian B., Yunxuan W., Xingzhou H. Surface modification of polyolefine by UV light/ozone treatment. Journal of Applied Polymer Science 1996; 60(13): 2397–2402.

[8] Ton-That C., Teare D. O. H., Campbell P. A., Bradley R. H. Surface characterisation of ultraviolet-ozone treated PET using atomic force microscopy and x-ray photoelectron spectroscopy. Surface Science 1999; 433: 278–282.

[9] Nie H.-Y., Walzak M. J., Berno B., McIntyre N. S. Atomic force microscopy study of polypropylene surfaces treated by UV and ozone exposure: modification of morphology and adhesion force. Applied Surface Science 1999; 114: 627–632.

[10] Teare D. O. H., Ton-That C., Bradley R. H. Surface characterization and ageing of ultraviolet–ozone-treated polymers using atomic force microscopy and x-ray photoelectron spectroscopy. Surface and Interface Analysis 2000; 29(4): 276–283.

[11] Efimenkoa K., Wallaceb W. E., Genzer J. Surface modification of Sylgard-184 poly(dimethyl siloxane) networks by ultraviolet and ultraviolet ozone treatment. Journal of Colloid and Interface Science 2002; 254(2): 306–315.

[12] Romero-Sanchez M. D., Pastor-Blas M. M., Martin-Martinez J. M., Walzak M. J. Addition of ozone in the UV radiation treatment of a synthetic styrene-butadiene-styrene (SBS) rubber. International Journal of Adhesion and Adhesives 2005; 25(4): 358–370.

[13] Koo G. H., Jang J. Surface modification of poly (lactic acid) by UV/Ozone irradiation. Fibers and Polymers 2008; 9(6): 674–678.

[14] Macmanus L. F., Walzak M. J., McIntyre N. S. Study of ultraviolet light and ozone surface modification of polypropylene. Journal of Polymer Science Part A: Polymer Chemistry 1999; 37(14): 2489–2501.

[15] McIntyre N. S., Walzak M. J. New UV/ozone treatment improves adhesiveness of polymer surfaces. Modern Plastics 1995; 25(3): 69–75.

[16] Sidouni F. Z., Nurdin N., Chabrecek P., Lohmann D., Vogt J., Xanthopoulos N., Mathieu H. J., Francois P., Vaudaux P., Descouts P. Surface properties of a specifically modified high-grade medical polyurethane. Surface Science 2001; 491(3): 355–369.

[17] Chen X., You B., Zhou S., Wu L. Surface and interface characterization of polyester-based polyurethane/nano-silica composites. Surface & Interface Analysis 2003; 35(4): 369–374.

[18] Owens D. K., Wendt R. Estimation of the surface free energy of polymers. Journal of Applied Polymer Science 1969; 13(8): 1741–1747.

[19] Butt H., Raiteri R. Surface characterization methods: principles, techniques, and applications. In: Milling A. J. (Ed.) Surface tension and surface stress of solids. Marcel Dekker, Inc., New York, 1999, p. 23.

[20] Sanchis M. R., Calvo O., Fenollar O., Garcia D., Balart R. Surface modification of a polyurethane film by low pressure glow discharge oxygen plasma treatment. Journal of Applied Polymer Science 2007; 105(3): 1077–1085.

[21] Lee J. H., Kuang P., Leung W., Kim Y. S., Park J. M., Kang H., Constant K., Ho K. M. Semicrystalline woodpile photonic crystals without complicated alignment via soft lithography. Applied Physics Letters 2010; 96(19): 193303.

[22] Peyvandi A., Abideen S. U., Huang Y., Lee I., Soroushian P., Lu J. Surface treatment of polymer microfibrillar structures for improved surface wettability and adhesion. Applied Surface Science 2014; 289: 586–591.

[23] Peyvandi A., Soroushian P., Lu J. A versatile perching mechanism employing shape memory wires and bio-inspired adhesives. Journal of Micro-bio Robotics 2014; 9: 1–13.

Chapter 5

Wetting and Navier-Stokes Equation — The Manufacture of Composite Materials

Mario Caccia, Antonio Camarano, Danilo Sergi, Alberto Ortona and Javier Narciso

Additional information is available at the end of the chapter

http://dx.doi.org/10.5772/61167

Abstract

It is well known that there are several processes to manufacture composite materials, a large part of which consist in the infiltration of a liquid (matrix) through a porous medium (reinforcement). To perform these processes, both thermodynamics (wetting) and kinetics (Navier-Stokes) must be considered if a good quality composite material is sought. Although wetting and the laws that govern it have been well known for over 200 years, dating back to the original works of Young and Laplace, this is not the case with the Navier-Stokes equation, which remains so far unsolved. Although the Navier-Stokes equation, which describes the motion of a fluid, has been solved for many particular cases, such as the motion of a fluid through a pipe, which has resulted in the well-known Poiseuille equation, or the motion of a fluid through a porous media, described by the Darcy's law (empirical law obtained by Darcy), its general solution remains one of the greatest challenges of mathematicians today. Therefore, the objective of this chapter is to present the resolution of the Navier-Stokes equation with the laws of wetting for different cases of interest in the manufacture of composite materials.

Keywords: Wettability, Infiltration, Navier-Stokes

1. Introduction

Composite materials are an important and oftentimes used type of engineering materials with tuneable properties which depend on the materials combined. They are usually made of two or more materials. The first phase, commonly referred to as "reinforcement" is constituted by a solid material, which can be employed in several morphologies such as fibers (short or continuous), particles, or as a co-continuous domain of a solid phase which occupies a fraction of the available volume (foams) [1-7]. The second phase, normally called "matrix," can be gaseous, liquid, or solid, and is in contact with the reinforcement through a proper interface. Composite materials are very common in nature, where a combination of fibers and matrix is often found. A good example is wood, which is composed of flexible fibers of cellulose embedded in a rigid matrix of lignin. Artificially produced composites are usually classified, according to the nature of their matrix, into metal, ceramic, or polymer matrix composites. Each type of composite presents its advantages and disadvantages, and together they cover a great scope of applications [1, 8-12].

Composite materials manufacturing can be accomplished through several techniques. Each technique is mainly the result of the trade-off between the target properties of the final product, the starting materials properties, and the final cost. In most of the cases, the matrix is in a liquid state during processing. A common action in composite materials manufacturing is "to infiltrate" the reinforcement with the matrix. When a liquid is used, the impregnation of the remaining free volume is highly dependent on how the first phase is placed in the space and on the physical properties of the liquid itself. In the case of polymeric matrix composites (PMC) techniques involving a liquid matrix are the only way to process them, as polymers are in a liquid state before their cross-linking in thermosets, or can be melted if they are thermoplastics. Also, ceramic matrix composites are processed through liquid routes (e.g., sol-gel techniques for oxide composites, or polymer infiltration and pyrolysis (PIP) [13-14] for ceramic matrix composites (CMC) produced by pre-ceramic precursors. In the latter case, where several impregnations are performed, the permeability of the porous materials decreases with the increasing of the final density of the composite under production). A widely utilized technique which stands among ceramic and metallic matrix composites (MMC) manufacturing is the so-called reactive silicon infiltration [15]. Its peculiarity is that, during processing, silicon is molten and by reacting with extra carbon present in the primary phase it yields silicon carbide. This method was not originally conceived to produce composite materials, but for the synthesis of bulk SiC, still, it proved to be a suitable methodology for ceramic matrix composites manufacture as well.

Some industrial products obtained with this infiltration technology are pantographs, connecting rods, loudspeakers, piston rings, heat sinks, electro-brushes, engine cylinders, brake discs, etc. In most of these examples, primary phase (reinforcement) content is above 50% in volume. For reinforcement contents of such magnitude, the most commonly used technique for manufacturing composites is infiltration [1, 5-7, 16-18].

Actually, infiltration processes are not exclusive of materials manufacture, but can be observed in many everyday life activities, such as making coffee, where hot water is infiltrated into a porous bed of coffee particles. Perhaps one of the most important cases where infiltration phenomenon is observed is in the penetration of rainwater into the ground, which ends up filling aquifers that guarantee water supply to population in dry regions where the surface runoff is scarce. Because of the great importance of infiltration in geological sciences, it was in this field where the deepest and most detailed studies of this phenomenon were performed. Many advances in this area of knowledge have been expanded and applied to other scientific fields such as materials science. The clearest example is Darcy's law which is widely used in the design of composite materials. This empirical law was obtained by Henry Darcy whose original work published in 1856 is entitled: *"Les Fontaines Publiques de la Ville de Dijon"* (The Public Fountains of the City of Dijon). However, Darcy's law is not the only example that can be found in the literature. The relationship between water saturation and organic compounds in soils of different origins has resulted in the semiempirical equations of Brooks and Corey [19] and Van Genuchten [20], both widely used in composite materials.

The aim of the present chapter is to introduce the reader to the basic concepts of wetting and infiltration, and to combine these concepts to solve the equations governing the infiltration process in several particular cases that often arise in the composite materials manufacturing industry.

2. Experimental

Although it will be discussed in detail in the upcoming sections of the chapter, it is important to clarify that infiltration, in most composite materials manufacturing processes, does not occur spontaneously (in the case of PMCs it does proceed spontaneously, but it is so slow that it is necessary to force it), so the liquid must be forced to infiltrate by applying an external pressure. From an engineering point of view, mainly two technologies have been developed for this purpose: the so-called gas-assisted pressure infiltration (GPI), where a gas is used to push the liquid; and another called squeeze-casting (SC) in which pressure is applied by means of a mechanical device. Each technique has its upsides and downsides, which will be discussed throughout this chapter. The main advantage of GPI over SC is that it allows a better control of physical parameters (pressure and temperature), and admits higher complexity in the manufactured parts. On the other hand, the main advantage of SC over GPI is that higher pressures can be achieved and the automatization of the process becomes easier.

First the GPI technique will be analyzed, and for this purpose, the example of a laboratory equipment developed at the University of Alicante will be used [10, 11, 21-29]. A scheme of the equipment is shown in Figure 1. The equipment consists of a pressure chamber (maximum admissible pressure at 300 °C of 5 MPa) equipped with a resistance furnace, which is thermally isolated from the chamber by the use of conventional refractory material (temperature of chamber wall is always below 100°C). The furnace has a power of 1500 W and can reach temperatures of 1000°C. For measurement and control of furnace temperature, K type thermocouples, connected to a temperature controller, are used. The pressure control is entirely performed by a system of pneumatic and electronic valves connected to a pressure controller. The pressure vessel is made entirely of stainless steel and consists of a cylindrical body with an internal diameter of 30 cm, an outer diameter of 34.6 cm, a height of 45 cm, and a wall thickness of 3 cm. The closing of the camera is made with 12 M16 mounting studs. Both, the electrical feed (to thermocouple input and power cables) and the gas inlet and outlet for pressurizing the chamber are positioned in the body. Additionally, the camera is connected to a primary vacuum pump, which allows the process to be performed in the absence of oxygen (low partial pressure).

The manufacturing process of a composite material using GPI can be described as follows: a preform, which consists of a porous bed of compacted particles, is placed into a quartz (or BN-coated graphite) crucible and a metal ingot is placed on top of it (see Figure 1b). The crucible is then introduced inside the pressure chamber and vacuum is applied during at least 30 min., until a pressure of about 100 Pa is reached. The chamber is heated under vacuum at a very low rate (3°C/min) up to 250°C, to remove humidity and gas adsorbed on particles. Subsequently, further heating is applied up to a temperature 50°C higher than the melting point of the metal at a rate of approximately 5°C/min. The maximum temperature is then maintained for 30 min followed by argon injection to reach the desired infiltration pressure. Once the preform has been infiltrated (approximately 2 min. under pressure) the crucible is pushed towards the bottom of the chamber, which is cold, and directional solidification occurs, ensuring a high-quality composite (null porosity). Figure 2 shows a composite material obtained at the University of Alicante with this process, where it is possible to observe the absence of porosity and an intimate contact between reinforcement and matrix.

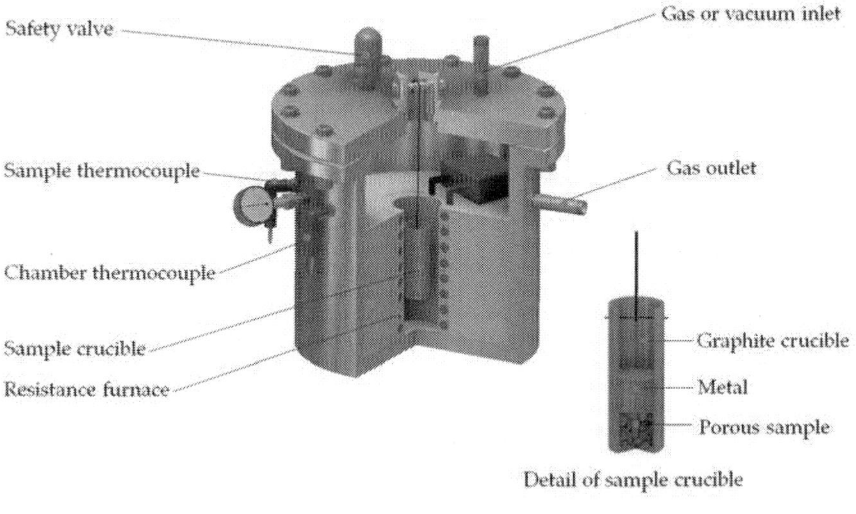

Figure 1. Schematic representation of a gas-assisted pressure infiltration (GPI) system developed at the University of Alicante and arrangement of the materials (reinforcement and matrix) inside the crucible.

At the University of Alicante, another equipment, in this case for squeeze-casting technique [30], was also developed. Figure 3 shows a schematic representation of the main body of this equipment and how it operates. The main feature of this apparatus is that it can work up to 450°C and at a maximum pressure of 150 MPa. To resist these operating conditions, special

(W720 Bölher) steel was used to build the body of the equipment. Embedded resistances in the central body heat the system, and temperature is measured and controlled by K type thermocouples connected to a temperature controller.

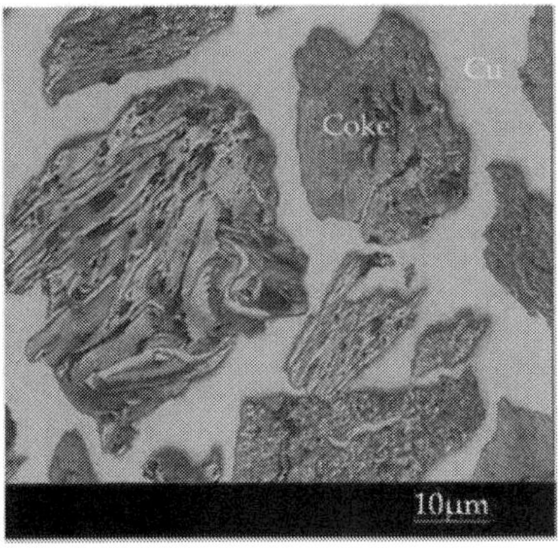

Figure 2. Polarized light optical microscopy image of a composite material obtained by GPI. The composite material consists of needle coke (reinforcement) and copper (matrix).

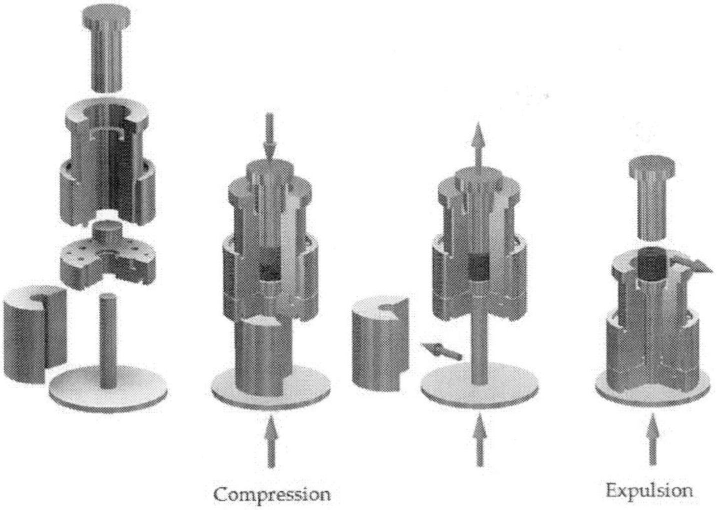

Figure 3. Schematic representation of a squeeze-casting (SC) system developed at the University of Alicante. The operating mode is indicated in the figure.

The manufacturing process of a composite material using SC can be described as follows: The system is first preheated to the operating temperature (between 350 and 450 °C depending on the alloy) to avoid early solidification of the metal matrix during infiltration. The preform is preheated (600 to 700°C) and the alloy is melted (150°C above the melting point) in two auxiliary furnaces. Thereafter, the preform is transferred to the main chamber, and the liquid metal is added. The desired pressure is applied as indicated in Figure 3. After infiltration, the system is cooled down and the manufactured material is removed as indicated in Figure 3.

The main problem with this technique is that the processing is complex because the times and temperatures must be very accurate to keep the metal from solidifying before penetrating into the preform. This disadvantage is also an advantage, since the high velocity of the process generates a finer microstructure and thus improves the mechanical properties of the final material.

Most importantly, it is possible to find in the market, both SC and GPI fully automated devices, but the basic operating principles are the same for the ones developed at University of Alicante with a scale factor between 2 and 10.

3. Wetting basic laws

One of the most relevant characteristics for the production of MMC through liquid processing routes is the wetting of the reinforcement by the liquid metals, which is quantified by measuring the contact angle between the reinforcement and matrix [31]. The contact angle between a flat substrate and a drop of liquid metal can be routinely measured by the method of the sessile drop. However, this measurement is not possible when the solid is in particulate form, in which case the wetting can be studied by means of liquid metal infiltration into compacted particulate samples [10, 11, 21-29].

Considering a solid, flat, non-deformable and chemically homogeneous surface in contact with a nonreactive liquid in the presence of a gaseous phase, if the liquid does not completely cover the solid, the liquid will form with the solid a contact angle θ, as depicted in Figure 4.

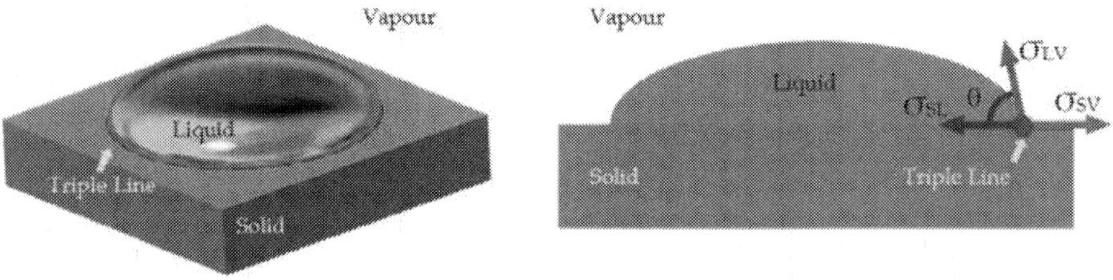

Figure 4. Configuration of a sessile drop on a substrate within which it forms a contact angle θ.

The equilibrium value of θ, which is used to define the wetting behavior of a fluid, can be written in terms of the balance of three energies present in the interfacial vapor-liquid-solid system, according to the well-known Young equation (1805):

$$\cos\theta = \frac{\sigma_{sv} - \sigma_{sl}}{\sigma_{lv}} \tag{1}$$

where σ_{ij} is the interfacial energy between the phases *i* and *j* (s, l, and v represent the solid, liquid, and vapor phases, respectively). A contact angle of less than 90° identifies a liquid that wets the substrate, while larger angles identify liquids that do not wet it. An ideal wetting liquid will have an assigned value of θ equal to 0°.

The interfacial energy σ_{lv}, also called surface tension (γ_{lv}), can be measured with sufficient accuracy by a large number of techniques, e.g., maximum bubble pressure [32, 33], pendant drop [34], or great crucible [35, 36]. However, the surface energy of the solid-vapor phase (σ_{sv}) and the solid-liquid (σ_{sl}) phase can only be estimated. Consequently, the two variables that are unknown in the Young's equation can be grouped into only one, called work of adhesion (W_a), which expresses the strength of the solid-liquid bonds and was introduced by Dupré (1869):

$$W_a = \sigma_{lv} + \sigma_{sv} - \sigma_{sl} \tag{2}$$

If now this equation is inserted in the Young's equation, the so-called Young-Dupré equation is obtained:

$$W_a = \gamma_{lv} \cdot (1 + \cos\theta) \tag{3}$$

All infiltration processes are driven by a differential pressure that allows the liquid matrix to penetrate into the porous reinforcement. This difference of pressure between the two fluids (liquid matrix and gas phase filling porosity) is defined as the capillary pressure (P_c) and described by the famous Young-Laplace equation for a straight circular capillary:

$$\Delta P_c = P_A - P_B = \frac{2 \cdot \gamma_{lv} \cdot \cos\theta}{r} \tag{4}$$

Where γ_{lv} is the liquid-vapour surface tension, θ the contact angle, and r the capillary radius. Figure 5 schematizes different infiltration situations based on the wetting or non-wetting behavior of the system.

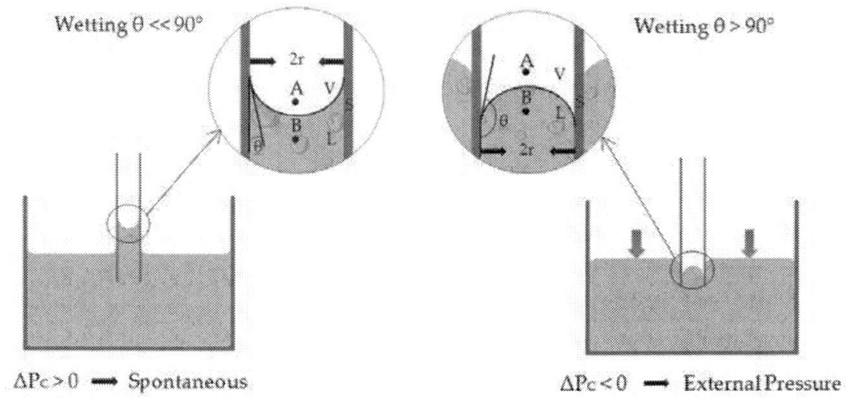

Figure 5. Scheme of infiltration process for different wetting behaving systems.

The energy shift that occurs when immersing the capillary into a liquid can be described in terms of work as:

$$W_i = \sigma_{sl} - \sigma_{sv} \tag{5}$$

where W_i is the work of immersion. For wetting systems (W_i<0), infiltration proceeds spontaneously without the necessity of applying an external pressure, as is the case of infiltration of silicon into carbon in the RBSC method. For non-wetting systems (W_i>0), infiltration does not occur unless an external pressure, superior to the threshold pressure, is applied. The threshold pressure can be written in terms of work of immersion as:

$$P_0 = S_i \cdot W_i \tag{6}$$

where S_i is the particle surface area per unit of volume of liquid matrix. Since $\sigma_{sv} = \sigma_{sl} - \gamma_{lv} \cdot cos\theta$, The threshold pressure can be related to the contact angle as:

$$P_0 = S_i \cdot \gamma_{lv} \cdot cos\theta \tag{7}$$

Since the measurement of the surface area of the particles becomes rather inaccurate when the particle diameter surpasses a certain value, and the surface roughness is very difficult to estimate, it is possible to assume a particular geometry for the porous media, e.g., cylindrical pores with a radius r as an approximation, in which case the surface area can be written as $S_i = \frac{2 \cdot \lambda \cdot (1-\phi)}{\phi \cdot r}$ and the threshold pressure can be written as:

http://dx.doi.org/10.5772/61167

$$P_0 = 2 \cdot \lambda \cdot \gamma_{lv} \cdot \cos\theta \cdot \frac{(1-\phi)}{\phi \cdot r} \tag{8}$$

Where λ is a geometrical factor introduced to describe deviation from the assumed geometry of the pores and the surface roughness of the material (it usually oscillates between 2 and 4 [10, 21-23, 25, 29], γ_{lv} is the surface tension, ϕ the porosity of the sample, r the mean pore radius, and θ is the contact angle at the triple line. It is important to emphasize that the contact angle in eq. 5 is the equilibrium contact angle that can be measured with high accuracy in sessile drop experiments.

4. The simplest case: Obtaining Darcy's law

As previously mentioned, Darcy's law predicts how a fluid moves through a porous system. This empirical law is based on the study of water movement in sand beds, and represents an ideal model in the manufacture of a composite material. Darcy's law can be derived from the resolution of the Navier-Stokes equation, considering the ideality of the system. In this section, the basis for describing the process of manufacturing of composite materials will be provided, and the most important factors that affect the process will be identified.

As aforementioned, the flow of viscous liquids is governed by the Navier-Stokes equation. This differential equation does not have a general solution. However, in some cases, it is possible to solve, e.g., for an incompressible fluid, in laminar flow (low Reynolds number), assuming that the time derivative of the fluid velocity is much smaller than its spatial derivatives, and that the effect of gravity is negligible. In this case, the equation can be reduced to:

$$\nabla P = \eta \cdot \nabla^2 v \tag{9}$$

where v is the speed of the fluid, η is the dynamic viscosity, and P the pressure. Assuming unidirectional flow (which is not so realistic in a standard porous system), it is possible to reduce the expression to the Darcy's law. This equation can be rewritten in the most commonly used form to analyze the flow of incompressible fluids through porous media:

$$v_0 = -\frac{K}{\eta} \cdot \frac{dP}{dz} \tag{10}$$

where z is the flow direction, v_0 the average superficial velocity of the fluid, dP/dz the pressure gradient along the infiltration front, and K the permeability of the porous media. The superficial velocity v_0 can be converted into infiltration velocity (dz/dt) in the porous medium by means of the porosity ϕ:

$$v_0 = \phi \cdot \frac{dz}{dt} \tag{11}$$

Combining both equations and integrating, the Darcy's law, which shows the relationship between infiltration distance (h), time (t) and the pressure drop in the infiltrating liquid ($\Delta P = P - Po$) is obtained.

$$h^2 = \frac{2 \cdot K \cdot t}{\eta \cdot \phi} \cdot \Delta P \tag{12}$$

It must be pointed out that the permeability of the porous media is proportional to the square of the average pore radius:

$$K = a \cdot r^2 \tag{13}$$

Darcy's law has been widely used in basic studies of manufacture of metal matrix composites. Despite its simplicity, it has proven to be very useful in metallurgical engineering. As an example, the results of infiltration of liquid aluminium into a porous bed of particles, obtained by Narciso et al., are shown in Figure 6 [10, 25]. Figure 6a provides the relationship between infiltration distance and applied pressure for liquid aluminium into a porous bed of alumina particles, while Figure 6b shows the results for a liquid infiltration of an aluminium alloy into preforms of graphite particles. As predicted by Darcy's law, the square of infiltration distance is proportional to the pressure drop (h^2 vs ΔP) in these systems.

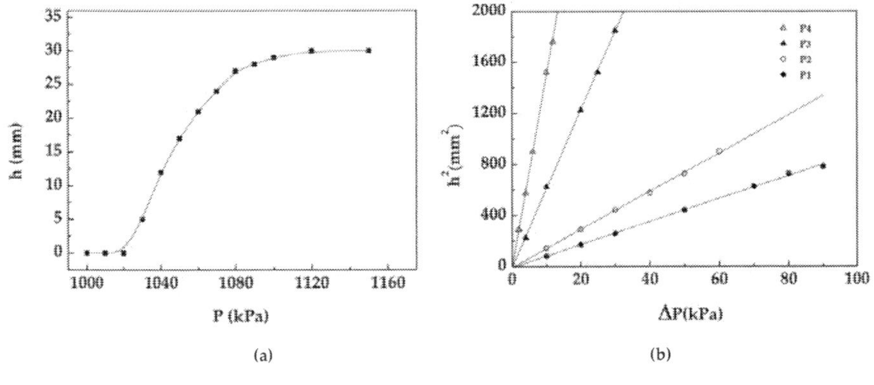

Figure 6. Evolution of the infiltrated height as a function of pressure for liquid aluminium on a preform of alumina particles (Alumina particles AA10, Vp = 59 %, Temperature = 700 °C) (a). Evolution of the square of the infiltrated height for a liquid aluminium alloy as a function of pressure drop on a preform of carbon particles at temperature of 700 °C (see the appendix for the characteristics of the graphite particles) (b).

As discussed in the preceding paragraphs, to solve the Navier-Stokes equation a full saturation and a laminar flow (Re < 2100) have been assumed. Besides, the effect of gravity has been considered to be negligible. Figure 6b shows a good correlation of the experimental data with Darcy's law, and thus, the assumed conditions should be valid. Reynolds number (Re) can be obtained from the following expression:

$$Re = \frac{\rho \cdot v \cdot D}{\eta} \tag{14}$$

Where ρ is the density of the liquid metal; D is the diameter of the channel, which in this case is the interparticle distance that can be assumed to be about 1/3 of the average size of the particles; v is the fluid velocity, which is comprised between 1 and 5 mm/s (in our equipments); and η the dynamic fluid viscosity. If the Reynolds number is calculated for the experiments shown in Figure 6, the values obtained are between 0.001 and 0.7, which means that the laminar flow assumption is indeed correct. The force exerted by gravity would be the metallostatic pressure of the melt, which is given by:

$$P = \rho \cdot g \cdot h \tag{15}$$

And considering that the liquid column is always less than 10 cm, in the case of aluminium for example, the pressure would be about 2 kPa, which is almost 3 orders of magnitude lower than the capillary forces, and therefore negligible.

In the following sections different real situations that can occur are considered and will be analyzed separately. The combination of all these particular cases simultaneously, although more realistic, is much more complex and goes beyond the scope of this chapter.

5. The effect of saturation

On systems where the liquid does not wet the solid, which is most common in MMCs, it must be taken into account that the permeability of the porous system is not constant [37-39] and that it depends on the saturation and this latter, and at the same time, on the applied pressure [21, 40-42].

The study of unsaturated flow is well known in geology and in 1931 it was described mathematically by Richards [34], whose partial differential equations are difficult to solve and usually solved numerically for each particular case. Numerous researchers have worked on this particular hydrodynamics and mathematical problem [44,45], and the books by Prof. Bear collect these study cases [46-48], which are essential reading for researchers working in this field. It is also noteworthy that Prof. Bear highlights in his books the goodness of Darcy's law, and that its application range is wider (more useful) than it was expected at first (i.e., unsaturated flow, high Reynolds number, etc.).

These studies were not limited to the case study of hydrology, but have crossed over to the field of metal matrix composites (MMC), performed especially by the group of Andreas Mortensen [40-42], but as mentioned above without analytical solution of the differential equations describing the phenomenon. The highlight to be found in the recent literature are two articles that show that close to the threshold pressure, infiltration is actually governed by percolation phenomena (instead of slug flow) [49,50].

Another aspect to consider is that pressure also depends on the position along the flow path, and is usually assumed that the pressure drop between the front (P_0) and head (applied pressure) is linear, which is reasonably sane. Nevertheless, in order for the differential equations to have an analytic solution, it will be assumed in this chapter that practically all the preform is at the same pressure, which in principle is something reasonable as working pressures oscillate between 1 and 5% of the threshold pressure (i.e., head=$1.05P_0$; front=P_0).

In the literature, several models showing the relationship between pressure and saturation have been described, but the most famous are the Brooks-Corey (BC) [19] and van Genuchten (VG) [20] models. In this section, the Navier-Stokes equation has been solved considering that the permeability is not constant, relating it to the saturation through the BC and VG models. Different authors have used these models to describe the infiltration process in MMCs [51]. However, previous works have shown that for low pressures (close to the threshold pressure) these models, especially the BC, do not describe the infiltration process correctly [52, 53].

Starting from reduced Navier-Stokes equation:

$$dz^2 = \frac{K \cdot t}{\eta \cdot \varnothing} \cdot dP \tag{16}$$

Where z is the infiltration distance, K the permeability, t the time of infiltration, η the dynamic viscosity of the fluid, $\varnothing$ the porosity of the porous body, and P the applied pressure. Permeability can be decomposed into two different terms, intrinsic permeability (K_i), which depends exclusively on the nature of the solid phase; and relative permeability (K_r), which depends on the degree of saturation (S) of the system:

$$K = K_i \cdot K_r \tag{17}$$

Relative permeability, depends on the degree of saturation according to:

$$K_r = S^n \tag{18}$$

Where n is a parameter that ranges from 1 to 3. In the case of composites fabricated by infiltration, it has been found that 1 is the most accurate value for n [51]. There are several models that relate saturation to the applied pressure (P) and the threshold pressure (P_0) of the system. One of the most used models of this kind is the Brooks-Corey equation:

$$S = 1 - \left(\frac{P_0}{P}\right)^{\lambda} \tag{19}$$

Where λ is a parameter that depends on the solid material. Replacing K in eq. 16 with eqs. 17, 18, and 19 yields a modified reduced Navier-Stokes equation:

$$dz^2 = \frac{K_i \cdot \left[1 - \left(\frac{P_0}{P}\right)^{\lambda}\right]^n \cdot t}{\eta \cdot \varnothing} \cdot dP \tag{20}$$

Solving this equation provides a model to estimate depth of infiltration as a function of applied pressure. In this form, there is no general analytical solution to the equation, so, to obtain an analytical model, a particular case must be considered. As discussed above, experimental data fit better when $n=1$, so to solve the equation, this particular case will be considered:

$$\frac{1}{2} \cdot h^2 = \frac{K_i \cdot t}{\eta \cdot \varnothing} \int \left(1 - \left(\frac{P_0}{P}\right)^{\lambda}\right) \cdot dP \tag{21}$$

$$h^2 = \frac{2 \cdot K_i \cdot t}{\eta \cdot \varnothing} \cdot \left(P + \frac{P \cdot \left(\frac{P_0}{P}\right)^{\lambda}}{\lambda - 1} + A \right) \tag{22}$$

Where A is the integration constant, and its value can be determined using initial conditions, i.e., for $P = P_0$ and $h = 0$

$$0 = \frac{2 \cdot K_i \cdot t}{\eta \cdot \varnothing} \cdot \left(P_0 + \frac{P_0 \cdot \left(\frac{P_0}{P_0}\right)^{\lambda}}{\lambda - 1} + A \right) \tag{23}$$

$$\begin{aligned} 0 &= P_0 + \frac{P_0}{\lambda - 1} + A \\ A &= -P_0 - \frac{P_0}{\lambda - 1} \\ & P_0 \cdot \left(-\frac{1}{\lambda - 1} - 1 \right) \end{aligned} \tag{24}$$

Then the final expression for *h(P)* is:

$$h^2 = \frac{2 \cdot K_i \cdot t}{\eta \cdot \varnothing} \cdot \left(P + \frac{P \cdot \left(\frac{P_0}{P} \right)^{\lambda}}{\lambda - 1} - P_0 \cdot \left(\frac{1}{\lambda - 1} + 1 \right) \right) \tag{25}$$

It must be noted that if λ acquires a very high value, the model tends to Darcy's law:

$$\lambda \to \infty$$

$$h^2 = \frac{2 \cdot K_i \cdot t}{\eta \cdot \varnothing} \cdot \left(P + 0 - P_0 \cdot (0 + 1) \right)$$

$$h^2 = \frac{2 \cdot K_i \cdot t}{\eta \cdot \varnothing} \cdot \left(P - P_0 \right)$$

If, instead of using the Brooks-Corey equation, the Van Genuchten model is applied:

$$S = 1 - \left[1 + (\alpha \cdot (P - P_0))^{n1} \right]^{-m} \tag{26}$$

$$m = 1 - \frac{1}{n1} \tag{27}$$

where α is a scale parameter inversely proportional to the mean pore diameter, and $n1$ (called n1 to differentiate from n in eq. 18) and m are shape parameters. The reduced Navier-Stokes equation results in:

$$dz^2 = \frac{K_i \cdot \left[1 - \left[1 + \alpha \cdot (P - P_0)^{n1} \right]^{-1 + \frac{1}{n1}} \right]^{n} \cdot t}{\eta \cdot \varnothing} \cdot dP \tag{28}$$

As in the case of the Brooks-Corey model, a general analytical solution is not available, so the case of n = 1 will be considered again:

$$dz^2 = \frac{K_i \cdot \left[1 - \left[1 + \alpha \cdot (P - P_0)^{n1} \right]^{-1 + \frac{1}{n1}} \right] \cdot t}{\eta \cdot \varnothing} \cdot dP \tag{29}$$

$$\frac{1}{2}\cdot h^2 = \frac{K_i \cdot t}{\eta \cdot \varnothing} \cdot \int \left[1 - \left[1 + (\alpha \cdot (P - P_0))^{n1} \right]^{-1 + \frac{1}{n1}} \right] \cdot dP \tag{30}$$

$$\frac{1}{2}\cdot h^2 = \frac{K_i \cdot t}{\eta \cdot \varnothing} \cdot \int 1 \cdot dP - \int \left[1 + (\alpha \cdot (P - P_0))^{n1} \right]^{-1 + \frac{1}{n1}} \cdot dP \tag{31}$$

$$\frac{1}{2}\cdot h^2 = \frac{K_i \cdot t}{\eta \cdot \varnothing} \cdot P - \int \left[1 + (\alpha \cdot (P - P_0))^{n1} \right]^{-1 + \frac{1}{n1}} dP \tag{32}$$

.

The second integral does not have an analytical solution either, so in order to obtain a solution as general as possible, the value of n1=2 is introduced [42, 51]:

$$h^2 = \frac{2 \cdot K_i \cdot t}{\eta \cdot \varnothing} \cdot \left[P - \int \left[1 + (\alpha \cdot (P - P_0))^2 \right]^{-1/2} dP \right] \tag{33}$$

$$\frac{1}{2}\cdot h^2 = \frac{2 \cdot K_i \cdot t}{\eta \cdot \varnothing} \cdot \left[P - \int \frac{1}{\sqrt{1 + (\alpha \cdot (P - P_0))^2}} dP \right] \tag{34}$$

Applying the substitution $u = \alpha \cdot (P - P_0)$, $du = \alpha \cdot dP$ results in:

$$h^2 = \frac{2 \cdot K_i \cdot t}{\eta \cdot \varnothing} \cdot \left[P - \int \frac{1}{\sqrt{1 + u^2}} \cdot \frac{1}{\alpha} \cdot du \right] \tag{35}$$

$$h^2 = \frac{2 \cdot K_i \cdot t}{\eta \cdot \varnothing} \cdot \left[P - \frac{1}{\alpha} \cdot \int \frac{1}{\sqrt{1 + u^2}} \cdot du \right] \tag{36}$$

$$h^2 = \frac{2 \cdot K_i \cdot t}{\eta \cdot \varnothing} \cdot \left[P - \frac{\operatorname{arcsinh}(u)}{\alpha} \cdot + C \right] \tag{37}$$

Replacing back $u = \alpha \cdot (P - P_0)$:

$$h^2 = \frac{2 \cdot K_i \cdot t}{\eta \cdot \varnothing} \cdot \left[P - \frac{\operatorname{arcsinh}\left(\alpha \cdot (P - P_0)\right)}{\alpha} + C \right] \tag{38}$$

To estimate the value of C, initial conditions must be applied $P = P_0$ and $h = 0$.

$$0 = \frac{2 \cdot K_i \cdot t}{\eta \cdot \varnothing} \cdot \left[P_0 - \frac{\operatorname{arcsinh}(0)}{\alpha} \cdot + C \right] \tag{39}$$

$$C = -P_0 \tag{40}$$

Then, the general solution for n1 = 2 is:

$$h^2 = \frac{2 \cdot K_i \cdot t}{\eta \cdot \varnothing} \cdot \left[(P - P_0) - \frac{\operatorname{arcsinh}\left(\alpha \cdot (P - P_0)\right)}{\alpha} \right] \tag{41}$$

In order to analyze the accuracy of the results, the evolution of the square of the infiltrated height as a function of pressure drop is plotted in Figures 7a and 7b. As shown, experimental results are perfectly framed between the line defined by Darcy's law (assuming that the liquid permeability is invariant and therefore permeability measured with polyethylene glycol is used) and the curve modified with the Van Genuchten equation.

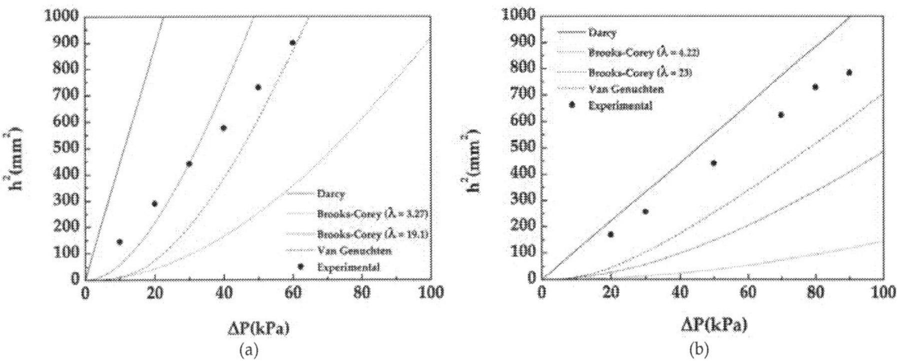

Figure 7. Evolution of the square of infiltrated height for liquid aluminium-silicon alloy as a function of pressure drop on a preform of graphite particles P1 at 700°C) (a). Evolution of the square of the infiltrated height for liquid aluminium-silicon alloy as a function of pressure drop on a preform of carbon particles P2 at 700°C) (b) Characteristics of the graphite particles are listed in the appendix. The coefficients for the BC model have been extracted from [53].

The results shown here clearly demonstrate that it is not necessary to solve the Richards equation (modified Darcy's law to take into account saturation), which has only numerical solutions to predict infiltration in composite materials unlike what was suggested by Mortensen et al. [54]. It is noteworthy that many authors in soil mechanics indicate that Darcy's law is more universal than it was originally supposed.

6. The effect of variable radius

Although the variation of the radius is not a very common phenomenon in the manufacture of composite materials, it is a process that can occur. The variation of the radius is a fundamental issue for example in the manufacture of SiC by the RBSC process or reactive infiltration and SiC derived CMCs. This is a rather complex situation since the variation of the radius affects the permeability of the porous media.

To solve the Navier-Stokes equation it has been assumed that the reduction of the radius, caused by a chemical reaction, is controlled by diffusion. The equations presented so far are valid for infiltration in nonreactive systems. However, they are not valid for reactive infiltration because the pore radius, in this kind of systems, is not constant but decreases with the generation of new solid. As discussed above, the threshold pressure and the permeability depend on the radius, so they are not invariant over time.

So for reactive systems, it is necessary to re-solve the reduced Navier-Stokes equation considering the variation of the radius:

$$\phi \cdot \frac{dz}{dt} = -\frac{K}{\eta} \cdot \frac{dP}{dz} \tag{42}$$

After integration, the following expression is obtained:

$$\frac{1}{2} h^2 = - \int \frac{K}{\varnothing \cdot \eta} dP \cdot dt \tag{43}$$

Usually in reactive systems, infiltration takes place spontaneously (without applying an external pressure), so $\Delta P = P_0$. Differentiating the expression for P_0 results in:

$$\mathrm{dP} = -2 \cdot \lambda \cdot \gamma_{lv} \cdot cos\theta \cdot \frac{(1-\phi)}{\Phi \cdot r^2} \cdot dr \tag{44}$$

Combining this expression with eq. 6 yields:

$$\frac{1}{2} \cdot h^2 = - \frac{a \cdot r^2}{\varnothing \cdot \eta} \left(-2 \cdot \lambda \cdot \gamma_{lv} \cdot cos\theta \cdot \frac{(1-\phi)}{\varnothing \cdot r^2} \right) \cdot dr \cdot dt \tag{45}$$

Simplifying and rearranging:

$$h^2 = 4 \cdot \lambda \cdot a \cdot \gamma_{lv} \cdot cos\theta \cdot \frac{(1-\phi)}{\eta \cdot \varnothing^2} \cdot \; dr \cdot dt \tag{46}$$

In principle, porosity of the system is not constant, but for a first approximation it can be considered so. To proceed with the solution of the equation, a relationship between radius and time is necessary. Assuming that the controlling step of the reaction is the diffusion of C through the SiC layer, then the model of decreasing nucleus developed in the decade of the 1950s can be used [55, 56]:

$$t = \frac{M_{Si} \cdot \rho_{SiC} \cdot r_o^2}{6 \cdot M_{SiC} . \mathcal{D}_{C/SiC} \cdot \rho_{Si}} \left[1 - 3 \cdot \left(\frac{r}{r_o} \right)^2 + 2 \cdot \left(\frac{r}{r_o} \right)^3 \right] \tag{47}$$

$$dt = \frac{M_{Si} \cdot \rho_{SiC} \cdot r_o^2}{6 \cdot M_{SiC} . \mathcal{D}_{C/SiC} \cdot \rho_{Si}} \cdot \left[-6 \cdot \left(\frac{r}{r_o^2} \right) + 6 \cdot \left(\frac{2r^2}{r_o^3} \right) \right] \cdot dr \tag{48}$$

Where ρ_{SiC} is the density of SiC, r_0 the initial pore radius, $D_{C/SiC}$ the diffusion coefficient of C atoms through SiC, and ρ_{Si} the density of Si atoms in kg/m^3, M_{SiC} and M_{Si} the molar weight of SiC and Si, respectively.

Introducing these relations into eq. 46 and defining:

$$G = 4 \cdot \lambda \cdot a \cdot \gamma_{lv} \cdot cos\theta \cdot \frac{(1-\phi)}{\eta \cdot \varnothing^2} \frac{M_{Si} . \rho_{SiC} \cdot r_o^2}{6 \cdot \mathcal{D}_{C,SiC} \cdot M_{SiC} \cdot \rho_{Si}} \tag{49}$$

Results in the following expression:

$$h^2 = G \cdot \left[-6 \cdot \left(\frac{r}{r_o^2} \right) + 6 \cdot \left(\frac{r^2}{r_o^3} \right) \right] dr \cdot dr \tag{50}$$

Solving the first integral yields:

$$h^2 = G \cdot \int_0^r \left[A - 3 \cdot \left(\frac{r}{r_o} \right)^2 + 2 \cdot \left(\frac{r}{r_o} \right)^3 \right] \cdot dr \tag{51}$$

Where A is an integration constant. Solving the second integral results in:

$$h^2 = G \cdot \left(A \cdot r - \left(\frac{r^3}{r_o^2} \right) + \frac{1}{2} \cdot \left(\frac{r^4}{r_o^3} \right) + B \right) \tag{52}$$

where B is an integration constant. In order to obtain the final solution, boundary conditions must be used to find the values of A and B:

a. $r = r_o \rightarrow h = 0 \rightarrow t = 0$

b. $r = 0 \rightarrow h = h_{MAX} \rightarrow t = \tau \rightarrow \frac{dh}{dr} = 0$

Where $\tau = \frac{M_{Si} \cdot \rho_{SiC} \cdot r_o^2}{6 \cdot M_{SiC} \cdot \mathcal{D}_{C/SiC} \cdot C_{Si}}$

By applying condition a), the next expression is obtained:

$$0 = G \cdot \left(A \cdot r_0 - \left(\frac{r_0^3}{r_o^2} \right) + \frac{1}{2} \cdot \left(\frac{r_o^4}{r_o^3} \right) + B \right) \tag{53}$$

Applying condition b) leads to:

$$h_{MAX}^2 = G \cdot B \tag{54}$$

and

$$\frac{dh}{dr} = 0 = \frac{G \cdot \left(A - 6 \cdot \left(\frac{r^2}{r_o^2} \right) + 2 \cdot \left(\frac{r^3}{r_o^3} \right) \right)}{2 \sqrt{G \cdot \left(A \cdot r - \left(\frac{r^3}{r_o^2} \right) + \frac{1}{2} \cdot \left(\frac{r^4}{r_o^3} \right) + B \right)}} \tag{55}$$

Solving eq. 55 leads to:

$$A = 0$$

Replacing the value of A in eq. 53 results in:

$$B = \frac{1}{2} \cdot r_0$$

Replacing the value of B in eq. 54 provides the expression for h_{MAX}:

$$h_{MAX} = \sqrt{\lambda \cdot a \cdot \gamma_{lv} \cdot cos\theta \cdot \frac{(1-\phi)}{\eta \cdot \varnothing^2} \cdot \frac{M_{Si} \cdot \rho_{SiC} \cdot r_o^3}{3 \cdot \mathcal{D}_{C,SiC} \cdot M_{SiC} \cdot \rho_{Si}}} \tag{56}$$

And after replacing the values of A and B, and rearranging the final expression h(r) is obtained:

$$h^2 = \lambda \cdot a \cdot \gamma_{lv} \cdot cos\theta \cdot \frac{(1-\phi)}{\eta \cdot \varnothing^2} \cdot \frac{\rho_{SiC} \cdot r_o^2 \cdot r}{3 \cdot \mathcal{D}_{C,SiC} \cdot \rho_{Si}} \cdot \left[\left(\frac{r}{r_0}\right)^3 - 2 \cdot \left(\frac{r}{r_0}\right)^2 + \frac{r_0}{r} \right] \tag{57}$$

$$h^2 = \frac{h_{MAX}^2}{r_0} \cdot r \cdot \left[\left(\frac{r}{r_0}\right)^3 - 2 \cdot \left(\frac{r}{r_0}\right)^2 + \frac{r_0}{r} \right] \tag{58}$$

$$h^2 = h_{MAX}^2 \cdot \left[\left(\frac{r}{r_0}\right)^4 - 2 \cdot \left(\frac{r}{r_0}\right)^3 + 1 \right] \tag{59}$$

It must be taken into consideration that this equation is valid only as long as the modification of the porosity is negligible. For evaluating h(t), eq. 59 and eq. 47 must be combined.

Figure 8 shows the calculated infiltration distance with time for a carbon preform with an initial pore radius of 30 micron and 50% open porosity, infiltrated with pure Si, using Darcy's law and the variable radius model for cylindrical and spherical pores. The variable radius models predict a slower rate of infiltration, which is closer to the experimental data obtained by several authors [12, 57-61].

Another approach is to consider that the pore closure is limited by the chemical reaction that originates it. In this case, the equations of the decreasing nucleus model to be used are:

$$t = \frac{M_{Si} \cdot \rho_{SiC} \cdot r_0}{k_s \cdot \rho_{Si} \cdot M_{SiC}} \cdot \left(1 - \frac{r}{r_0}\right) \tag{60}$$

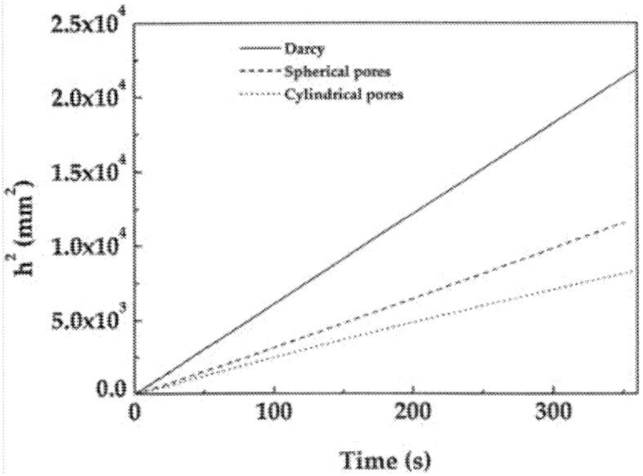

Figure 8. Comparison of the square distance of infiltration (h^2) vs. time, calculated using Darcy's law and variable radius models for a carbon with an initial pore radius of 30 micron, and 50% porosity, infiltrated with pure Si. For the variable radius model, the pore closure is assumed to be controlled by diffusion, and two different pore geometries are considered. Infiltration is simulated until the pore radius decreases by 30%.

$$dt = -\frac{\rho_{SiC}}{k_s \cdot \rho_{Si}} \cdot dr \tag{61}$$

Introducing this equation into the reduced Navier-Stokes equation and solving the integrals results in:

$$h^2 = -4 \cdot \lambda \cdot a \cdot \gamma_{lv} \cdot cos\theta \cdot \frac{(1-\phi)}{\eta \cdot \varnothing^2} \cdot \frac{M_{Si} \cdot \rho_{SiC}}{k_s \cdot \rho_{Si} \cdot M_{SiC}} \cdot \int (r + A) \cdot dr \tag{62}$$

$$h^2 = -4 \cdot \lambda \cdot a \cdot \gamma_{lv} \cdot cos\theta \cdot \frac{(1-\phi)}{\eta \cdot \varnothing^2} \cdot \frac{M_{Si} \cdot \rho_{SiC}}{k_s \cdot \rho_{Si} \cdot M_{SiC}} \cdot \left(\frac{r^2}{2} + A \cdot r + B \right) \tag{63}$$

The integration constants A and B are calculated using contour conditions:

c. $r = r_o \rightarrow h = 0 \rightarrow t = 0$

d. $r = 0 \rightarrow h = h_{MAX} \rightarrow t = \tau \rightarrow dh/dr = 0$

a. For condition a):

$$0 = -4 \cdot \lambda \cdot a \cdot \gamma_{lv} \cdot cos\theta \cdot \frac{(1-\phi)}{\eta \cdot \varnothing^2} \cdot \frac{M_{Si} \cdot \rho_{SiC}}{k_s \cdot \rho_{Si} \cdot M_{SiC}} \cdot \left(\frac{r_0^2}{2} + A \cdot r_0 + B \right) \tag{64}$$

For condition b):

$$\frac{dh}{dr} = 0 = \frac{-4\cdot\lambda\cdot a\cdot\gamma_{lv}\cdot cos\theta\cdot\frac{(1-\phi)}{\eta\cdot\varnothing^2}\cdot\frac{M_{Si}\cdot\rho_{SiC}}{k_s\cdot\rho_{Si}\cdot M_{SiC}}\cdot(0+A)}{2\cdot\sqrt{-4\cdot\lambda\cdot a\cdot\gamma_{lv}\cdot cos\theta\cdot\frac{(1-\phi)}{\eta\cdot\varnothing^2}\cdot\frac{M_{Si}\cdot\rho_{SiC}}{k_s\cdot\rho_{Si}\cdot M_{SiC}}\cdot\left(\frac{0}{2}+A\cdot 0+B\right)}} \quad (65)$$

$$A = 0 \quad (66)$$

and replacing in condition a):

$$0 = -4\cdot\lambda\cdot a\cdot\gamma_{lv}\cdot cos\theta\cdot\frac{(1-\phi)}{\eta\cdot\varnothing^2}\cdot\frac{M_{Si}\cdot\rho_{SiC}}{k_s\cdot\rho_{Si}\cdot M_{SiC}}\cdot\left(\frac{r_0^2}{2}+0\cdot r_0+B\right) \quad (67)$$

$$B = -\frac{r_0^2}{2} \quad (68)$$

and

$$h_{MAX} = \sqrt{\lambda\cdot a\cdot\gamma_{lv}\cdot cos\theta\cdot\frac{(1-\phi)}{\eta\cdot\varnothing^2}\cdot\frac{M_{Si}\cdot\rho_{SiC}}{k_s\cdot\rho_{Si}\cdot M_{SiC}}\cdot r_0^2} \quad (69)$$

Replacing the values of A, B and h_{MAX}:

$$h^2 = 4\cdot\lambda\cdot a\cdot\gamma_{lv}\cdot cos\theta\cdot\frac{(1-\phi)}{\eta\cdot\varnothing^2}\cdot\frac{M_{Si}\cdot\rho_{SiC}}{k_s\cdot\rho_{Si}\cdot M_{SiC}}\cdot\left(\frac{r_0^2-r^2}{2}\right) \quad (70)$$

Figure 9 shows the square of the infiltration distance as a function of time for a carbon preform with an initial pore radius of 30 micron and 50% open porosity infiltrated with pure Si, using Darcy's law and the variable radius model for a chemically controlled pore closure. As seen, infiltration rate decreases with time, just as in the previous case (diffusion controlled pore closure). However, the experimental data [12, 42, 59-66] show that the infiltration rate is lower than the one predicted by these models; furthermore, infiltrated height is proportional to t instead of to $t^{(1/2)}$. Experiments indicate that infiltration is controlled by the reaction at the triple line and not by the viscous phenomena described here. However, both interpretations are compatible. It is probable that in the first stage of the process, infiltration is controlled by the reaction at the triple line, and after the SiC interface has been formed, the viscous flow becomes the controlling process. A completely different approach to the process of infiltration is performed by Sergi et al. [67, 68], in which they use Lattice-Bolztman models to describe infiltration in a single capillary and predict that infiltration distance is proportional to time when considering pore narrowing of the capillary due to the reaction. However, their predictions have not been validated experimentally up to date. It is possible to conclude that a deeper understanding of the reactive infiltration phenomenon should be gained in order to discern what is happening in the overall process.

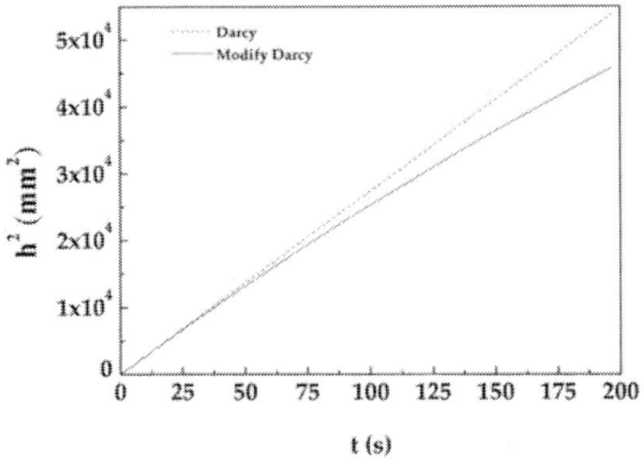

Figure 9. Comparison of the square distance of infiltration (h^2) as a function of time, calculated using Darcy's law and the variable radius models for a carbon with an initial pore radius of 30 micron, and 50% porosity, infiltrated with pure Si. For the variable radius model, pore closure is assumed to be controlled by the chemical reaction. Infiltration is simulated until the pore radius decreases by 30%.

7. Effect of variable viscosity: Curing polymeric resins

The properties of PMCs depend strongly on the type of matrix used. Among the possible binders, thermoset resins present a singularity: their viscosity is time dependent [69, 70]. These kinds of polymers present an initial low viscosity that grows exponentially with time. During this process, the material undergoes a series of chemical reactions that cause the cross-linking of polymeric chains yielding a 3D structure with excellent thermal stability and solvent resistance. The process of cross-linking that takes place in these resins is known as curing. In polymer matrix composites in which the matrix cures with time, thus hardening and losing fluidity, viscosity becomes a limiting factor in infiltration.

The evolution of viscosity is usually expressed with the following equation:

$$\eta = \eta_0 \cdot e^{k_n \cdot t} \tag{71}$$

Where η_0 is the initial viscosity and k_n is a strongly temperature-dependent rheological kinetic constant [70]. Starting from the reduced Navier-Stokes equation:

$$\phi \cdot \frac{dz}{dt} = -\frac{K}{\eta} \cdot \frac{dP}{dz} \tag{72}$$

After integration, the following expression is obtained:

$$\frac{1}{2\cdot} h^2 = - \frac{K}{\varnothing \cdot \eta} dP \cdot dt \tag{73}$$

Introducing eq. 71 and solving the integral yields a model for predicting infiltration distance as a function of time for curing polymeric resins:

$$\frac{1}{2\cdot} h^2 = - \frac{K}{\varnothing \cdot \eta_0 \cdot e^{k_n \cdot t}} dP \cdot dt \tag{74}$$

$$\frac{1}{2\cdot} h^2 = - \frac{K}{\varnothing \cdot \eta_0} \cdot \frac{dP}{e^{k_n \cdot t}} \cdot dt \tag{75}$$

$$h^2 = -\frac{2 \cdot K \cdot \Delta P}{\varnothing \cdot \eta_0} \cdot \int e^{-k_n \cdot t} \cdot dt \tag{76}$$

$$h^2 = \frac{2 \cdot K \cdot \Delta P}{\varnothing \cdot \eta_0} \cdot \frac{e^{-k_n \cdot t}}{k_n} + A \cdot t + B \tag{77}$$

To calculate the values of integration constants A and B, initial conditions are applied:

a. $t = 0; h = 0$

b. $t = \infty; \frac{dh}{dt} = 0$

For condition a):

$$\begin{gathered} 0 = \frac{2 \cdot K \cdot \Delta P}{\varnothing \cdot \eta_0} \cdot \frac{1}{e^{k_n \cdot 0}} + A \cdot 0 + B \\ B = -\frac{2 \cdot K \cdot \Delta P}{\varnothing \cdot \eta_0 \cdot k_n} \end{gathered} \tag{78}$$

for condition b):

$$\frac{dh}{dt} = 0 = \frac{-\frac{2 \cdot K \cdot \Delta P}{\varnothing} \cdot e^{-k_n \cdot t} + A}{2 \cdot \sqrt{\cdot \frac{e^{-k_n \cdot t}}{-k_n} + A \cdot t + B}} \qquad (79)$$

$$A = 0$$

Then the final equation results in:

$$h^2 = \frac{2 \cdot K \cdot \Delta P}{\varnothing \cdot \eta_0 \cdot k_n} \cdot e^{-k_n \cdot t} - \frac{2 \cdot K \cdot \Delta P}{\varnothing \cdot \eta_0 \cdot k_n} \qquad (80)$$

$$h^2 = -\frac{2 \cdot K \cdot \Delta P}{\varnothing \cdot \eta_0 \cdot k_n} \cdot (1 - e^{-k_n \cdot t}) \qquad (81)$$

It must be highlighted that the negative sign is only indicative of the direction of flow. Figure 10 shows the evolution of the square of infiltrated height for a preform infiltrated with curing resins with an initial viscosity of 100 Pa s at different temperatures. The drastic effect of temperature on infiltration distance becomes evident, since increasing temperature to 50°C reduces infiltrated distance by one order of magnitude.

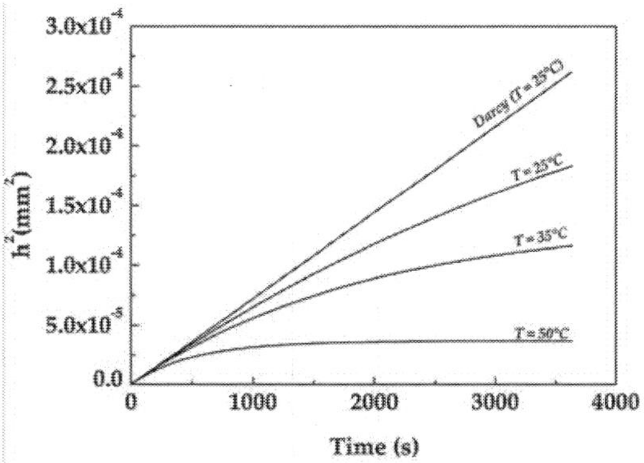

Figure 10. Evolution of the square of infiltration height with time for curing resins into porous preforms at different temperatures.

8. Concluding remarks

The present chapter has introduced the reader to the processes involved in the manufacture of different types of composite materials (PMCs, MMCs, CMCs) via infiltration. Over 90% of the world production of composite materials is carried out by infiltration, especially in the case of PMCs where it is almost the only available processing technique, while in case of metal matrix, production via infiltration represents only 70% [7]. In order to accurately control the properties of the final material and to be able to optimize the production, it is key to identify and understand the different processes involved in the manufacture of these materials. For this purpose, in this chapter, different models have been elucidated, starting from the differential equation of Navier-Stokes and taking into account the basic laws of wetting, by applying different real conditions often found in industry. For PMCs, the main discussed issue was the change in viscosity that thermoset matrixes experiment during infiltration. By introducing this phenomenon in the infiltration equations, it becomes clear that it can be a limiting factor during production since it shows a great deviation from Darcy's law. In the case of MMCs, the main phenomenon analyzed was the effect of saturation on infiltration. It became clear that variable saturation not only affects in a huge manner infiltration kinetics, but also changes the dominating phenomena. The last publications in this area have identified percolation as probably the leading phenomenon on infiltration in unsaturated systems at very low overpressure. Finally, for CMCs, the role of pore narrowing, an issue that has been widely discussed in the literature over the past 10 years was revised. This phenomenon is especially relevant in processing of ceramics and ceramic matrix composites via reactive techniques like the RBSC method used for SiC production. Even though the research on this area is vast, there is still not a full understanding of the effect of pore reduction on infiltration. Most models proposed still yield higher infiltration rates than the ones observed experimentally, because they do not consider the effect of chemical reactivity.

Appendix

Table 1 shows the main characteristic of the carbon particles used in this work.

Particle	D(90)	D(50)	D(10)	D(4,3)	D(3,2)	Span	V_p
P1	25.8	13.4	7.0	15.1	11.6	1.39	0.511
P2	41.0	25.8	15.5	27.2	23.0	0.99	0.515
P3	95.1	62.0	36.8	64.0	53.0	0.94	0.518
P4	173.6	117.6	56.6	124.1	110.4	0.99	0.482

Table 1. Main characteristics of the carbon particles used. Span is defined as D(90) – D(10)/D(50), where D(x) represents the diameter below which x% of particles are encountered. V_p is the particle volume fraction.

Table 2 shows the value of the parameters required to assess h(t) for the several cases exposed in this work. All parameters are in SI units.

Symbol	Magnitude	Value	SI units	Ref.
λ	Geometrical factor	2-4	-	[10, 21-23,25, 29]
a	Proportionality constant	$4\cdot10^{-4}$		[61]
γ_{lv}	Liquid-vapor surface tension of Si (at melting point)	0.750	N/m	[34-36]
θ	Contact angle of Si/C (at Si melting point)	40	°	[31]
η	Dynamic viscosity of molten Si (at melting point)	$0.605\cdot10^{-3}$	Pa·s	[34]
ϱ_{SIC}	Density of SiC	3210	kg/m^3	[9, 12]
M_{SiC}	Molar weight of SiC	$40\cdot10^{-3}$	kg/mol	[9, 12]
M_{Si}	Molar weight of Si	$28\cdot10^{-3}$	kg/mol	[9, 12]
C/,SiC	Diffusion coefficient of C through SiC	$2\cdot10^{-14}$	m^2/s	[62]
ϱ_{SI}	Density of Silicon (at melting point)	2570	kg/m^3	[65]
k_s	kinetic constant in the SiC formation reaction	$4\cdot10^{-8}$	m/s	[71, 72]

Table 2. Value of the main parameters used in this work in SI units.

Acknowledgements

The authors would like to thank the University of Alicante for Danilo Sergi's scholarship under the program "Support for short stays of foreign researchers at the University of Alicante".

Financial support from the Generalitat Valenciana (PROMETEO II/2014/004- FEDER, Master grant "Santiago Grisolia" and PhD grant Vali+d), and the European Union's Seventh Framework Programme (FP7/2007-2013) under the HELM project (grant agreement no. 280464) is gratefully acknowledged.

Author details

Mario Caccia[1], Antonio Camarano[1], Danilo Sergi[2], Alberto Ortona[2] and Javier Narciso[1]

*Address all correspondence to: narciso@ua.es

1 University of Alicante, Department of Inorganic Chemistry, Apdo. Alicante, Spain

2 University of Applied Sciences SUPSI, The iCIMSI Research Institute, Manno, Switzerland

References

[1] Hull D, Clyne TW. An introduction to Composites Materials (2nd edn.). In: Clarke DR, Suresh S, Ward IM, (eds.), *Cambridge Solid State Series*. Cambridge: Cambridge University Press, 1996. p. 1 – 36. ISBN: 0 521 38190 8.

[2] Chung D. Composite Materials - science and applications (2nd edn.). In: Derby B (ed.), *Engineering Materials and Processes*. New York: Springer, 2010. p. 1 – 34. ISBN: 978-1-84882-830-8.

[3] Callister W Jr. Materials science and engineering: an introduction (6th edn.). In: Anderson W, Santor K, (eds.). *Wiley General Materials Science*. New York: Wiley & Son, 2000. p. 577 – 619. ISBN: 0-471-22471-5

[4] Chawla N, Chawla K. *Metal Matrix Composites*. In: Springer (ed.). Springer Science +Business Media New York: Springer-Verlag, 2006. p. 1 – 63. ISBN: 10 0-387-23306-7.

[5] Mallick P, Newman S. *Composites Materials Technology. Process and Properties*. In: Mallick P, Newman S (eds.). Hansel Publisher. Munich: Carl Hanser Verlag, 1990. p. 1 – 9. ISBN: 3-446-15684-4.

[6] Miracle D, Donaldson S. *ASM HandBook Volume 21 – Composites*. In: Moosbrugger Ch, Kinson J (eds.). ASM International. Materials Park (Ohio): ASM International Committee, 2001. p. 39 – 483. ISBN: 0-87170-703-9.

[7] Evans A, Marchi CS, Mortensen A. *Metal Matrix Composites in Industry: An Introduction and a Survey*. In: Springer group (ed.). Springer Science+Business Media. Dordrecht (Netherland): Kluwer Academic Publisher, 2003. p. 1 - 79 ISBN: 978-1-4020-7521-6

[8] Chawla K. *Composite Materials: Science and Engineering (Materials Research and Engineering)*. In: Springer (ed.). Materials Research and Engineering. New York: Springer, 2013. p. 1 – 176. ISBN-10: 0387743642.

[9] Narciso-Romero FJ, Rodríguez-Reinoso F, Díez MA. Influence of the carbon material on the synthesis of silicon carbide. *Carbon* 1999; 37(1):1771-1778. DOI: 10.1016/S0008-6223(99)00045-7

[10] Rodríguez-Guerrero A, Sánchez SA, Narciso J, Louis E, Rodríguez-Reinoso F. Pressure infiltration of Al–12wt.% Si–X (X= Cu, Ti, Mg) alloys into graphite particle preforms. *Acta Mater* 2006; 54(7):1821-1831. DOI: 10.1016/j.actamat.2005.11.041

[11] Narciso J, Alonso A, Pamies A, Garcia-Cordovilla C, Louis E. Factors affecting pressure infiltration of packed SiC particulates by liquid aluminum. *Metall Mater Trans A* 1995; 26(4): 983-990. DOI: 10.1007/BF02649095

[12] Calderon NR, Martinez-Escandell M, Narciso J, Rodríguez-Reinoso F. The role of carbon biotemplate density in mechanical properties of biomorphic SiC. *J Eur Ceram Soc* 2009; 29(3): 465-472. DOI: 10.1016/j.jeurceramsoc.2008.05.049

[13] Ortona A, Donato A, Filacchioni G, De Angelis U, La Barbera A, Nannetti CA, Riccardi B, Yeatman J. SiC–SiC f CMC manufacturing by hybrid CVI–PIP techniques: process optimization. *Fusion Eng Des* 2000; 51: 159-163. DOI:10.1016/S0920-3796(00)00310-0

[14] Nannetti CA, Ortona A, Pinto DA, Riccardi B. Manufacturing SiC-fiber-reinforced SiC matrix composites by improved CVI/slurry infiltration/polymer impregnation and pyrolysis. *J Am Ceram Soc* 2004; 87(7): 1205-1209. DOI: 10.1111/j.1551-2916.2004.tb20093.x

[15] Ortona A. Ceramic matrix composites: reaction bonded. In: Wiley (ed.). *Wiley Encyclopedia of Composites*. Manno (Switzerland): Wiley, 2011. p. 1 – 4. DOI: 10.1002/9781118097298.weoc031

[16] Kaczmara J, Pietrzakb K, Wlosinski W. The production and application of metal matrix composite materials. *J Mater Process Tech* 2000; 106: 58-67. doi:10.1016/S0924-0136(00)00639-7

[17] Harrigan Jr W. Commercial processing of metal matrix composites. *Mat Sci Eng A-Struct* 1998; 244: 75-79. DOI: 10.1016/S0921-5093(97)00828-9

[18] Miracle D. Metal matrix composites – from science to technological significance. *Compos Sci Technol* 2005; 65: 2526-2540. doi:10.1016/j.compscitech.2005.05.027

[19] Brooks RH, Corey AT. *Hydraulic Properties of Porous Media*. Hydrology papers Colorado State University. Fort Collins: Colorado State University; 1964; 5: 1-27.

[20] Van Genuchten M. A closed-form equation for predicting the hydraulic conductivity of unsaturated soils. *Soil Sci Soc Am J* 1980; 44(5): 892-898. doi:10.2136/sssaj1980.03615995004400050002x

[21] Garcia-Cordovilla C, Louis E, Narciso J. Pressure infiltration of packed ceramic particulates by liquid metals. *Acta Mater* 1999; 47(18): 4461-4479. DOI: 10.1016/S1359-6454(99)00318-3

[22] Alonso A, Pamies A, Narciso J, Garcia-Cordovilla C, Louis E. Evaluation of the wettability of liquid aluminum with ceramic particulates (SiC, TiC, AI2O3) by means of pressure infiltration. *Metall Trans A* 1993; 24 (6): 1423-1432. DOI: 10.1007/BF02668210

[23] Molina JM, Saravanan RA, Arpón R, Garcıa-Cordovilla C, Louis E, Narciso J. Pressure infiltration of liquid aluminium into packed SiC particulate with a bimodal size distribution. *Acta Mater* 2002; 50(2): 247-257. DOI: 10.1016/S1359-6454(01)00348-2

[24] Prieto R, Molina JM, Narciso J, Louis E. Fabrication and properties of graphite flakes/metal composites for thermal management applications. *Scripta Mater* 2008; 59(1): 11-14. DOI: 10.1016/j.scriptamat.2008.02.026

[25] Molina JM, Piñero E, Narciso J, García-Cordovilla C, Louis E. Liquid metal infiltration into ceramic particle preforms with bimodal size distributions. *Curr Opin Solid St M* 2005; 9(4): 202-210. DOI: 10.1016/j.cossms.2006.02.005

[26] Narciso J, Alonso A, Pamies A, Garcia-Cordovilla C, Louis E. Wettability of binary and ternary alloys of the system Al-Si-Mg with SiC particulates. *Scripta Metall Mater* 1994; 31(11): 1495-1500. DOI: 10.1016/0956-716X(94)90063-9

[27] Tian J, Piñero E, Narciso J, Louis E. Effects of temperature on pressure infiltration of liquid Al and Al–12wt.% Si alloy into packed SiC particles. *Scripta Mater* 2005; 53(12): 1483-1488. DOI: 10.1016/j.scriptamat.2005.07.038

[28] Molina JM, Arpón R, Saravanan RA, Garcıa-Cordovilla C, Louis E, Narciso J. Threshold pressure for infiltration and particle specific surface area of particle compacts with bimodal size distributions. *Scripta Mater* 2004; 51(6): 623-627. DOI: 10.1016/j.scriptamat.2004.05.009

[29] Piñero E, Molina JM, Narciso J, García-Cordovilla C, Louis E. The intrinsic permeability of packed SiC particles with monomodal and bimodal size distributions. *J Compos Mater* 2008; 42: 2795-2804. DOI:10.1177/0021998308096502

[30] Sánchez AS, Narciso J, Rodríguez-Reinoso F, Bernard D, Watson IG, Lee PD, Dashwood RJ. Characterization of lightweight graphite based composites using X-ray microtomography. *Adv Eng Mater* 2006; 8(6): 491-495. DOI: 10.1002/adem.200600101

[31] Eustathopoulos N, Nicholas MG, Drevet B. Wettability at High Temperatures. In: Cahn RW (ed.). *Pergamon Materials Series*. Cambridge: Pergamon; 1999. ISBN: 978-0-08-042146-9

[32] Saravanan RA, Molina JM, Narciso J, García-Cordovilla C, Louis E. Effects of nitrogen on the surface tension of pure aluminium at high temperatures. *Scripta Mater* 2001: 44 (6): 965–970. DOI: 1016/S1359-6462(00)00688-6

[33] Pamies A, Garcia-Cordovilla C, Louis E. The measurement of surface tension of liquid aluminium by means of the maximum bubble pressure method: the effect of surface oxidation. *Scripta Metall* 1984; 18 (9): 869-872. DOI: 10.1016/0036-9748(84)90251-5

[34] Amore S, Giuranno D, Novakovic R, Ricci E, Nowak R, Sobczak N. Thermodynamic and surface properties of liquid Ge–Si alloys. *Calphad*. 2014; 44: 95-101. DOI:10.1016/j.calphad.2013.07.014

[35] Novakovic R, Ricci E, Gnecco F, Giuranno D, Borzone G. Surface and transport properties of Au–Sn liquid alloys. *Surf Sci* 2005; 599: 230-247. DOI:10.1016/j.susc.2005.10.009.

[36] Egry I, Ricci E, Novakovic R, Ozawa S. Surface tension of liquid metals and alloys — recent developments. *Adv Colloid Interface Sci* 2010;159: 198-212. DOI:10.1016/j.cis.2010.06.009

[37] Bahraini M, Molina JM, Kida M, Weber L, Narciso J, Mortensen A. Measuring and tailoring capillary forces during liquid metal infiltration. *Curr Opin Solid St M* 2005; 9(4): 196-201. DOI: 10.1016/j.cossms.2006.02.007

[38] M Bahraini M, Weber L, Narciso J, Mortensen A. Wetting in infiltration of alumina particle preforms with molten copper. *J Mater Sci* 2005; 40(9-10): 2487-2491. DOI: 10.1007/s10853-005-1980-1

[39] Molina M, Rodriguez-Guerrero A, Bahraini M, Weber L, Narciso J, Rodriguez-Reinoso F, Louis E, Mortensen A. Infiltration of graphite preforms with Al–Si eutectic alloy and mercury. *Scripta Mater* 2007; 56(11): 991-994. DOI: 10.1016/j.scriptamat.2007.01.042

[40] Mortensen A, Masur LJ, Cornie JA, Flemings MC. Infiltration of fibrous preforms by pure metal. 1. Theory. *Metall Mater Trans A* 1989; 20(11): 2535-2547. DOI: 10.1007/BF02666688

[41] Masur LJ, Mortensen A, Cornie JA, Flemings MC. Infiltration of fibrous preforms by pure metal. 2. Experiment. *Metall Mater Trans A* 1989; 20(11): 2549-2557. DOI: 10.1007/BF02666689

[42] Michaud V, Mortensen A. Infiltration processing of fibre reinforced composites: governing phenomena. *Compos Part A-Appl S* 2001; 32(8): 981-996. DOI: 10.1016/S1359-835X(01)00015-X

[43] Richards LA. Capillary conduction of liquids through porous mediums. *Physics 1*. 1931; 1: 318-333. DOI: 10.1007/s10853-005-1950-7

[44] Benavente D. Locks P, García-del-Cura MA, Ordoñez S. Predicting the capillary imbibition of porous rocks from microstructure. *Transp Porous Med* 2002; 49: 59–76. DOI: 10.1023/A:1016047122877

[45] Szymkiewicz A. Modelling water flow in unsaturated porous media. In: Rowínski P (ed.), *GeoPlanet: Earth and Planetary Sciences*. Berlin: Springer: 2013. p. 9 -170 DOI: 10.1007/978-3-642-23559-7_2

[46] Bear, Jacob, Cheng, Alexander HD. Modeling Groundwater Flow and Contaminant Transport. In: Bear J, editor. Theory and Applications of Transport in Porous Media. New York: Springer: 2010. p. 109-247. DOI: 10.1007/978-1-4020-6682-5

[47] Bear J, Bachmat Y. Introduction to modeling of transport phenomena in porous media. In: Bear J (ed.), *Theory and Applications of Transport in Porous Media*. London: Kluwer Academic Publishers: 1990. p. 263 - 285. DOl: 10.1007/978-94-009-1926-6

[48] Bear J. Dynamics of fluids in porous media. In: Bear J (ed.) *Dovers Books on Physics and Chemistry*. New York: Dovers Publications: 1972. p. 119 – 186. ISBN:0-486-65675-6

[49] Louis E, Miralles JA, Molina JM. High temperature infiltration at low overpressures: Darcy's law, the slug-flow hypothesis and percolation. *J Mater Sci* 2015; 50: 1655-1665. DOI: 10.1007/s10853-014-8726-x

[50] Léger A, Molina-Jorda JM, Weber L, Mortensen A. Percolation and universal scaling in composite infiltration processing. *Mater Res Lett* 2015; 3(1): 7-15. DOI: 10.1080/21663831.2014.948692

[51] Dopler T, Modaressi A, Michauds V. Simulation of metal-matrix composite isothermal infiltration processing. *Metall Mater Trans B* 2000; 31(2): 225-234. DOI: 10.1007/s11663-000-0041-z

[52] Molina JM, Narciso J, Louis E. On the triple line in infiltration of liquid metals into porous preforms. *Scripta Mater* 2010; 62(12): 961-965. DOI: 10.1016/j.scriptamat.2010.03.015

[53] Rodríguez-Guerrero A, Molina JM, Rodrıguez-Reinoso F, Narciso J, Louis E. Pore filling in graphite particle compacts infiltrated with Al–12 wt.%Si and Al–12 wt.%Si–1 wt.%Cu alloys. *Mat Sci Eng A* 2008; 495: 276-281. DOI: doi:10.1016/j.msea.2008.01.071

[54] Michauds V. Mortensen A. On measuring wettability in infiltration processing. *Scripta Mater* 2007; 56(10); 859-862. doi:10.1016/j.scriptamat.2007.02.002

[55] Levenspiel O. Ingeniería de las Reacciones (2nd edn). Wiley (ed.). Barcelona: Revertre; 1978. ISBN: 84-291-7325-0

[56] Levenspiel O. El Omnilibro de los Reactores Químicos. Wiley (ed.). Barcelona: Revertre; 1986. ISBN: 84-291-7336-6

[57] Calderon NR, Martínez-Escandell M, Narciso J, Rodríguez-Reinoso F. The combined effect of porosity and reactivity of the carbon preforms on the properties of SiC produced by reactive infiltration with liquid Si. *Carbon* 2009; 47(9): 2200-2210. DOI: 10.1016/j.carbon.2009.04.002

[58] Voytovych R, Bougiouri V, Calderon NR, Narciso J, Eustathopoulos N. Reactive infiltration of porous graphite by NiSi alloys. *Acta Mater* 2008; 56(10): 2237-2246. DOI: 10.1016/j.actamat.2008.01.011

[59] Calderon NR, Voytovych R, Narciso J, Eustathopoulos N. Wetting dynamics versus interfacial reactivity of AlSi alloys on carbon. *J Mater Sci* 2010; 45(8): 2150-2156. DOI: 10.1007/s10853-009-3909-6

[60] Bougiouri V, Voytovych R, Rojo-Calderon N, Narciso J, Eustathopoulos N. The role of the chemical reaction in the infiltration of porous carbon by NiSi alloys. *Scripta Mater* 2006; 54(11): 1875-1878. DOI: 10.1016/j.scriptamat.2006.02.015

[61] Calderon NR, Voytovych R, Narciso J, Eustathopoulos N. Pressureless infiltration versus wetting in AlSi/graphite system. *J Mater Sci* 2010; 45(16): 4345-4350. DOI: 10.1007/s10853-010-4358-y

[62] Mortensen A, Drevet B, Eustathopoulos N. Kinetics of diffusion-limited spreading of sessile drops in reactive wetting. *Scripta Mater* 1997; 36(6): 654-651.

[63] Landry K, Eustathopoulos N. Dynamics of wetting in reactive metal/ceramic systems: linear spreading. *Acta Mater* 1996; 44(10): 3923-3932. DOI: 10.1016/S1359-6454(96)00052-3

[64] Dezellus O, Hodaj F, Eustathopoulos N. Chemical reaction-limited spreading: the triple line velocity versus contact angle relation. *Acta Mater* 2002; 50(19): 4741-4753. DOI: 10.1016/S1359-6454(02)00309-9

[65] Dezellus O, Jacques S, Hodaj F, Eustathopoulos N. Wetting and infiltration of carbon by liquid silicon. *J Mater Sci* 2005; 40(10): 2307-2311. DOI: 10.1007/s10853-005-1950-7

[66] Bougiouri V, Voytovych R, Dezellus O, Eustathopoulos N. Wetting and reactivity in Ni-Si/C system: experiments versus model predictions. *J Mater Sci* 2007; 42(6): 2016-2023. DOI: 10.1007/s10853-006-1483-8

[67] Sergi D, Grossi L, Leidi T, Ortona A. Surface growth effects on reactive capillary-driven flow: Lattice Boltzman investigation. *Eng Appl Comp Fluid* 2014; 8(4): 549-561. DOI: 10.1080/19942060.2014.11083306

[68] Sergi D, Grossi L, Leidi T, Ortona A. Lattice Boltzmann simulations on the role of channel structure for reactive capillary infiltration. *Cond Mat Soft* 2014: arXiv: 1409.0954

[69] Bowman JM, Buffenbarger R, Clark J, Embrekts L, Goldberg S, Lambert S. Advanced materials by design. In: Nichols RW (ed.). *U.S. Congress, Office of Technology Assessment*. Washington, DC: U.S. Government Printing Office, 1988. NTIS order #PB88-243548

[70] Cañamero-Martínez P, Fernández-García M, Fuente JL. Rheological cure characterization of a polyfunctional epoxy acrylic resin. *React Funct Polym* 2010; 70: 761-766. doi:10.1016/j.reactfunctpolym.2010.07.010

[71] Fitzer E, Gadow R. Investigations of the reactivity of different carbons with liquid silicon. In: *Proceedings of International Symposium on Ceramic Components for Engines; 1983, Japan*. Tokyo, Japan: KTK Scientific publishers; 1984. p. 561-572.

[72] Messner RP, Chiang Y. Liquid-phase reaction-bonding of silicon carbide using alloyed silicon-molybdenum melts. J Am Ceram Soc 1990; 73(151): 1193-1200. DOI: 10.1111/j.1151-2916.1990.tb05179.x

Chapter 6

Modification of Surface Energy and Wetting of Textile Fibers

Franco Ferrero and Monica Periolatto

Additional information is available at the end of the chapter

http://dx.doi.org/10.5772/60812

Abstract

The modification of the surface energy of textile fibers to improve functional properties such as the wettability was reviewed. This modification can be achieved by physical or chemical methods or by the combination of both. Applications of plasma treatment to improve the wettability of natural and synthetic fibers were considered and some methods of wettability measurement were mentioned. Subsequently the methods aimed to confer water and oil repellency were discussed and the treatment by UV curing of fluorochemicals was explained in detail. Finally the sol-gel techniques useful to modify the surface properties of textiles were introduced and the results of water and oil repellency achievable by sol-gel were presented.

Keywords: Surface energy, Wetting, Textile fibers, Plasma, Contact angle, Water repellency, Oil repellency, Fluorochemicals, UV-curing, Sol-gel

1. Introduction

The modification of the surface energy of textile fibers is pursued with the aim of improving their own hydrophilicity, wettability, and dyeability or of conferring functional properties such as hydro and oil repellency, soil release, adhesion improvement, and antistatic performances. If the modification is confined to a thin surface layer of the fibers, the bulk properties of a textile material, such as mechanical strength, flexibility, breathability, and softness, should not be compromised.

The wettability of a surface depends on the surface tension of the liquid that goes into contact with the same. Water has a very high surface tension (72.8 mN/m), so it tends to wet only surfaces bearing highly polar groups; otherwise, it forms spherical drops with contact angles higher than 90°. Instead, apolar liquids of lower surface tension get drops flatter than those of water. Hydrophobic or oleophobic surfaces are difficult to wet by water or apolar liquids, respectively, and are called low-energy surfaces. Wetting, in reality, is much more complex than as described by classical laws. This primarily comes from the non-ideality of solid substrates that are both rough and chemically heterogeneous [1]. Such a situation commonly occurs in the case of textile materials and can be evidenced by the hysteresis of the contact angle [2].

The surface modification of textile fibers can be achieved by physical or chemical methods or by the combination of both. Plasma treatments and exposure to radiations are mainly representative of physical methods, although their effects are often accomplished in the presence of reactive gases or after impregnation with suitable chemicals. However, chemical treatments can generally be carried out with oxidants or other finishing agents, followed by thermal treatment. Among the latter methods, sol–gel techniques are the more promising for surface modification.

Among the textile fibers, there are many differences regarding wettability and surface structure. Natural fibers have cellular structures that are more complex than those of man-made fibers obtained by chemical spinning. Wool and fine animal fibers are mainly composed of keratin, but show the structure of composite material formed by an assembly of cuticle cells in the form of scales and cortical cells surrounded by a keratin cell membrane and held together by an intercellular cement. The cell membrane consists of a chemically resistant protein layer and a lipidic layer that constitutes a hydrophobic barrier to the transfer of water and dye molecules from an aqueous solution [3]. Therefore, a wettability improvement of these fibers can be obtained by the removal of the hydrophobic lipidic layer and by the introduction of polar groups on the surface, both performed by chemical and/or physical means. A chemical treatment is usually made by chlorination, which modifies the scale edges of wool and increases the critical surface tension of the fibers. In such a manner, the chlorinated fibers are made more wettable and dyeable and can be coated with polymers, conferring the shrink-resist effect. However, many research works are carried out to find alternative processes that avoid chlorination in order to remove the problem of the formation of absorbable organohalogen compounds in wastewaters [4]. To this aim, plasma treatments have extensively been studied, but even other cheaper eco-friendly processes have been experimented, in particular for dyeability improvement, such as treatment with enzymes or ultraviolet (UV) irradiation. A list of selected references is reported in Table 1.

Cotton and cellulose fibers are more homogeneous than wool and are richer in hydroxyl groups ththat confer higher hydrophilicity; hence, their surface modifications are mainly directed to confer hydro and oil repellency. This topic will be treated in paragraphs 3 and 4. Moreover, the exhaust dyeing of cotton with anionic dyes, i.e., direct and reactive, request a high concentration of electrolytes in dyebath to reduce negative charges on the fiber surface and to promote the exhaustion of dyes. This requirement creates environmental issues due to the removal of high concentrations of salts and dyes from wastewaters. To overcome such

Process type	Topic	Reference
LPP	Surface characterization	Kan 2004 [5]
LPP	Surface composition	Kan 2005 [6]
LPP	Review	Kan 2007a [7]
LPP	Effect on wool dyeing	Kan 2007b [8]
APP	Wettability improvement	Wang 2007 [9]
LPP	Application of HMDSO coating	Rombaldoni 2008 [10]
LPP	Characterization of wool coated	Rombaldoni 2009 [11]
APP	Ageing effect	Naebe 2011 [12]
Enzyme	Effect on dyeing kinetics	Riva 2002 [13]
Enzyme	Effect on wool dyeing	Onar 2005 [14]
Enzyme	Dyeability, antifelting	Cardamone 2006 [15]
Enzyme	Effect on wool dyeing	Parvinzadeh 2007 [16]
Enzyme	Effect on wool dyeing	Cui 2009 [17]
Enzyme	Low temperature dyeing	Periolatto 2010 [18]
Enzyme	Influence on dyeing	Periolatto 2011 [19]
Enzyme	Influence of lipids on dyeing	Ferrero 2015 [3]
UV irradiation	Printability improvement	Shao 2001 [20]
UV irradiation	Surface modification, dyeing	Xin 2002 [21]
UV irradiation	Improving properties	El-Zaher 2002 [22]
UV irradiation	Dyeability improvement	Micheal 2003 [23]
UV irradiation	Multifunctional finishing	Periolatto 2013 [24]
UV irradiation	Differential dyeing	Migliavacca 2014 [25]
UV irradiation	Low temperature dyeing	Periolatto 2014 [26]

APP: Atmospheric pressure plasma; LPP: low-pressure plasma.

Table 1. Processes of wool surface modifications.

problems, many treatments with plasma, chemicals, and polymers were experimented to impart a cationic character to the cotton fiber surface [27]. Chitosan, 2-amino-2-deoxy-(1 → 4)-β-D-glucopyranan, derived from the deacetylation of the chitin component of the shells of crustaceans, is undoubtedly one of the more promising multifunctional polymers for surface modification of textiles [28]. It is a biopolymer with unique properties such as biodegradability, nontoxicity, and antimicrobial activity; hence, it was mainly applied to textiles as an antimicrobial finishing agent. Moreover, cotton treated with chitosan shows improved absorption of anionic dyes due to electrostatic attraction arising from the cationized amino groups of chitosan in an acidic medium. However, chitosan should stably be bonded to the fiber surface to ensure the fastness of treatment to washing. To this aim, crosslinking agents such as dialdehydes are used, although these are toxic chemicals. On the contrary, an eco-friendly grafting of chitosan onto cotton can be carried out by radical UV curing in the presence of a low concentration of a photoinitiator [29, 30].

A list of selected references on the surface modification of cotton and cellulose fibers is reported in Table 2, whereas a recent review was published by Kalia et al. [31].

Process type	Topic	Reference
LPP, APP	Penetration into textile structures	Poll 2001 [32]
LPP	Dyeability of fabrics	Ozdogan 2002 [33]
LPP	Repellent coating	Allan 2002 [34]
LPP	Fluorination	McCord 2003 [35]
APP	Bleaching and dyeing	Prabaharan 2005 [36]
APP	Hydrophobic coating	Kim 2006 [37]
APP	Dyeability of fabrics by acid dyes	Karahan 2008 [38]
LPP	Hydrophilicity improvement	Pandiyaraj 2008 [39]
LPP	Dyeability of Tencel	Mak 2006 [40]
Chitosan	Crosslinking for antimicrobial cotton	El-tahlawy 2005 [41]
Chitosan	Crosslinking on UV-irradiated fibers	Alonso 2009 [42]
Chitosan	UV curing for antimicrobial textiles	Ferrero 2012 [29]
Chitosan	UV curing for antimicrobial textiles	Periolatto 2012 [30]
Chitosan	UV curing for antimicrobial cotton	Ferrero 2013 [43]

APP: atmospheric pressure plasma; LPP: low-pressure plasma.

Table 2. Processes of surface modifications of cotton and cellulose fibers.

Synthetic fibers such as polyamide and polyester have a compact structure with a low content of polar groups, whereas polypropylene is lacking of them; these are substantially hydrophobic and can be subjected to surface etching by plasma and oxidation to increase porosity and wettability, but the latter purpose is often pursued by grafting of polar monomers or application of hydrophilic coating.

In a similar manner, hydrophobicity and oleophobicity can be increased with a suitable coating obtained by plasma. A selection of articles published in the last 15 years is reported in Table 3.

Fiber	Plasma	Topic	Reference
PET	LPP	Surface modification by silane	Negulescu 2000 [44]
Cotton, PET, silk	LPP	Surface modification by air	Riccardi 2001 [45]
PET	LPP	Surface modification by SF_6	Riccardi 2003 [46]
PET	LPP	Wettability and dyeability by SF_6	Barni 2005 [47]
PET, PA	LPP	Surface modification by acrylic acid	Cireli 2007 [48]
PET	LPP	Hydrophilicity by O_2 and NH_3	Calvimontes 2011 [49]
PET	APP	Penetration depth of modification	Wang 2008 [50]
PET	APP	Fluorocarbon nano-coating	Leroux 2008 [51]
PET	APP	Surface oxidation and silicon resin	Leroux 2009 [52]
Cotton, PET, PA	APP	Water and oil absorbency	Samanta 2009 [53]
PET, PA	APP	Antistatic effect	Samanta 2010 [54]
PET	APP	Surface modification	Gotoh 2010 [55]
Acrylic	LPP	Hydrorepellent coating	Pane 2001 [56]
Acrylic	APP	Water and oil repellency	Ceria 2010 [57]
PET, acrylic	LPP	Wettability by capillary rise	Ferrero 2003 [58]
PET, PA, PP	LPP	Dyeability by acrylic acid	Ferrero 2004 [59]
PP	LPP	Wettability by CF_4	Kwon 2002 [60]
PP	APP	Surface treatment of nonwoven fabrics	Väänänen 2010 [61]

APP: Atmospheric pressure plasma; LPP: low-pressure plasma; PA: polyamide; PET: polyethylene terephtalate; PP: polypropylene.

Table 3. Selected articles on the surface modification of synthetic fibers by plasma treatment.

2. Wettability by plasma treatments

Plasma is commonly defined as an ionized gas in a neutral state with an equal density of positive and negative charges. It is often referred to as the "fourth state of matter," which can be reached in a wide range of temperatures and pressures. From a chemical point of view, plasma mostly consists of a mix of activated gaseous particles comprising free electrons, radicals, ions, UV radiations, and various highly excited neutral and charged species. It is considered to be a very interesting medium for material processing technologies, with particular regard to surface engineering. In fact, a large variety of treatments aimed at surface modification of different substrates can be carried out, including etching, cleaning, activation, and coating, as detailed below:

- *Etching*: the removal of the bulk substrate material, occurs when the interaction between the solid surface and plasma generates gaseous byproducts, including atoms or molecules, carried away from the surface substrate, removing minimal fractions of bulk material.

- *Cleaning*: the removal of contamination in the form of etching, but with very high selectivity. Only the unwanted surface contaminant is volatilized and removed, whereas the substrate remains unaffected by the process.
- *Activation*: the enhancement of the substrate energy, which generates chemically reactive sites on a previously nonreactive surface.
- *Coating*: the deposition of a functional thin film, occurs if the plasma–solid surface interaction creates a solid-phase material. This process is sometimes called plasma enhanced chemical vapor deposition or plasma polymerization.

Plasmas can be classified by taking into account the employed generation technology (power supply and pressure range) into two major categories: thermal plasmas (very high temperature, not suitable for heat-sensitive materials) and non-thermal plasmas (close to room temperature and suitable for treating textiles). Non-thermal plasmas are also known as low-temperature plasmas and can be subdivided into many different technologies considering the electrical power supply, the operating pressure (low or atmospheric), and the geometrical arrangements.

The laws of plasma physics would make it easier to generate large-volume plasmas at reduced pressure rather than at atmospheric pressure. Moreover, in a closed system under low pressure, it is easier to control the characteristics and composition of the gas atmosphere from which the plasma is generated and, hence, the process chemistry rather than in a system at atmospheric pressure open to ambient air. Nevertheless, a closed system is not easily adaptable to a continuous process of fabric treatment. Therefore, although most of the plasma applications on polymeric materials, including textiles, have been studied using low-pressure plasma (LPP), the atmospheric pressure plasma (APP) technique has demonstrated to be the most interesting tool for large-scale applications on textiles.

Plasma treatments are able to modify the fiber surface, leaving the bulk properties unaffected, and this characteristic is very important for the modification of textile fibers that should not lose their mechanical and chemical properties after treatment. Moreover, the finishing of textile fabrics by plasma technologies can advantageously replace some wet chemical applications as environmental friendly processes, since they do not require water and a high amount of chemicals. By controlling the plasma variables, such as the nature of gas, discharge power, pressure, and exposure time, a great variety of surface properties can be improved, mainly cleaning, wettability, hydrophobic and oleophobic properties, soil release, adhesion of coatings, dyeability, printability, and flame resistance.

An exhaustive book on the applications of plasma technologies to textiles was published by Shishoo [62], whereas the surface modifications by plasma treatments were reviewed by Radu et al. [63] and Morent et al. [64]. Reviews on atmospheric plasma treatments were published by Kale et al. [65] and, more recently, by Wolf [66]. In particular, the plasma pre-treatments to improve dyeability were considered by Deshmukh and Bhat [67], whereas Hossain and Hegemann studied the deposition of thin coatings on synthetic fibers to confer a substrate independent dyeability [68].

The modification induced by plasma treatment on a polymeric surface as a film is easily measured by contact angle determination with the sessile drop method. In the case of textile materials, instead, the measure is strongly affected by the heterogeneous surface structure; therefore, the porosity of the fabrics can often determine a suction effect on the water drop, preventing the contact angle determination. Another technique consists of weight variation measurement by a Wilhelmy balance during capillary wicking. The variety of techniques commonly used to measure contact angles has recently been highlighted by Yuan et al. [69].

Moreover, the hydrophilicity of plasma-treated fabrics can be tested according to the BS 4554:1970 method, known as the drop test, in which a drop of 100.0 μL of deionized water is placed on the surface of the specimen. The time required for the droplet to completely penetrate the fabric was measured by means of a stopwatch.

Poll et al. [32] measured the hydrophilization effect induced by plasma through the suction test. A capillary is filled with a colored test liquid and positioned onto the surface of the fabric layer to be checked. The liquid is absorbed by the fabric to form a colored circle. The diameter of the circle formed after an exposure time of 20 s is a measure of the hydrophilization effect.

Another wettability test can be performed on a fabric strip that is kept vertical, with the lower end immersed in water–dye liquor. A spontaneous wicking occurs due to capillary forces. The absorption height *h* is recorded as a function of time, and the absorption rate is calculated (capillary rise method, as shown in Figure 1; reprinted with the kind permission of the author in [70]).

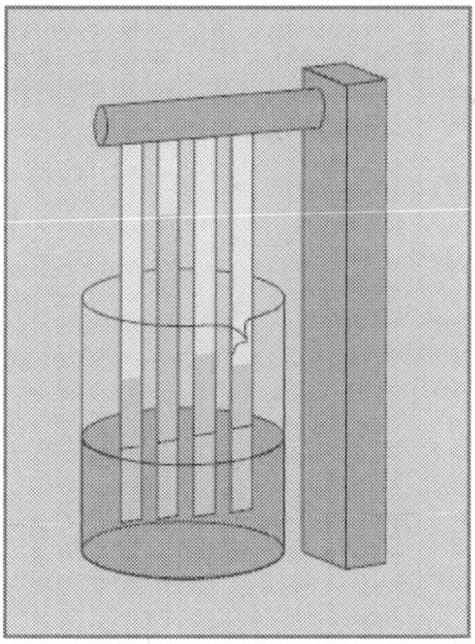

Figure 1. Capillary rise method.

Ferrero [58] applied this method to the wettability measurements on plasma-treated synthetic fabrics (polyester and acrylic). The processing of capillary rise data proposed in this work allowed the assessment of wettability improvement by plasma treatment carried out in different gases (nitrogen, air, and oxygen). Wettability is strongly enhanced by plasma treatment, and after some time, the height reaches an equilibrium value h_{eq}, as shown, for example, in Figure 2, where the wicking curves of untreated and plasma-treated PETs are compared. It is evident that wettability is strongly improved by a mild nitrogen plasma treatment (25 W, 30 s, 65-Pa pressure in a Plasmod apparatus).

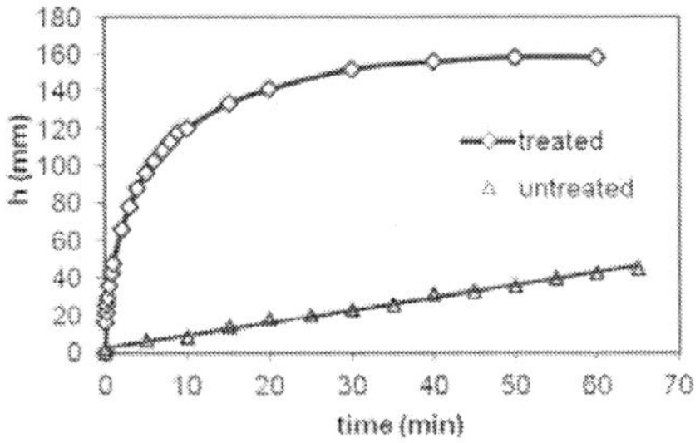

Figure 2. Comparison between the wicking curves of untreated and plasma-treated PET fabrics.

The maximum equilibrium height h_{eq} can be related to the equilibrium static contact angle θ_{eq}, generally smaller than the dynamic one, by equation (1):

$$h_{eq} = \frac{2\gamma \cos\theta_{eq}}{\rho g R_S} \tag{1}$$

where γ and ρ are the surface tension and density of the liquid, respectively, R_S is the mean static radius of pores, and g is the gravity acceleration. In the early stages of the process, the hydrostatic pressure in equation (1) can be neglected; hence, h can be related to θ, advancing contact angle of the liquid on the solid, by Washburn's equation (2):

$$h^2 = \frac{r\gamma \cos\theta}{2\eta} t \tag{2}$$

where r in fiber networks means an equivalent radius of the capillary porous structure, and η is the viscosity of the liquid.

Hence, h^2 values plotted against time show a straight line according to equation (3):

$$h^2 = D \cdot t \tag{3}$$

where the slope D is a capillary diffusion coefficient related to the size of the capillaries r and to the physicochemical characteristics of the liquid. Therefore, a surface treatment of a fabric that modifies r and the contact angle, such as plasma treatment, causes variation of this diffusion coefficient. It was observed that D is affected by the nature of the gas plasma, power, exposure time, and aging. On polyester fabric, nitrogen plasma induces higher wettability than air and oxygen, probably mainly due to surface etching, whereas the wettability of acrylic fabric slightly increases in air plasma with respect to nitrogen.

http://dx.doi.org/10.5772/60812

3. Water and oil repellency by UV curing

Cotton has always been the principal fiber for clothing fabrics due to its attractive characteristics such as softness, comfort, warmness, biodegradability, and low cost. However, the high concentration of hydroxyl groups on the cotton surface makes the fabric water absorbent and easily stained by liquids. Therefore, additional finishes are required to impart hydrophobicity and self-cleaning properties to cotton fabrics. The same finishes are applied to other fabrics, although based on less hydrophilic fibers such as wool, silk, polyamides, and polyester.

In general, water repellency of a fabric can be defined as the fabric's ability to withstand wetting or penetration by water under test conditions. It is important to distinguish between the terms "water repellent" and "waterproof". A fabric is made water repellent by the surface modification or deposition of hydrophobic material on the fibers. Water repellent fabrics have open pores and are permeable to air and water vapor. They are resistant to wetting by rain drops, water spreading, and wicking. Waterproofing involves filling the pores in the fabric with a material that is impermeable to water and, usually, to air as well. Water-repellent, but not waterproof, fabrics allow passage of water once the hydrostatic pressure is sufficiently high. Waterproof and water-repellent finishes are required, in particular, for cotton fabrics.

Polysiloxanes are widely used for textile finishing to impart desirable properties such as softness, crease resistance, and water repellency. However, the specific properties conferred by siloxanes depend on the nature of organic functional groups that are incorporated in the polymer structure [71]. Moreover, water and oil repellency is required for protective clothes and is currently achieved by thermal polymerization of fluorinated monomers, which enable a strong increase of water and oil contact angles on the treated fabrics [72]. However, the application of a polymeric coating to a cotton fabric in the form of a thin film ensures good homogeneity of the conferred properties, but the fabric could lose comfort characteristics, such as handling and breathability. Therefore, an alternative method that allows the uniform adsorption of monomers onto each fiber and the formation of polymer chains inside the fibers should be preferred, since the interpenetration of components and uniform distribution of monomers, even at a low concentration, contribute to obtaining textile materials with modified surface properties without a high add-on of polymer. This result can be achieved by a radiation curing method.

Radiation processes have several commercial applications for the coating of metals, plastics, and glass in printing, wood finishing, film and plastic crosslinking, and in adhesives and electrical insulations. The advantages of this technology are well known: energy savings (low-temperature process), low environmental impact (no solvent emissions), simple, cheap, small equipment, and high treatment speed. Despite these advantages, there have been few applications of radiation curing in the textile industry, such as nonwoven fabric bonding, fabric coating, pigment printing, silk grafting, and surface modification of cotton and synthetic fibers [73, 74]. In fact, in textile finishing processes, the conventional thermal curing technique is still used, regardless of energy consumption and cost. Among the textile finishing processes by radiation curing, pigment printing of fabrics has received much attention [75], whereas coatings for shrink-resistant wool, flame-retardant fabrics, and durable press finishes have also been investigated. Recently, studies on the effects of radiations on textile dyeing have been reviewed by Bhatti et al. [76], whereas the application of ultraviolet irradiation to wool dyeing processes has been experimented [24–26].

Water-repellent fabrics have been obtained by γ-radiation grafting of poly(vinyl methyl siloxane) or methyl hydrogen silicone on hydrophilic substrates. In industrial applications, UV light from a mercury vapor lamp is preferred for thin coatings because of its high efficiency of energy absorption and low equipment cost. In UV curing, radical or cationic species are generated by the interaction of UV light with a suitable photoinitiator, which quickly induces the curing reaction of reactive monomers and oligomers at low temperature, with lower environmental impact and lower process cost than the thermal process. If a monomer and photoinitiator mixture is adsorbed onto the fibers and subsequently UV cured, the polymeric chains can form inside the textile structure, which can be also involved in the formation of graft bonds, making the treatment solid and water resistant.

Ferrero et al. [77] proposed the water-repellent finishing of cotton fabrics by radical UV curing of silicone and urethane-acrylates with different formulations. The results of contact angle, wettability, and moisture adsorption showed that water repellency is already significant at a low resin add-on, whereas the treated fabric maintains its own breathability. SEM analysis confirmed that UV curing yields a coating layer onto each single fiber than a film on the fabric surface.

Polyester and nylon fabrics were made superhydrophobic by the UV curing of a polydimethylsiloxane-containing polyurethane oligomer that was synthesized on purpose. The UV-curable system helps the super hydro-repellent polydimethylsiloxane moiety to anchor onto textile surface, improving the washing stability of the treatment [78].

Moreover, photografting as a surface modification method to provide permanent wettability and wicking performance to deep-groove polypropylene fibers was proposed by Zhu and Hirt [79]. In this case, polyacrylamide and polyacrylic acid were grafted on the fibers by UV irradiation; then, the advancing water contact angle on single fibers decreased from 100° to 55°, and spontaneous wicking of water was observed after surface modification.

On the other hand, a number of research papers have been published on the production and application of different types of fluorochemicals for textile finishing. Fluorochemicals are organic compounds consisting of perfluorinated carbon chains with more fluorine than hydrogens attached to carbon, having thermal and chemical stability. These chains, evenly distributed on the fiber with proper orientation, present an essentially fluorinated surface, which imparts water and oil repellency. In fact, the critical surface tension for fluorocarbon surfaces is in the range of 6 mN/m (-CF_3) to 28 mN/m, whereas for bleached cotton, it is 44 mN/m. One of the most successful ways of obtaining this condition is the incorporation of the fluorinated groups into polymer molecules in which perfluoro groups constitute the side chains [80]. The fluorochemicals used nowadays are based on C_6 carbon chains, which have substituted the C_8 fluorocarbons that release perfluorooctanesulfonate and perfluorooctanoic acid, higly hazardous and toxic substances. Selected articles regarding the surface modification of fibers achieved by fluorination are listed in Table 4.

http://dx.doi.org/10.5772/60812

Fiber	Topic	Reference
Wool, cotton, PET	Fluoropolymer latexes coating	Castelvetro 2001 [81]
FR PET, FR rayon	Padding and thermal curing	Shekar 2001 [82]
FR PET, aramid+viscose	Padding and thermal curing	Kasturiya 2003 [83]
Cotton	Perfluoro-alkyl-epoxy	Shao 2004 [84]
Nylon 6	Padding and thermal curing	De 2005 [72]
Nylon 6,6 and PET	Superhydrophobic rough surface	Lee 2007 [85]
Cotton	Crosslinking of perfluorinated acrylate	Li 2008 [86]
Various	Waterproof breathable fabrics	Mukhopadyai 2008 [87]
Cotton, PA, PET	Perfluoro-alkyl-polyacrylates UV-cured	Ferrero 2011 [74]
Cotton	Perfluoro-acrylates UV and thermal cured	Ferrero 2012 [88]
Cotton	Fluorocarbon and sodium CMC	Dhiman 2014 [89]
Cotton	Direct fluorination and polymerization	Maity 2010 [90]
Cotton, PET	Fluorination by SF_6 plasma treatment	Selli 2001 [91]
Cotton	Plasma sputtering	Wi 2010 [92]
Cotton	Thermal, UV curing, plasma	Udrescu 2011 [93]

FR: Flame retardant; PET: polyethyleneterephtalate.

Table 4. Selected articles on the surface modification of textile fibers by fluorination.

Fluorochemical finishings are commercially available as water emulsions and are applied to fabrics by the pad–dry–cure method, i.e., bath impregnation followed by squeezing, drying in air at 80–100 °C, and final curing at 150–175 °C in hot flue for some minutes. Fluorochemicals give water-repellent and soil-release finishes in conjunction with other water repellents, called extenders, which are able to yield fiber coatings with good resistance to washing. Castelvetro et al. [81] studied the performance of fluoropolymer latexes applied by padding to wool, cotton, and polyester fabrics. They evaluated the performances of the fabrics by means of technological standard test methods whose results correlated well with static and dynamic contact angle measurements.

Alternative fluorination methods have been proposed. Maity et al. [90] experimented the direct fluorination of cotton using elemental fluorine and admicellar polymerization, with a surfactant and fluoromonomer system. Selli et al. [91] used a SF6 plasma to confer water and oil repellency to cotton and PET, whereas plasma sputtering was used by Wi et al. to obtain a water-repellent PTFE coating on cotton fibers [92].

On the other hand, Ferrero et al. [88] extended the study of the UV-curing method to the use of perfluoro-alkyl-polyacrylate resins that are able to impart water as well as oil repellency to cotton fabrics, and the results obtained by UV curing were compared with those obtained by conventional thermal polymerization. This study was focused on the use of commercial finishes for thermal application, Repellan EPF and NFC, by Pulcra Chemicals, and Oleophobol CP-C, by Huntsman, supplied in water emulsions (about 17% solid content, dispersible in cold water in all ratios). Darocur 1173 (2-hydroxy-2-methyl-1-phenylpropan-1-one supplied by Ciba Specialty Chemicals) as a radical photoinitiator was added in an amount of 2% weight on the resin, enough to obtain a film by UV curing with each formulation considered. Water was added as a diluent to the mixtures, continuously stirred until complete homogeneity, in order to enable uniform spread of the liquids on cotton. The formulations were applied by dipping onto strips of fabric that were subsequently dried in an oven. The amount of resin put on the fabrics was adjusted according to the desired final weight add-on and the emulsion concentration. Weight percentages of 3% and 5% on the weight fiber were usually applied in order to obtain the desired properties without loss of fabric handling.

The surface-coated fabrics were exposed to UV radiation using a medium-pressure mercury lamp with a light irradiance on the fabric of about 20 mW/cm^2, in a small box equipped with a quartz window under nitrogen atmosphere, since oxygen interferes with the formation of radicals. The required radiation dose was obtained by adjusting the distance of the textile from the lamp and the exposure time, which was assessed between 40 and 60 s. Instead, thermal curing was carried out in 2–3 min at 140 °C or 150 °C according to the indications of the producer.

Resin emulsion is adsorbed by the fibers, so the polymerized product does not form films onto the fabric surface, but penetrate inside. To test if UV curing was effective even inside the fabric compared with the thermal treatment, the polymerization yield was evaluated by the determination of the unpolymerized resin extracted by chloroform at room temperature from the cured fabrics. Repellan EPF shows the highest yields after UV curing and is about similar to those reached with the thermal treatment (93–96%), whereas Repellan NFC shows lower, although acceptable, yields in UV curing (80–81%) than in the thermal one (98%). With these finishes, the yields remained unaffected by exposure time and resin add-on. Oleophobol CP-C gives lower yields either in thermal or UV curing, with a marked dependence on add-on and exposure time; hence, 60 s is needed to obtain a good yield for a 3% add-on (91%).

The surface properties of coated and uncoated textiles were tested with optical measurements of static and dynamic contact angles of water and oil drops on the textile. The measuring liquids were HPLC grade water (72.8 mN/m at 25 °C) and olive oil (32.0 mN/m). The contact angle values should be higher, as the hydrophobic or oil repellency behavior of the textile is greater. On cotton samples finished with both curing methods, water and oil repellency fastness to domestic washing was evaluated after five washings according to UNI-EN ISO 105-C01.

In Figure 3, the results of dynamic contact angle measurements of water on cotton finished by Repellans before and after washing are compared. In this evaluation, the analysis of advancing and receding contact angles and the resulting difference, i.e. the hysteresis, can give information on the influence of surface roughness and chemical heterogeneity on fabric wettability [81]. A $\Delta\theta>0$ is typical of most real surfaces, as confirmed by all the results obtained. With both resins, the advancing contact angles were slightly reduced after five washing cycles, and this proved the good wash fastness of water repellency, regardless of the curing type and polymer add-on. The hysteresis values generally decreased, indicating a lower surface heterogeneity, probably due to the washing effect. Oleophobol gave slightly lower contact angles but was practically unaffected by repeated washings.

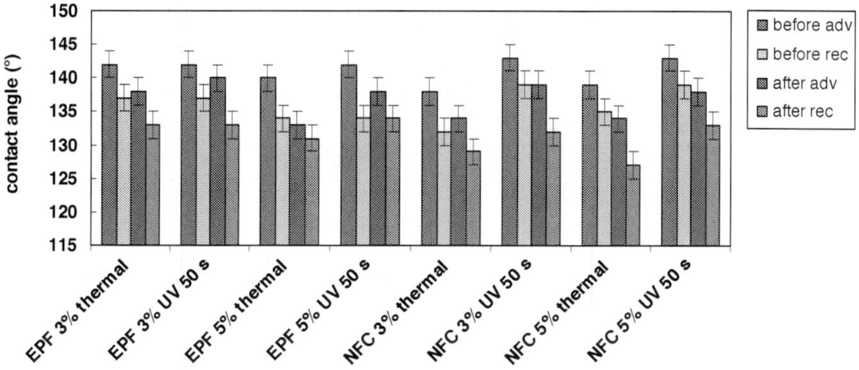

Figure 3. Dynamic contact angles of water before and after washing on cotton fabrics finished with Repellan EPF and NFC (adv: advancing; rec: receding) [with kind permission from Springer Science+Business Media: Ferrero F., Periolatto M., Udrescu C. Water- and oil-repellent coatings of perfluoro-polyacrylate resins on cotton fibers: UV curing in comparison with thermal polymerization. Fibers and Polymers 2012; 13 (2), 191–198, p. 195, Fig. 1].

Static contact angles of oil before and after washing are compared in Figure 4, and the results confirmed the satisfactory wash fastness of oil repellency, in particular with Repellan NFC, without differences between the thermal and UV curing methods.

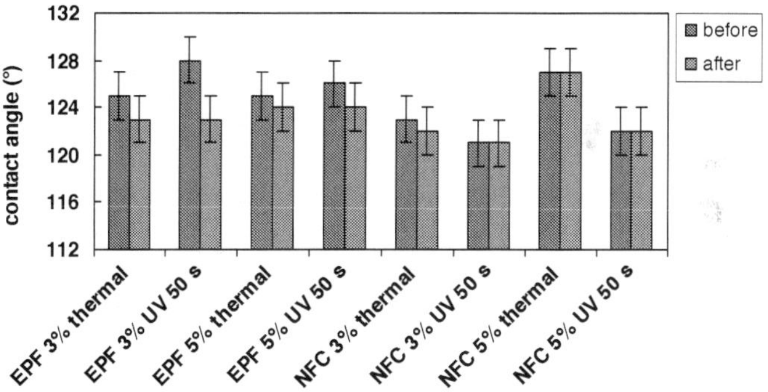

Figure 4. Static contact angles of oil before and after washing on cotton fabrics finished with Repellan EPF and NFC. [with kind permission from Springer Science+Business Media: Ferrero F., Periolatto M., Udrescu C. Water and oil-repellent coatings of perfluoro-polyacrylate resins on cotton fibers: UV curing in comparison with thermal polymerization. Fibers and Polymers 2012; 13(2), 191–198, p. 195, Fig. 2].

The contact angle value on untreated cotton, both with water and oil, was 0° due to the immediate absorption of the drops. It is evident of the high water and oil repellency conferred by the treatment. Measurements on 10 different points of the same sample surface are in good agreement (average values estimated with a confidence interval of ±2° at a 95% confidence level), showing a good uniformity of the coating. The results of thermal and UV curing are very close and poorly affected by weight gain and UV curing time, suggesting that a low polymer add-on is enough to modify the fiber surface.

With each finishing type, water contact angles are higher than with oil, in agreement with the results reported in the literature with other finishes on cotton [84, 94], although the values are lower than 150°, which is considered the lower limit for super hydrophobic surfaces showing the so-called Lotus effect. However, the UV-cured resins yielded oil contact angles mostly higher than 120°, denoting super oil-repellent surfaces.

X-ray photoelectron spectroscopy (XPS) analysis gives the chemical composition of the fabric surface and provides useful information on the fiber coating. Table 5 shows the relative peak intensities of C_{1s}, O_{1s}, F_{1s}, and Cl_{2p} in XPS measurements of untreated and finished cotton fabrics. For the untreated cotton, only two peaks corresponding to C and O are observed. F_{1s} intensity was found to be about the same for samples that were thermally or UV cured, whereas Repellan NFC showed the lowest values.

Resin	Curing	C_{1s} (%)	O_{1s} (%)	F_{1s} (%)	Cl_{2p} (%)
Untreated cotton	–	60.6	39.4	–	–
Repellan EPF	thermal	42.0	6.1	51.8	–
	UV	43.1	5.7	51.1	–
Repellan NFC	thermal	44.7	10.7	43.3	1.3
	UV	46.0	8.7	43.5	1.7
Oleophobol CP-C	thermal	46.4	7.7	44.7	1.3
	UV	48.0	9.2	41.9	0.9

Table 5. Relative intensities in the XPS spectra of untreated and resin-treated cotton fabrics (3% polymer add-on and 60-s UV curing time).

Information on how fluorine binds to the polymer surface can be obtained from the high-resolution C_{1s} signals. According to Selli et al. [91], the C_{1s} spectrum was resolved into six components corresponding to the groups reported in Table 6 with the relative peak areas. For each resin, small differences arise from the comparison between thermal and UV curing, whereas higher differences can be observed between coatings of the different fluorocarbons. With Oleophobol CP-C, the coatings yielded the lowest percentage of the $-CF_2-$ groups and, conversely, the highest for the –CO- groups, although these differences did not affect water and oil repellency. However, in any case, a much lower concentration of the $-CF_3$ groups was found. Such considerations suggest that the lower content of fluorine groups yielded by Repellan NFC coating is enough to confer the requested surface properties to cotton.

In conclusion, UV curing of cotton with commercial perfluoro-alkyl-polyacrylates applied in water emulsion yielded water- and oil-repellent cotton fabrics like the thermal process. The polymerization yields as well as the contact angles with water and oil were of the same order of those obtained with thermal curing, even at low-resin add-ons. Moreover, the UV-cured resins yielded mostly super oil-repellent surfaces, whereas water and oil repellency was adequately maintained after washing.

Group	Binding energy (eV)	Relative peak area (%)					
		Repellan EPF		Repellan NFC		Oleophobol CP-C	
		Thermal	UV cured	Thermal	UV cured	Thermal	UV cured
-CH-	284.5	35.4	32.1	24.7	30.8	28.9	25.9
-C-O-	285.9	11.1	15.6	37.3	30.5	45.3	41.8
-C=O,-C-CF_x	287.3	-	-	2.9	3.5	5.4	5.0
-COO-,CHF-	288.5	11.9	14.1	8.7	7.7	1.6	7.4
-CF_2-	290.9	37.2	33.3	23.0	24.5	15.8	17.2
-CF_3	293.6	4.4	4.9	3.5	3.1	2.0	2.5

Table 6. High-resolution C_{1s} spectra for resin-treated cotton fabrics (3% polymer add-on and 60-s UV curing time).

XPS analyses showed small differences between thermal and UV-cured coatings with each resin, whereas lower percentages of fluorine groups were observed in the case of Repellan NFC coatings without worsening of water and oil repellency, suggesting that such properties can be obtained with a low polymer add-on (3%) and with the lowest fluorine content.

Therefore, UV curing can be indicated as a valid alternative and environment-friendly method to confer water-resistant hydro and oil repellency to cotton fabrics. A comparison with plasma polymerization on cotton of the same perfluoro-alkyl-polyacrylates [93] confirmed that UV curing yields similar results, but with a simpler apparatus that can easily be introduced in the production lines of continuous fabric finishing.

4. Water and oil repellency by sol–gel techniques

There have been many articles in the literature on the improvement of hydrophobic properties of several kinds of fabrics using nanostructures achieved by nanotechnology. It was demonstrated that superhydrophobicity depends not only on surface chemistry but also on surface topology. Two theoretical models (Wenzel and Cassie–Baxter) have inspired how to engineer superhydrophobic surfaces by either roughening the same through microstructures or nanostructures or lowering the surface-free energy due to waxy materials applied on top of the rough structures, or both. An example is a microprocessing technique for producing rough surface and subsequent chemical treatment with silane- or fluorine-containing polymers to reduce the surface-free energy.

Roughened surfaces have commonly been obtained by the introduction of nano-size particles onto the pristine surface, and the sol–gel technique has been reported as a promising tool for the preparation of water-repellent coatings that is especially versatile for applications on glass, paper, and textile [95–100]. An exhaustive review on the application of sol–gel techniques to textiles has been published by Mahltig and Textor [101], and a series of selected articles is reported in Table 7.

Fiber	Topic	Reference
Nylon, PET/cotton	Hydrorepellency	Mahltig 2003 [102]
Nylon	Finishing of carpeting	Satoh 2004 [103]
Cotton	Superhydrophobicity	Daoud 2004 [104]
Cotton	Superhydrophobicity	Yu 2007 [105]
Wool, cotton, PET	Superhydrophobicity	Wang 2008 [106]
Cotton	Antimicrobial and repellency	Tomšič 2008 [107]
Cotton	Superhydrophobicity	Bae 2009 [108]
Cotton	Superhydrophobicity	Erasmus 2009 [109]
Cotton	Durable hydrophobic finishing	Roe 2009 [110]
Cotton, PET	Hydrorepellency	Gao 2009 [111]
PET, PET/cotton	Hydrorepellency and antistaticity	Textor 2010 [112]
Cotton	Superhydrophobicity	Liu 2011 [113]
Cotton	Water and oil repellency, antimicrobial	Simončič 2012 [114]
Cotton	Superhydrophobicity	Shi 2012 [115]
Cotton	Superhydrophobicity and UV blocking	Pan 2012 [116]
Cotton	Super hydro-oleophobicity, self-cleaning	Vasiljević 2013 [117]
Cotton	Hydrorepellency	Periolatto 2013 [118]
Cotton	Hydrorepellency and oil repellency	Ferrero 2013 [119]
Cotton, PET	Hydrorepellency improved by plasma	Montarsolo 2013 [120]

Table 7. Selected articles on the surface modification of fabrics by sol–gel techniques.

In many research works, sol–gel formulations of fluoroalkylsilanes in combination with other silanes to obtain co-condensates are used. The solvents are mostly alcohols, but some water-based systems have been described. In these nanocomposites, the organic and the inorganic networks are covalently bound and homogeneously intermingled at the nanometer scale so that the resulting coatings show enhanced mechanical stability [96].

These materials have a pronounced gradient structure, with a high concentration of fluoroalkyl groups at the coating–air interface so that only a small amount (1.7 mol%) of fluoroalkyl silane is necessary to obtain an effective repellency. Moreover, it accounts for an excellent adhesion of the coatings on various substrates such as glass, metals, and polymers. The gradient is due to the accumulation of surface-active fluorosilanol molecules and condensates at the interface.

Employing organically modified alkoxysilanes containing long-chained aliphatic or highly fluorinated groups, sol–gel offers far-reaching possibilities to prepare water- as well as oil-repellent textiles. A low required add-on is of great interest for textile applications; in fact, it keeps the typical hand and breathability of fabrics uncompromised. Furthermore, most fluorinated materials are very expensive and may often cause serious risks to the human health in case of skin contact and for the environment. Therefore, it is necessary to minimize the use of such substances.

Periolatto et al. [118] obtained highly hydrophobic and oil-repellent cotton fabrics by a one-step deposition of a modified silica-based coatings by sol–gel prepared by co-hydrolysis and condensation in weakly acid medium of TEOS-based sols with low amounts of hydrophobic additives such as hexadecyltrimethoxysilane or fluorooctyltriethoxysilane. This work was further developed [119] with the aim of comparing the effect of the laboratory-grade fluorinated reagent 1H,1H,2H,2H-Fluorooctyltriethoxy-silane (FOS) with that of a commercial product (Fluorolink S10). During the acid-catalyzed hydrolysis of TEOS or fluorinated alkoxysilanes, labile silanol groups are formed, which can first promote the silane adsorption onto the OH-rich cellulose structure of cotton fibers through hydrogen bonding. Successively, during the thermal curing step (120° for 1 h), the condensation reactions reported in Figure 5 can occur.

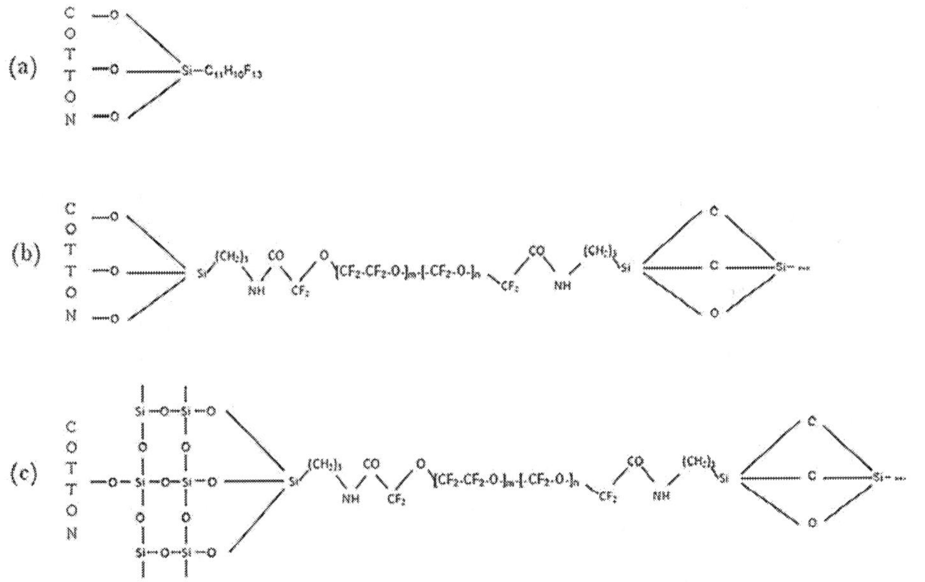

Figure 5. Grafting reactions of fluoromonomers on cotton [reprinted with permission from Elsevier: Ferrero F., Periolatto M. Application of fluorinated compounds to cotton fabrics via sol–gel. Applied Surface Science 2013; 275, 201–207]

The procedure of preparation of the nanosols is illustrated in Figure 6.

The cotton samples were subjected to contact angle measurements by a Krüss DSA20E "Easydrop standard" drop shape analysis tensiometer using the sessile drop method for fitting. Measuring liquid drops were deposited from a glass syringe on the fabric's surface by means of software-controlled dosing. The contact angles were the average of at least five measurements for each sample, with a standard deviation of about 2–3%. The contact angles on untreated cotton were 0°C, whereas the drops are immediately absorbed. Moreover, the time necessary for the total absorption of both water and oil drops was measured. The results are summarized in Figure 7.

On samples finished with an impregnation time of 24 h, higher values of contact angles were measured (169°), denoting the importance of a deep penetration of the finishing agent inside the fibers.

A better behavior of Fluorolink-treated samples, with respect to the FOS-treated ones, was found: contact angles higher than 150 °C were measured, typical of super hydro- and oil-repellent surfaces. Absorption times higher than 2 h were measured with both water and oil drops, whereas on FOS-treated samples, the oil drop is absorbed in about 15 min, a good result but worse than Fluorolink's performance. This can be due to the molecular structure of Fluorolink, which is longer and more complex than FOS. The presence of TEOS seems to be ineffective.

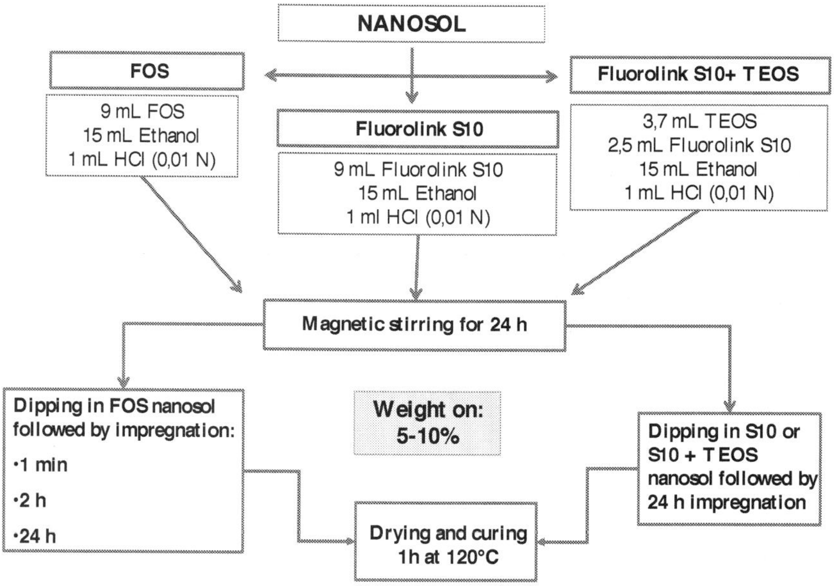

Figure 6. Procedure of sol–gel preparation and its application to cotton fabrics to confer hydro and oil repellency.

The same measurements were made after five repeated washing cycles (at 40 °C for 30 min using 5-g/l ECE detergent according to ISO 105 C01 standard) to assess the durability of the treatments to laundering. The results are summarized in Figure 8.

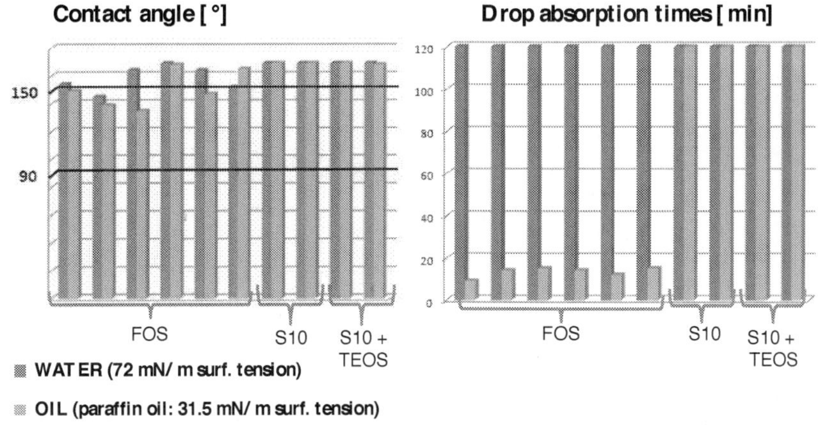

Figure 7. Results of contact angle measurements and drop absorption time on cotton fabrics treated with sol–gel finishes. From left to right; for FOS-finished samples add-on and impregnation time: 5% 1 min, 10% 1 min, 5% 2 h, 10% 2h, 5% 24 h, and 10% 24 h; for the other finishes: 5% 24 h and 10% 24 h.

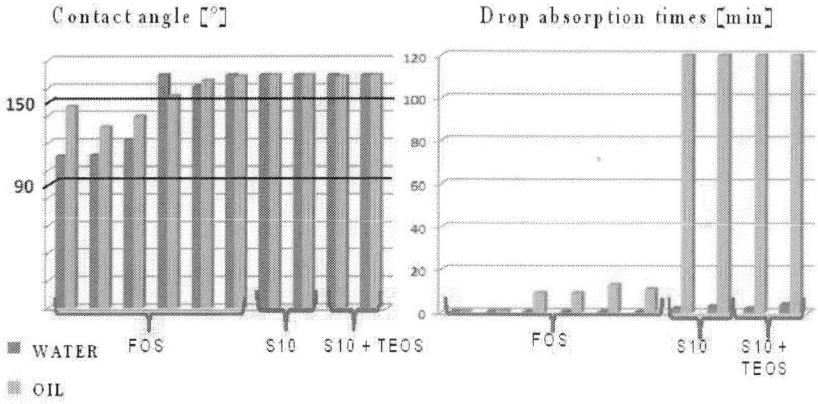

Figure 8. Contact angle and drop absorption time measurements after five washing cycles (description of samples as shown in Figure 7).

On FOS samples after washing, a decrease in contact angles was observed, but the behavior was better with 24 h of impregnation, showing that a longer contact time enables better interpenetration of the finishing agent inside the cotton fibers. Instead, the samples finished by Fluorolink S10 showed the best performance, without the influence of TEOS. After washing, the drop absorption times of water strongly decreased, whereas the oil drop absorption time of 2 h was maintained by samples finished with Fluorolink S10.

The strong loss of hydrophobicity after washing can be due to a rearrangement of the fluorinated chains with an orientation toward the internal part of the fibers. In fact, it is well known that fluorine-containing polymers are usually quite susceptible to rapid rearrangement when the polymer surface is contacted with water, in particular with short perfluorinated-side chains to minimize the interfacial free-energy response to the environmental media. This was confirmed by the increase in contact angle and water drop absorption time of the washed samples after ironing.

The results of XPS analysis confirmed the presence of finishing agents on the surface as evidenced by the content of F and Si. CF_3 groups are present in the structure of FOS, whereas CF_2 groups are present in FOS as well as in Fluorolink. In fact, the F content was higher on the FOS-treated samples (58.3% on the cotton finished with 10% FOS) than that finished with 10% Fluorolink (38.9%). These values were significantly reduced after washing (53.2% and 28.3%, respectively).

In conclusion, the application of a fluorinated alkoxysilane to cotton textiles by sol–gel is a promising textile finishing process to confer durable hydro and oil repellency. In fact, high contact angles and drop absorption time values were measured on treated cotton with both water and oil. Low add-ons (5%) are enough to confer the properties, unaffecting the fabric's characteristics. A prolonged impregnation time (24 h) significantly improves repellency and fastness to washing, whereas ironing of the washed samples can partially restore the hydro and oil repellency lost after washing.

The best performances were obtained with a commercial product (Fluorolink S10). This can allow the application of the treatment at the industrial level, taking into account that the sol–gel process can immediately be implemented on existing production lines of fabric finishing.

5. Conclusions

The modification of surface energy and wettability of textile fibers can be achieved by several techniques, such as plasma treatments, thermal or UV curing of suitable monomers and oligomers, and, finally, nanotechnology based on sol–gel processes.

LPPs are versatile, enabling various treatment types: etching, grafting of groups onto the surface, coating with polymers produced *in situ* from gaseous monomers or liquid oligomers previously impregnated, sputtering, and so on. Despite this flexibility, the apparatus is complex, cumbersome, expensive, and unsuitable for continuous fabric finishing. APP processes are preferable to this aim, but are limited with regard to utilizable gases and plasma polymerization. In any case, the major drawback of plasma activation aimed at improving the following processes is because most of the free radicals remaining on the treated fiber surface is extinguished when exposed to air oxygen; therefore, the time lapse between the plasma treatment and the exploitation of the effects should be as short as possible.

UV curing of suitable monomers or oligomers can be a valid alternative to traditional thermal curing for fabric water and oil repellency and other surface properties. It is a low-cost and environment-friendly process that can easily be introduced in the finishing processes.

Finally, sol–gel processes applied to fabric finishings allow for obtaining engineered surfaces with a great variety of applications arising from the properties of the nanostructures. If low-cost precursors will be found and the need for organic solvents will be reduced to a minimum amount, these techniques will show a significant development in the next years.

Author details

Franco Ferrero* and Monica Periolatto

*Address all correspondence to: franco.ferrero@polito.it

Department of Applied Science and Technology, Politecnico di Torino, Torino, Italy

References

[1] Quéré D., Rough ideas on wetting, Physica A 2002 (1–2); 313 32–46.

[2] He B., Lee J., Patankar N. A. Contact angle hysteresis on rough hydrophobic surfaces. Colloids and Surfaces A: Physicochemical and Engineering Aspects 2004; 248 (1–3) 101–104.

[3] Ferrero F., Mossotti R., Innocenti R., Coppa F., Periolatto M. Enzyme-aided wool dyeing: influence of internal lipids. Fibers and Polymers 2015; 16 (2) 363–369.

[4] Holme I. Innovative technologies for high performance textiles. Coloration Technology 2007; 123 59–73.

[5] Kan C. W., Kwon C., Yuen C. W. M. Surface characterization of low temperature plasma treated wool fiber—the effect of the nature of gas. Fibers and Polymers 2004; 5 (1), 52–58.

[6] Kan C. W., Chan K., Yuen C. W. M. Influence of plasma gas on surface composition of low-temperature plasma-treated wool fibre. Indian Journal of Fibre and Textile Research 2005; 30 (1) 60–67.

[7] Kan C. W., Yuen C. W. M., Plasma technology in wool. Textile Progress 2007; 39 (3) 121–187.

[8] Kan C. W. Effect of low temperature plasma on different wool dyeing systems. AUTEX Research Journal 2007; 8 (4) 255–263.

[9] Wang C. X., Qiu Y. P. Two sided modification of wool fabrics by atmospheric pressure plasma jet: influence of processing parameters on plasma penetration. Surface and Coatings Technology 2007; 201 (14) 6273–6277.

[10] Rombaldoni F., Mossotti R., Montarsolo A., Innocenti R., Mazzuchetti G., Vassallo E. Application of HMDSO plasma-aided coating to wool fabrics and characterization of their chemical, surface and physical properties. Proceedings of the 8th AUTEX 2008 Word Textile Conference, 24–26 June, Biella, Italy, CD-ROM version.

[11] Rombaldoni F., Mossotti R., Montarsolo A., Innocenti R., Mazzuchetti G., Vassallo E., Characterization of plasma-coated wool fabrics. Textile Research Journal 2009; 79 (9) 853–861.

[12] Naebe M., Denning R., Huson M., Cookson P. G., Wang X. Ageing effect of plasma-treated wool. Journal of the Textile Institute 2011; 102 (12) 1086–1093.

[13] Riva A., Algaba I., Prieto R. Dyeing kinetics of wool fabrics pretreated with a protease. Coloration Technology 2002; 118 (2) 59–63.

[14] Onar N., Sarıışık M. Use of enzymes and chitosan biopolymer in wool dyeing. Fibers and Textiles in Eastern Europe 2005; 13 (1) 54–59.

[15] Cardamone J. M., Damert W. C. Low-temperature dyeing of wool processed for shrinkage control. Textile Research Journal 2006 (1); 76 78–85.

[16] Parvinzadeh M., Effect of proteolytic enzyme on dyeing of wool with madder. Enzyme and Microbial Technology 2007; 40 (7) 1719–1722.

[17] Cui L., Yu Y., Fan X., Wang P., Wang Q. Effect of protease treatment on dyeing properties of wool fabrics for single bath. Engineering in Life Sciences 2009; 9 (2) 135–139.

[18] Periolatto M., Ferrero F., Giansetti M., Mossotti M., Innocenti R. Enzyme-aided wool dyeing with a neutral protease at reduced temperatures. Engineering in Life Sciences 2010; 10 (5) 474–479.

[19] Periolatto M., Ferrero F., Giansetti M., Mossotti M., Innocenti R. Influence of protease on dyeing of wool with acid dyes. Central European Journal of Chemistry 2011; 9 (1) 157–164.

[20] Shao, J., Liu, J., Carr, C. M. Investigation into the synergistic effect between UV/ozone exposure and peroxide pad—batch bleaching on the printability of wool. Coloration Technology 2001; 117 (5) 270–275.

[21] Xin, J. H., Zhu, R., Hua, J., Shen, J. Surface modification and low temperature dyeing properties of wool treated by UV radiation. Coloration Technology 2002; 118 (4) 169–173.

[22] El-Zaher, N. A., Micheal, M. N. Time optimization of ultraviolet-ozone pretreatment for improving wool fabrics properties. Journal of Applied Polymer Science 2002; 85 (7) 1469–1476.

[23] Micheal, M. N., El-Zaher, N. A. Efficiency of ultraviolet/ozone treatments in the improvement of the dyeability and light fastness of wool. Journal of Applied Polymer Science 2003; 90 (13) 3668–3675.

[24] Periolatto M., Ferrero F., Vineis C., Rombaldoni F. Multifunctional finishing of wool fabrics by chitosan UV-grafting: an approach. Carbohydrate Polymers 2013; 98 (1) 624–629.

[25] Migliavacca G., Ferrero F., Periolatto M. Differential dyeing of wool fabric with metal-complex dyes after ultraviolet irradiation. Coloration Technology 2014; 130 (5) 327–333.

[26] Periolatto M., Ferrero F., Migliavacca G. Low temperature dyeing of wool fabric by acid dye after UV irradiation. The Journal of The Textile Institute 2014; 105 (10) 1058–1064.

[27] Bashar M. M., Khan M. A. An overview on surface modification of cotton fiber for apparel use. Journal of Polymers and the Environment 2013; 21 (1) 181–190.

[28] Enescu D. Use of chitosan in surface modification of textile materials. Romanian Biotechnological Letters 2008; 13 (6) 4037–4048.

[29] Ferrero F., Periolatto M. Antimicrobial finish of textiles of textiles by chitosan UV-curing. Journal of Nanoscience and Nanotechnology 2012; 12 (6) 4803–4810.

[30] Periolatto M., Ferrero F., Vineis C. Antimicrobial chitosan finish of cotton and silk fabrics by UV-curing with 2-hydroxy-2-methylphenylpropane-1-one. Carbohydrate Polymers 2012; 88 (1) 201–205.

[31] Kalia S., Thakur K., Celli A., Kiechel M. A., Schauer C. L. Surface modification of plant fibers using environment friendly methods for their application in polymer composites, textile industry and antimicrobial activities: a review. Journal of Environmental Chemical Engineering 2013; 1 (3) 97–112.

[32] Poll H. U., Schladitz U., Schreiter S. Penetration of plasma effects into textile structures. Surface and Coatings Technology 2001; 142–144 489–483.

[33] Ozdogan E., Saber R., Ayhan H., Seventekin N. A new approach for dyeability of cotton fabrics by different plasma polymerization methods. Coloration Technology 2002; 118 (3) 100–103.

[34] Allan G., Fotheringham A., Weedall P. The use of plasma and neural modeling to optimize the application of a repellent coating to disposable surgical garment. AUTEX Research Journal 2002; 2 (2) 64–68.

[35] McCord M. G., Hwang Y. J., Qiu Y., Hughes L. K., Bourham M. A. Surface analysis of cotton fabrics fluorinated in radio-frequency plasma. Journal of Applied Polymer Science 2003; 88 (8) 2038–2047.

[36] Prabaharan M., Carneiro N. Effect of low-temperature plasma on cotton fabric and its application to bleaching and dyeing. Indian Journal of Fibre and Textile Research 2005; 30 (1) 68–74.

[37] Kim J. H., Liu G., Kim S. H. Deposition of stable hydrophobic coatings with in-line CH_4 atmospheric RF plasma. Journal of Materials Chemistry 2006; 16 (10) 977–981.

[38] Karahan H. A., Ozdoğan E., Demir A., Ayhan H., Seventekin N. Effects of atmospheric plasma treatment on the dyeability of cotton fabrics by acid dyes. Coloration Technology 2008; 124 (2) 106–110.

[39] Pandiyaraj K. N., Selvarajan V. Non-thermal plasma treatment for hydrophilicity improvement of grey cotton fabrics. Journal of Materials Processing Technology 2008; 199 (1–3) 130–139.

[40] Mak C. M., Yuen C. W. M., Ku S. K. A., Kan C. W. Low-temperature plasma treatment of Tencel. Journal of The Textile Institute 2006; 97 (6) 533–540.

[41] El-tahlawy, K. F., El-Bendary, M. A., Elhendawy, A. G., Hudson, S. M. The antimicrobial activity of cotton fabrics treated with different crosslinking agents and chitosan. Carbohydrate Polymers 2005; 60 (4) 421–430.

[42] Alonso, D., Gimeno, M., Olayo, R., Vázquez-Torres, H., Sepúlveda-Sánchez, J. D., Shirai, K. Cross-linking chitosan into UV-irradiated cellulose fibers for the preparation of antimicrobial-finished textiles. Carbohydrate Polymers 2009; 77 (3) 536–543.

[43] Ferrero F., Periolatto M., Ferrario S. Sustainable antimicrobial finishing of cotton fabrics by chitosan UV-grafting: from laboratory experiments to semi industrial scale-up. Journal of Cleaner Production 2015; 96, 244-252.

[44] Negulescu I. I., Despa S., Chen J., Collier B. J., Despa M., Denes A., Sarmadi M., Denes F. S., Characterizing polyester fabrics treated in electrical discharges of radio-frequency plasma. Textile Research Journal 2000; 70 (1) 1–7.

[45] Riccardi C., Barni R., Fontanesi M., Marcandalli B., Massafra M., Selli E., Mazzone G. A SF_6 RF plasma reactor for research on textile treatment. Plasma Sources Science and Technology 2001; 10 (1) 92–98.

[46] Riccardi C., Barni R., Selli E., Mazzone G., Massafra M. R., Marcandalli B., Poletti G. Surface modification of poly(ethylene terephthalate) fibers induced by radio frequency air plasma treatment. Applied Surface Science 2003; 211 (1–4) 386–397.

[47] Barni R., Riccardi C., Selli E., Massafra M. R., Marcandalli B., Orsini F., Poletti G., Meda L. Wettability and dyeability modulation of poly(ethylene terephthalate) fibers through cold SF6 plasma treatment. Plasma Processes and Polymers 2005; 2 (1) 64–72.

[48] Cireli A., Kutlu B., Mutlu M., Surface modification of polyester and polyamide fabrics by low frequency plasma polymerization of acrylic acid. Journal of Applied Polymer Science 2007; 104 (4) 2318–2322.

[49] Calvimontes A., Saha R., Dutschk V. Topological effects of O_2- and NH_3-plasma treatment on woven plain polyester fabric in adjusting hydrophilicity. AUTEX Research Journal 2011; 11 (1) 24–30.

[50] Wang C. X., Liu Y., Xu H. L., Ren Y., Qiu Y. P. Influence of atmospheric pressure plasma treatment time on penetration depth of surface modification into fabric. Applied Surface Science 2008; 254 (8) 2499–2505.

[51] Leroux F., Campagne C., Perwuelz A., Gengembre L. Fluorocarbon nano-coating of polyester fabrics by atmospheric air plasma with aerosol. Applied Surface Science 2008; 254 (13) 3902–3908.

[52] Leroux F., Campagne C., Perwuelz A., Gengembre L., Atmospheric air plasma treatment of polyester textile materials. Textile structure influence on surface oxidation and silicon resin adhesion. Surface and Coatings Technology 2009; 203 (20–21) 3178–3183.

[53] Samanta K. K., Jassal M., Agrawal A., Improvement in water and oil absorbency of textile substrate by atmospheric pressure cold plasma treatment. Surface and Coatings Technology 2009; 203 (10–11) 1336-1342.

[54] Samanta K. K., Jassal M., Agrawal A. K. Antistatic effect of atmospheric pressure glow discharge cold plasma treatment on textile substrates. Fibers and Polymers 2010; 11 (3) 431–437.

[55] Gotoh K., Yasukawa A. Atmospheric pressure plasma modification of polyester fabric for improvement of textile-specific properties. Textile Research Journal 2010; 81 (4) 368–378.

[56] Pane S., Tedesco R., Greger R. Acrylic fabrics treated with plasma for outdoor applications. Journal of Industrial Textiles 2001; 31 (2) 135–145.

[57] Ceria A., Hauser P. J. Atmospheric plasma treatment to improve durability of a water and oil repellent finishing for acrylic fabrics. Surface and Coatings Technology 2010; 204 (9–10) 1535–1541.

[58] Ferrero F. Wettability measurements on plasma treated synthetic fabrics by capillary rise method. Polymer Testing 2003; 22 (5) 571–578.

[59] Ferrero F., Tonin C., Peila R., Ramella Pollone F. Improving the dyeability of synthetic fabrics with basic dyes using in situ plasma polymerization of acrylic acid. Coloration Technology 2004; 120 (1) 30–34.

[60] Kwon Y. A. Influence of the CF_4 plasma treatments on the wettability of polypropylene fabrics. Fibers and Polymers 2002; 3 (4) 174–178.

[61] Väänänen R., Heikkilä P., Tuominen M., Kuusipalo J., Harlin A. Fast and efficient surface treatment for nonwoven materials by atmospheric pressure plasma. AUTEX Research Journal 2010; 10 (1) 8–13.

[62] Shishoo R. (ed.) Plasma Technologies for Textiles. Cambridge: Woodhead Publishing Ltd: 2007.

[63] Radu C. D., Kiekens P., Verschuren J. Surface modification of textiles by plasma treatments. In: Pastore C. M., Kiekens P. (eds.). Surface Characteristics of Fibers and Textiles. New York: Marcel Dekker: 2001. p203–218.

[64] Morent R., De Geyter N., Verschuren J., De Clerk K., Kiekens P., Leys C. Non-thermal plasma treatment of textiles. Surface and Coatings Technology 2008; 202 (14) 3427–3449.

[65] Kale K. H., Desai A. N. Atmospheric pressure plasma treatment of textiles using non-polymerising gases. Indian Journal of Fibre and Textile Research 2011; 36 (3) 289–299.

[66] Wolf R. A. Atmospheric plasma modification of textile surfaces. In: Atmospheric Pressure Plasma for Surface Modification. Somerset: John Wiley & Sons: 2012. p139–154.

[67] Deshmukh R. R., Bhat N. V. Pretreatments of textiles prior to dyeing: plasma processing. In: Hauser P. (ed.). Textile Dyeing. Rijeka: InTech: 2011. p34–56. Available from: http://www.intechopen.com/books/textile-dyeing/pretreatments-of-textiles-prior-to-dyeing-plasma-processing (accessed 10 February 2014).

[68] Hossain M. M., Hegemann D. Substrate independent dyeing of synthetic textiles treated with low-pressure plasmas. In: Hauser P. (ed.). Textile Dyeing. Rijeka: InTech: 2011. p173–194. Available from: http://www.intechopen.com/books/textile-dyeing/substrate-independent-dyeing-of-synthetic-textiles-treated-with-low-pressure-plasmas (accessed 10 February 2014).

[69] Yuan Y., Lee T. R. Contact angle and wetting properties. In: Bracco G., Holst B. (eds.). Surface Science Techniques. Berlin, Heidelberg: Springer-Verlag: 2013. p3–33.

[70] Hegemann D. Plasma Functionalization of Textiles. 2nd ACTECO Training Day 2007, 18 April, Turin, Italy.

[71] El Naggar, A. M.; Zhody, M. H.; Mohammed, S. S.; Alam, E. A. Water resistance and surface morphology of synthetic fabrics covered by polysiloxane/acrylate followed by electron beam irradiation. Nuclear Instruments and Methods in Physics Research B 2003; 201 (4), 595–603.

[72] De P., Sanke M. D., Chaudhari S. S., Mathur M. R. UV-resist, water-repellent breathable fabric as protective textiles. Journal of Industrial Textiles 2005; 34 (4) 209–222.

[73] Ferrero F., Periolatto M., Bianchetto Songia M. Silk grafting with methacrylic and epoxy monomers: thermal process in comparison with ultraviolet curing. Journal of Applied Polymer Science 2008; 110 (2) 1019–1027.

[74] Ferrero F., Periolatto M. Ultraviolet curing for surface modification of textile fabrics. Journal of Nanoscience and Nanotechnology 2011; 11 (10) 8663–8669.

[75] Neral B., Šostar-Turk S., Vončina B. Properties of UV-cured pigment prints on textile fabric. Dyes and Pigments 2006; 68 (2–3) 143–150.

[76] Bhatti I. A., Adeel S., Abbas M. Effect of Radiation in Textile Dyeing. In: Hauser P. (ed.). Textile Dyeing. Rijeka: InTech: 2011. p1–17. Available from: http://www.intechopen.com/books/textile-dyeing/effect-of-radiation-on-textile-dyeing (accessed 10 February 2014).

[77] Ferrero F., Periolatto M., Sangermano M., Bianchetto Songia M. Water-repellent finishing of cotton fabrics by ultraviolet curing. Journal of Applied Polymer Science 2008; 107 (2) 810–818.

[78] Chen W.-H., Chen P.-C., Wang S.-C., Yeh J.-T., Huang C.-Y., Chen K.-N. UV-curable PDMS-containing PU system for hydrophobic textile surface treatment. Journal of Polymer Research 2009; 16 (5) 601–610.

[79] Zhu S., Hirt D.E. Improving the wettability of deep-groove polypropylene fibers by photografting. Textile Research Journal 2009; 79 (6) 534–547.

[80] Kasturiya N., Bhargava G. S. Liquid repellency and durability assessment: a quick technique. Journal of Industrial Textiles 2003; 32 (3) 187–222.

[81] Castelvetro V., Francini G., Ciardelli G., Ceccato M. Evaluating fluorinated acrylic latices as textile water and oil repellent finishes. Textile Research Journal 2001; 71 (5), 399–406.

[82] Shekar R. I., Kasturiya N., Raj H., Mathur G. N. Studies on effect of water repellent treatment on flame retardant properties of fabric. Journal of Industrial Textiles 2001; 30 (3) 222–254.

[83] Kasturiya N., Katiyar P., Bhargava G. S., Nishkam A. Effect of fluorochemicals on flame-retardant fibers. Journal of Industrial Textiles 2003; 32 (4) 245–254.

[84] Shao H., Sun J. Y., Meng W.-D., Qing F.-L. Water and oil repellent and durable press finishes for cotton based on a perfluoroalkyl-containing multi-epoxy compound and citric acid. Textile Research Journal 2004; 74 (10) 851–855.

[85] Lee H. J., Michielsen S. Preparation of a superhydrophobic rough surface. Journal of Polymer Science Part B: Polymer Physics 2007; 45 (3) 253–261.

[86] Li Z.-R., Fu K.-J., Wang L.-J., Liu F. Synthesis of a novel perfluorinated acrylate copolymer containing hydroxyethyl sulfone as crosslinking group and its application on cotton fabrics. Journal of Materials Processing Technology 2008; 205 (1) 243–248.

[87] Mukhopadhyay A., Midha V. K. A review on designing the waterproof breathable fabrics part I: fundamental principles and designing aspects of breathable fabrics. Journal of Industrial Textiles 2008; 37 (3) 225–262.

[88] Ferrero F., Periolatto M., Udrescu C. Water and oil-repellent coatings of perfluoropolyacrylate resins on cotton fibers: UV curing in comparison with thermal polymerization. Fibers and Polymers 2012; 13 (2) 191–198.

[89] Dhiman G., Chakraborty J. N. Soil release performance of cotton finished with oleophobol CPR and CMC–Na salt. Fashion and Textiles 2014; 1 23. Available from: http://link.springer.com/article/10.1186/s40691-014-0023-4

[90] Maity J., Kothary P., O'Rear E. A., Jacob C. Preparation and comparison of hydrophobic cotton fabric obtained by direct fluorination and admicellar polymerization of fluoromonomers. Industrial Engineering Chemistry Research 2010; 49 (13) 6075–6079.

[91] Selli E., Mazzone G., Oliva C., Martini F., Riccardi C., Barni R., Marcandalli B., Massafra M. R. Characterisation of poly(ethylene terephthalate) and cotton fibres after cold SF_6 plasma treatment. Journal of Materials Chemistry 2001; 11 (8) 1985–1991.

[92] Wi D.-Y., Kim I. W., Kim J. Water repellent cotton fabrics prepared by PTFE RF sputtering. Fibers and Polymers 2010; 10 (1) 98–101.

[93] Udrescu C., Periolatto M., Ferrero F. Water-repellent cotton fabrics by ultraviolet curing and plasma treatment. 11th World Textile Conference AUTEX 2011, Mulhouse (France), 8–10 June 2011. p273–276.

[94] Černe L., Simončič B., Željko M. The influence of repellent coatings on surface free energy of glass plate and cotton fabric. Applied Surface Science 2008; 254 (20) 6467–6477.

[95] Latthe S. S., Hirashima H., Rao A. V. TEOS based water repellent silica films obtained by a co-precursor sol–gel method. Smart Materials and Structures 2009; 18 095017 1–6.

[96] Pilotek S., Schmidt H. K. Wettability of microstructured hydrophobic sol–gel coatings. Journal of Sol-Gel Science and Technology 2003; 26 (1–3) 789–792.

[97] Cunha A. G., Gandini A. Turning polysaccharides into hydrophobic materials: a critical review part 1: cellulose. Cellulose 2010; 17 (5) 875–889.

[98] Cappelletto E., Callone E., Campostrini R., Girardi F., Maggini S., della Volpe C., Siboni S., Di Maggio R. Hydrophobic siloxane paper coatings: the effect of increasing methyl substitution. Journal of Sol-Gel Science and Technology 2012; 62 (3) 441–452.

[99] Gowri S., Amorim T., Carneiro N., Souto A. P., Esteves M. F. Polymer nanocomposites for multifunctional finishing of textiles—a review. Textile Research Journal 2010; 80 (13) 1290–1306.

[100] Mahltig B., Haufe H., Bottcher H. Functionalization of textiles by inorganic sol–gel coatings. Journal of Materials Chemistry 2005; 15 (41) 4385–4398.

[101] Textor T., Mahltig B. Nanosols and textiles. Singapore: World Scientific Publishing Co. Pte. Ltd.: 2008.

[102] Mahltig B., Böttcher H. Modified silica sol coatings for water repellent textiles. Journal of Sol-Gel Science and Technology 2003; 27 (1) 43–52.

[103] Satoh K., Nazakumi H., Morita M. Novel fluorinated inorganic–organic finishing materials for nylon carpeting. Textile Research Journal 2004; 74 (12) 1079–1084.

[104] Daoud W. A., Xin J. H., Tao X. Superhydrophobic silica nanocomposite coating by a low-temperature process. Journal of the American Ceramic Society 2004; 87 (9) 1782–1784.

[105] Yu M., Gu G., Meng W.-D., Qing F.-L. Superhydrophobic cotton fabric coating based on a complex layer of silica nanoparticles and perfluorooctylated quaternary ammonium silane coupling agent. Applied Surface Science 2007; 253 (7) 3669–3673.

[106] Wang H., Fang J., Cheng T., Ding J., Qu L., Dai L., Lin T. One-step coating of fluoro-containing silica nanoparticles for universal generation of surface superhydrophobicity. Chemical Communications 2008; 7 877–879.

[107] Tomšič B., Simončič B., Orel B., Černe L., Forte Tavčer P., Zorko M., Jerman I., Vilčnik A., Kovač J. Sol–gel coating of cellulose fibres with antimicrobial and repellent properties. Journal of Sol-Gel Science and Technology 2008; 47 (1) 44–57.

[108] Bae G. Y., Min B. G., Jeong Y. G., Lee S. C., Jang J. H., Koo G. H. Superhydrophobicity of cotton fabrics treated with silica nanoparticles and water-repellent agent. Journal of Colloid and Interface Science 2009; 337 (1) 170–175.

[109] Erasmus E., Barkhuysen F.A. Superhydrophobic cotton by fluorosilane modification. Indian Journal of Fibre and Textile Research 2009; 34(4), 377–379.

[110] Roe B., Zhang X. Durable hydrophobic textile fabric finishing using silica nanoparticles and mixed silanes. Textile Research Journal 2009; 79 (12) 1115–1122.

[111] Gao Q., Zhu Q., Guo Y., Yang C. Q. Formation of highly hydrophobic surfaces on cotton and polyester fabrics using silica sol nanoparticles and nonfluorinated alkylsilane. Industrial Engineering Chemistry Research 2009; 48 (22) 9797–9803.

[112] Textor T., Mahltig B. A sol–gel based surface treatment for preparation of water repellent antistatic textiles. Applied Surface Science 2010; 256 (6) 1668–1674.

[113] Liu J., Huang W., Xing Y., Li R., Dai J. Preparation of durable superhydrophobic surface by sol–gel method with water glass and citric acid. Journal of Sol-Gel Science and Technology 2011; 58 (1) 18–23.

[114] Simončič B., Tomšič B., Cerne L., Orel B., Jerman I., Kovac J., Zerjav M., Simončič A. Multifunctional water and oil repellent and antimicrobial properties of finished cotton: influence of sol–gel finishing procedure. Journal of Sol-Gel Science and Technology 2012; 61 (2) 340–354.

[115] Shi Y., Wang Y., Feng X., Yue G., Yang W. Fabrication of superhydrophobicity on cotton fabric by sol–gel. Applied Surface Science 2012; 258 (20) 8134–8138.

[116] Pan C., Shen L., Shang S., Xing Y. Preparation of superhydrophobic and UV blocking cotton fabric via sol–gel method and self-assembly. Applied Surface Science 2012; 259 110–117.

[117] Vasiljević J., Gorjanc M., Tomšič B., Orel B., Jerman I., Mozetič M., Vesel A., Simončič B. The surface modification of cellulose fibres to create super-hydrophobic, oleophobic and self-cleaning properties. Cellulose 2013; 20 (1) 277–289.

[118] Periolatto M., Ferrero F., Mossotti R., Montarsolo A. Hydrorepellent finishing of cotton fabrics by chemically modified TEOS based nanosol. Cellulose 2013; 20 (1) 355–364.

[119] Ferrero F., Periolatto M., Application of fluorinated compounds to cotton fabrics via sol–gel. Applied Surface Science 2013; 275 201–207.

[120] Montarsolo A., Periolatto M., Zerbola M., Mossotti R., Ferrero F. Hydrophobic sol-gel finishing for textiles: improvement by plasma pre-treatment. Textile Research Journal 2013; 83 (11) 1190–1200.

Chapter 7

Surface Energy and Wetting in Island Films

Sergei Dukarov, Aleksandr Kryshtal and
Vladimir Sukhov

Additional information is available at the end of the chapter

http://dx.doi.org/10.5772/60900

Abstract

The chapter describes the fundamental aspects of the effects of scale on surface phenomena in condensed films. Experimental and theoretical data for the size and temperature dependencies of the surface energy (including the solid phase); wetting of solid surfaces and free thin films by small metal particles are discussed. Several modern methods of contact angle measurement in small-sized systems based on the optical and electron microscopy methods are described.

Keywords: surface energy, small particles, thin films, wetting, size effects

1. Introduction

Surface energy is one of the most important characteristics of condensed matter. While methods available for the liquid phase enable to determine reliably not only the value but also the temperature dependence of surface energy [1–3], for the solid phase the accuracy of existing methods, as a rule, does not allow to trace its temperature dependence [4, 5]. Therefore, the following approach is justified: what information can be obtained about surface energy of the condensed matters on studies of various properties and processes in small size samples [6–9]. This article considers mainly the investigations of temperature and size dependence of surface energy of condensed matter based on the analysis of surface phenomena and phase transitions in nano-sized systems.

Wetting of solid surfaces with a liquid as well as spreading of a liquid over solid surfaces as a manifestation of interaction between the solid and liquid phases is one of the universal phenomena and covers a wide variety of both fundamental and technological processes. Despite their crucial importance, these processes are still unclear [1, 2, 10]. Therefore, this paper focuses on effects of wetting in nanodispersed systems and considers various physical and chemical factors affecting it. Such statement of the problems seems actual since these details are important to describe a wide range of processes and phenomena, whereas the available data are disembodies and often ambiguous or even lacking.

2. The surface energy of nanoparticles

2.1. The size dependence of the surface energy of nanoparticles

In the framework of the Gibbs thermodynamics of heterophase systems, the size dependence of surface energy is due to the curvature of the phase interface. The Gibbs method applied to interfaces with a small curvature radius was first developed by Tolman [11], who derived the equation relating the surface energy σ of a spherical particle with its radius R

$$\frac{1}{\sigma}\frac{d\sigma}{dR}=\frac{\left(2\delta/R^2\right)\left(1+\delta/R+\delta^2/3R^2\right)}{1+\left(2\delta/R\right)\left(1+\delta/R+\delta^2/3R^2\right)} \tag{1}$$

where δ is the difference between the radii of equimolecular surface and surface of tension. Since the function $\delta = \delta(R)$ is unknown, at $\delta << R$ the variable δ can be considered constant and equal to the value for plane interface, that is, $\delta \approx \delta_\infty$. Under this assumption approximate solution to equation (1) was obtained in [11] known in literature as the Tolman formula:

$$\sigma/\sigma_\infty = 1/\left(1+2\delta_\infty/R\right)$$

For particles with $R >> \delta_\infty$ the first term of the expansion will be sufficient

$$\sigma = \sigma_\infty\left(1-\alpha/R\right). \tag{2}$$

In this approach the parameter α = $2\delta_\infty$ has a definite physical meaning as a width of the respective phase interface for any condensed phases. Character of the dependence $\sigma(R)$ is determined by the sign of the parameter δ_∞. In the case when $\delta_\infty > 0$, surface energy of microparticles will decrease, and at $\delta_\infty < 0$, conversely, increase of σ with decrease of R. Since the sign and value of parameter δ_∞ cannot be obtained based on thermodynamic prerequisites, determination of the type of the dependence $\sigma(R)$ calls for the use of the model of the structure of the transition layer for real systems or experimental results. In [3] this problem is solved for a simple idealized fluid using correlation function methods and it is shown that in this system the transition layer has a layered structure corresponding to some orderliness of particles therein. This is essential for many phenomena in surface physics.

Calculations of δ_∞ for plane liquid–vapor interfaces made by statistical methods [12] has shown that the quantity $\delta_\infty > 0$ and has a value of approximately a few tenth of nanometer (e.g., according to [12] for argon at 90 K $\delta_\infty = 0.36$ nm). Further research with use of a computer simulation [13–17], electron theory of surface energy of metals [18, 19], the thermodynamic perturbation theory [20, 21] is consistent with the results in terms of the type of the dependence $\sigma(R)$.

Qualitatively, the decrease in the surface energy of small particles can be explained as follows. For the condensed phase being in equilibrium with its own vapor, the interface surface energy at first approximation is proportional to the difference between the number of atoms (molecules) per unit volume of the condensed and vapor phases. With decreasing particle size of the condensed phase vapor pressure increases, and, consequently, its density increases, which causes decrease of the surface energy of the particle – saturated vapor interface approximately in inverse proportion to the particle radius.

In this way, theoretical studies suggest the existence of the size dependence of surface energy in the nanodispersed systems. According to estimates made using different methods, dependencies of σ on size for particles and films are manifested as a monotone decrease with decreasing size starting from a radius of less than 20 nm for particles and a thickness of < 5 nm for films.

2.2. Experimental foundations for the determination of surface energy of nanoparticles

Experimental determination of surface energy of solid bodies is a challenging task. Experimental methods available to scientists today offer the measurement of values of surface energy of liquid-phase matters with a reasonable degree of accuracy and in a broad temperature band, which is not the case for the crystalline phase. Known experimental methods for the determination of the surface energy for the crystalline phase are limited, and, as a rule, have a very narrow range of pre-melting temperatures and provide precision of not more than 10–20% [4]. This is largely due to the fact that surface energy is not a directly measurable value, but in most cases it is estimated as an adjustable parameter in various processes such as, for example, wetting, spreading, melting, crystallization, dissolution, analysis of high-temperature creep, electronic work function, etc. Among the best-known methods are the following: the crystal cleavage method, the dispersed powder dissolution method, the "neutral" droplet method, the multiphase equilibrium method, the growth and evaporation steps method, the "healing" scratch method, and, finally, the zero creep method [4]. Surface energy may also be evaluated by the measurement of electronic work function [18, 19]. However, the analysis of these methods shows that they are not applicable to the measurement of surface energies of small particles.

2.2.1. Kinetics of evaporation of small particles and surface energy

Surface energy of small particles can be determined by kinetics of evaporation in vacuum at a constant temperature [6]. The method is based on the concepts of the molecular-kinetic theory that supposes that the rate of evaporation from a unit of free surface in vacuum is defined by the expression

$$\frac{dM}{dt} = \left(\frac{m}{2\pi kT}\right)^{1/2} P(T),$$

where m is mass of atom (molecule), k is Boltzmann constant, P is pressure of saturated vapor at temperature T. For the particle with the radius R saturated vapor pressure is linked to the vapor pressure over a flat surface $P_\infty(T)$ with the Kelvin equation

$$P(T,R) = P_\infty(T)\exp\left(\frac{2v_a}{kT}\frac{\sigma}{R}\right). \quad (3)$$

(v_a is atomic volume). The evaporation rate will be equal to

$$\frac{dM}{dt} = \left(\frac{m}{2\pi kT}\right)^{1/2} P_\infty(T)\exp\left(\frac{2v_a}{kT}\frac{\sigma}{R}\right). \quad (4)$$

For an array of particles on the substrate it is more practicable to measure not the evaporation rate dM/dt, but the dependence of the particle radius on its evaporation time t at a constant temperature. From the relation of R to t one can determine the variation of the particle radius dR/dt at different R and, consequently, find σ. Indeed, as it follows from (4)

$$\left|\frac{dR}{dt}\right| = \frac{1}{\rho}\frac{dM}{dt} = \frac{1}{\rho}\left(\frac{m}{2\pi kT}\right)^{1/2} P_\infty \exp\left(\frac{2v_a}{kT}\frac{\sigma}{R}\right) \text{ or } \ln\left|\frac{dR}{dt}\right| = \ln A + \frac{B\sigma}{R}, \quad (5)$$

where

$$A = \frac{1}{\rho}\left(\frac{m}{2\pi kT}\right)^{1/2} P_\infty(T) \text{ and } B = \frac{2v_a}{kT}. \quad (6)$$

According to (5) and (6), knowing the temperature, particle size reduction rate dR/dt, and $P_\infty(T)$, one can determine the value of σ. These expressions adequately describe evaporation kinetics of liquid Pb particles and crystalline Ag particles [6] at values of σ close to handbook ones.

Electron microscope investigation of the kinetics of particle evaporation was later used to register the melting temperature of small crystalline Au particles by breaks in dependencies $R(t)$ [7]. This effect is due to the difference in evaporation rates for crystalline and liquid states.

The melting temperature lowering data obtained for small Au particles from their evaporation kinetics [7] correlate well with similar results established later using electron diffraction analysis [22]. Investigating the kinetics of evaporation of silver particles on carbon substrates has shown that the observed sublimation temperature generally decreased with decreasing particle size [23], in agreement with the predictions from the Kelvin equation. However, sublimation of smaller nanoparticles was often observed to occur in discrete steps, which led to faceting of the nanoparticles.

This method was used to determine the surface energy of small particles in Bi, Pb, and Au island films [8, 9]. The sample film was heated in the electron microscope by electron beam up to the onset temperatures of evaporation. The temperature and, hence, evaporation rate was controlled with beam density. The particles radius variation rate $\Delta R/\Delta t$ during evaporation was established by the analysis of a series of successive electron micrographs taken at fixed time intervals. These data allowed to establish the temperature of particles heated by the electron beam. For this purpose, expression (6) at $R \to \infty$ should be represented in the form

$$\lg P_{\infty}(T) = \lg C + 1/2 \ \lg T, \tag{7}$$

Where $C = \rho \left(\frac{2\pi k}{m} \right)^{1/2} (dR/dt)_{R \to \infty}$.

Tabular data are available for the function $P_{\infty}(T)$, while the value $(dR/dt)_{R\to\infty}$ is to be determined experimentally from the results of change of the radius of the particles during their evaporation represented in accordance with (5) and (6) in the coordinates "ln $|\Delta R/\Delta t|$ — $1/R$." Equation (7) can be solved graphically for T and, thus, the temperature of the observed object can be found.

Figure 1a presents an example of a series of successive micrographs of Bi island films obtained in the process of their evaporation with the time interval of 15 s, and Figure 1b, c present the results of analysis of evaporation of Au island films. These data were used to find values of ΔR for different-sized particles at fixed time intervals Δt. Resulting dependencies for the ensemble of particles in Au island films (the size range of 10–50 nm) are presented in Figure 1c in the coordinates "ln $|\Delta R/\Delta t|$ — $1/R$." Since these dependencies are linear, according to (5) and (6) they allow us to establish σ and values of $(dR/dt)_{R\to\infty}$. The range of particle sizes in experiments [8, 9] made 10–150 nm. For Pb and Bi temperatures of particle evaporation were higher than melting temperatures, whereas for Au island films data on evaporation rates both in liquid and crystalline state were obtained.

Values of surface energies σ for Au, Pb, and Bi found as a result of the preceding experiments are presented in Table 1, which also presents available literature data for σ at similar temperatures. Comparison of values of σ obtained by kinetics of evaporation of small particles with available data for bulk samples shows their satisfactory fit.

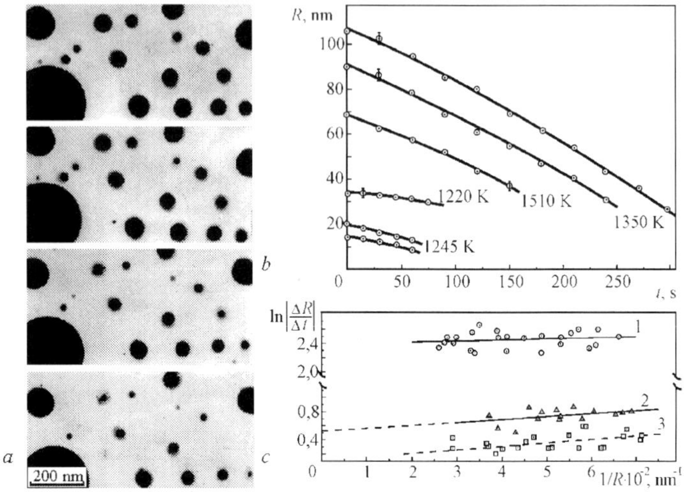

Figure 1. Electron micrographs of successive stages of evaporation of Bi island films on amorphous Si films (a); change of particle size in the process of evaporation (temperatures are given in the charts) (b), and relations of evaporation rate to reciprocal size of particles for gold island films on carbon substrates (c)

Me	Evaporation kinetics [8, 9]		Literature data	
	T, K	σ, mJ/m²	*T*, K	σ, mJ/m²
Au	1245	1410±20	1240	1410 [5]
	1260	1430	1176–1306	1390±80 [5]
	1310	1320±100	1297	1137 [6]
	1350	1230±100	1348	1135 [6]
	1510	1160		
Pb	670	385	557–589	560 [5]
	720	484	735	438 [7]
	740	452	730	439 [7]
	750	450	748	436 [7]
	770	447		
Bi	650	386	509–518	501 [5]

Table 1. Comparison of surface energy values for Au, Pb, and Bi [8]

2.2.2. *Dependence of surface energy on particle size*

Available experimental data, for example, [8, 9, 24, 25], offer contradictory conclusions regarding the sign of the size dependence of the surface energy of small particles.

The preceding paragraph demonstrates that the surface energy of small particles can be directly determined using the kinetics of their evaporation in vacuum. Table 1 presents the results of such experiments for nanoparticles over 20 nm in size. At the same time, the kinetics of evaporation of Pb and Au nanoparticles with the size below 20 nm both in liquid and crystalline state was investigated in works [6, 7]. The authors of these studies used these results to test the applicability of the Kelvin equation (3) and to estimate values of σ at temperatures at which the evaporation of particles is observed. However, the authors [6, 7] did not analyze the dependence of particle evaporation rates on particle size. Such analysis was offered by the authors in [8, 9], where they demonstrate that for particles with a size of less than 10 nm their surface energy decreases. Figure 2a presents an example of the plot of $R(t)$ for Pb particles at different temperatures, and Figure 2b presents particle size reduction rate in the coordinates "$\ln|\Delta R/\Delta t| - 1/R$" plotted using these data.

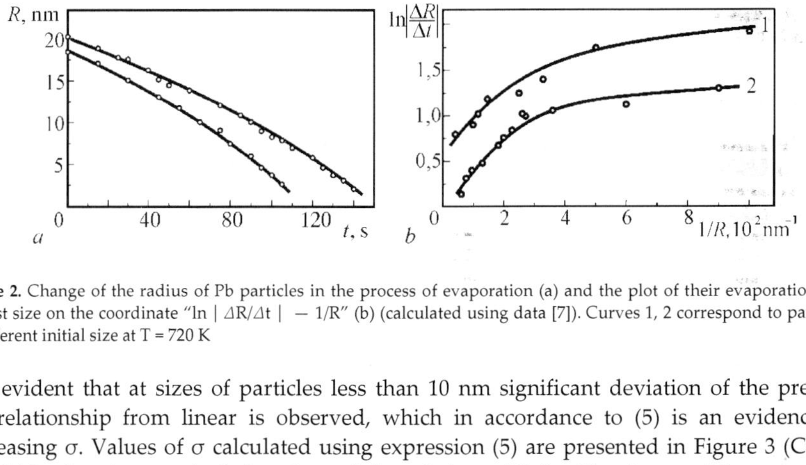

Figure 2. Change of the radius of Pb particles in the process of evaporation (a) and the plot of their evaporation rate against size on the coordinate "ln | ΔR/Δt | — 1/R" (b) (calculated using data [7]). Curves 1, 2 correspond to particles of different initial size at T = 720 K

It is evident that at sizes of particles less than 10 nm significant deviation of the preceding relationship from linear is observed, which in accordance to (5) is an evidence of decreasing σ. Values of σ calculated using expression (5) are presented in Figure 3 (Curve 5), which also shows calculation data of the relation $\sigma(R)$ for Pb microparticles using the Tolman equation (2) with the parameter α = 0.29 nm, which value is determined from the empirical relation $\alpha = 0.916 v_a^{1/3}$ [26] (Curve 4). The same figure presents the results of calculation of values of gold and lead nanoparticles surface energy at different temperatures (Curves 2, 3, 5) obtained from the data of analysis of particle evaporation kinetics given in [6, 7].

Comparison of these dependencies produces qualitatively the same result, that is, the surface energy of small particles decreases with decrease of their size, but nanoparticles evaporation experiments suggest a stronger relationship $\sigma(R)$.

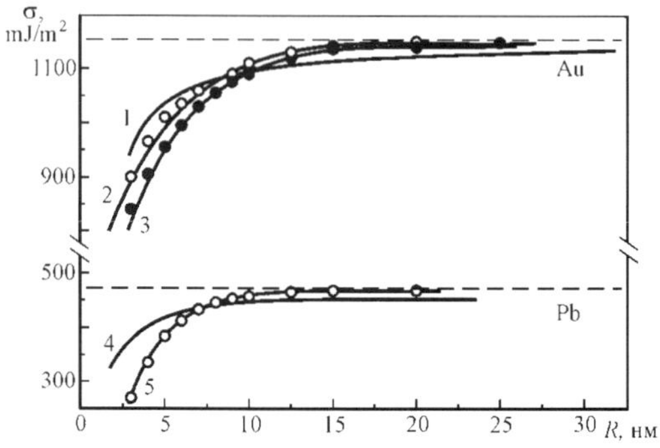

Figure 3. The plot of surface energy against microparticles size: a – Au (1 – calculation using the Tolman equation at α = 0.24 nm, 2 and 3 – using microparticles evaporation data at T = 1348 K and T = 1316 K); b – Pb (4 – calculation using the Tolman equation at α =0.29 nm, 5 – particle evaporation data at T = 735 K)

2.2.3. Decrease in small particles melting temperature and surface energy

It is common knowledge that melting temperature of small particles, thin metal, and alloy films is a function of size [6, 7, 22, 27–36]. When considered in terms of thermodynamics, there exist several models to describe the size dependence of melting temperature of small particles [27]; however, as the quantitative analysis of experimental data shows, the triple point model proves to be the most feasible. Within the framework of this model the problem of the melting temperature of the small particle was first solved by Pavlov [37], who obtained expression for the size dependence of melting temperature

$$\frac{T_S - T_R}{T_S} = \frac{3}{\lambda R}\left(\sigma_S - \sigma_l\left(\rho_l / \rho_s\right)^{1/3}\right), \tag{8}$$

where T_S and T_R are the melting temperatures of a bulk sample and a particle with the radius R, λ is melting heat, σ and ρ are surface energies and densities of crystalline (s) and liquid (l) phases, respectively. As it is seen from (8) using experimental data $T_R(R)$, one can establish the difference of surface energies of solid and liquid phases, that is, $\Delta\Omega=\sigma_S-\sigma_l\ (\rho_l/\rho_S)^{1/3}$, and given known values of σ_l find σ_S. The possibility of determining surface energy in solid phase and its temperature dependence using experimental data $T_R(R)$ are detailed in work [9]. It is possible since the difference $\Delta\Omega$ is not a constant value, but changes in accordance to variation of σ_S and σ_l. Considering this, the expression for the melting temperature of small particles can be represented in the following form

$$\frac{T_s - T_r}{T_s} = \frac{3\Delta\Omega_0}{\lambda\left(R - \Delta R\right)}, \tag{9}$$

http://dx.doi.org/10.5772/60900

where $\Delta\Omega_0 = \sigma_s^{\,0} - \sigma_l^{\,0}(1 + \delta_V / 3)$ is the variation of surface energy during melting at temperature T_s; $\Delta R = 3T_s[\gamma_l(1 + \delta_V/3) - \gamma_s]/\Lambda\rho_s$, $\gamma_l = \partial\sigma_l/\partial T$ is the temperature coefficient of liquid phase surface energy; $\gamma_s = \partial\sigma_s/\partial T$ is the temperature coefficient of solid phase surface energy, σ^0 is surface energy of relevant phase at bulk sample melting temperature T_s, δ_V is relative variation of volume during melting.

It follows from (9) that using the experimental relation $T_R(R)$ one can calculate $\Delta\Omega$ at different temperatures and, provided known type of temperature dependence of liquid phase surface energy $\sigma_l(T)$ below T_s, find dependence $\sigma_s(T)$. This approach allowed to evaluate relations $\Delta\Omega(T)$ and $\sigma_s(T)$ over a broad temperature interval using experimental data of $T_R(R)$ for Sn and In [31, 32]. The validity of linear extrapolation of values of liquid phase surface energy σ_l to the region of significant supercooling was substantiated in works [9, 38].

Considering the preceding, we calculated values of surface energies for a number of metals (In, Sn, Bi, Pb, Al, Au) in crystalline state over the temperature interval of $(0.6–1)T_s$. The obtained results are presented in Figure 4. These values agree well with available literature data on σ_s for bulk samples obtained by other methods.

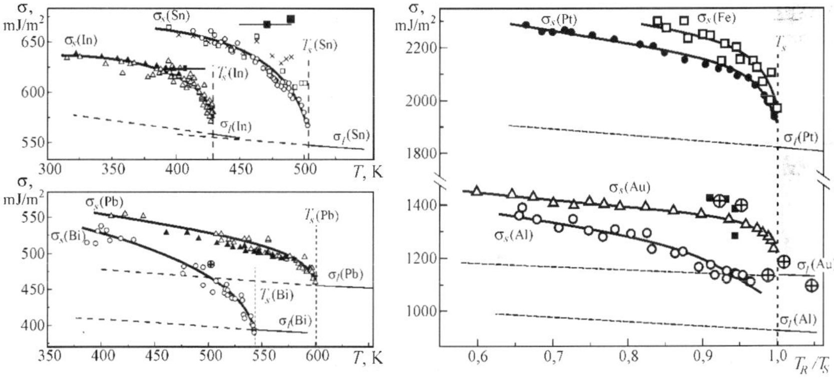

Figure 4. Surface energy temperature dependence for different metals according to the $T_R(R)$ data [6, 7, 22, 31–36] (the dotted lines are extrapolation of dependencies $\sigma_l(T)$ to the area of supercooling state) ■ – creep data [5], ⊕ – small particles evaporation data [7], □ – data for Fe [39]

It is evident that there is a common tendency observed for all of the preceding metals manifested in the fact that the values of σ_s have nonlinear temperature dependence. With relative temperatures below $T/T_s \leq (0.85–0.9)$ the temperature coefficient for these metals becomes approximately constant and makes module (0.3–0.4) mJ/m²K (Figure 4). Nonlinear decrease in σ_s at $T \rightarrow T_s$ is probably common for metals. The result of work [39] supports this assumption. In this work, surface energy of macroscopic iron samples is determined using the method of wetted solid surface deformation over the range of temperatures (1580–1790) K (Figure 4).

The nonlinear increment effect $|\partial\sigma_S/\partial T|$ at $T \rightarrow T_S$ was considered for In and Sn and supported by case study calculations of σ_S, made for Bi, Pb, Al, Au in work [9]. A detailed analysis of nonlinear relation of the temperature coefficient $\partial\sigma_S/\partial T$ in premelting temperature band showed its vacancy nature [9]. Based on the analysis of data on $\sigma_S(T)$ in the coordinates "$\ln(\Delta\sigma/\sigma) - 1/T$" the paper also estimated the vacancy formation energy E_V that yielded the following values: In – 0.5 eV, Sn – 0.62 eV and Pb – 0.6 eV. Close values of the values of E_V support vacancy mechanism of nonlinear temperature dependence of solid phase surface energy in the premelting temperature band.

3. Wetting in condensed films

Information on values of surface energy and interfacial energy of contacting phases can be obtained when studying wetting in solid–liquid systems. Analysis of known methods for the determination of the wetting contact angles θ shows that the use of traditional methods [1, 40] for studying wetting in ultradispersed systems is quite limited. In view of these, new methods [41, 42] were developed that allowed to investigate wetting in ultradispersed systems with different types of contact interaction (i.e., applicable both at $\theta < 90°$, and for $\theta > 90°$), with typical phase size changing over a broad range.

Test samples were island films of various metals condensed in vacuum by vapor–liquid method on solid substrates, which, as a rule, were prepared using vacuum condensation as well [9, 42–44]. The substance substrate was deposited on the NaCl (or KCl) cleavages in a vacuum of 10^{-7}–10^{-9} mm Hg. After that the investigated metal was condensed at a substrate temperature that ensured condensation of the metal into liquid phase. The obtained films were cooled in vacuum to room temperature and the crystallized particles were further analyzed using the methods of optical, scanning, and transmission electron microscopy. According to the estimates and data of experimental research [42–44] (Figure 5) contact angle measurement error due to changing droplet volume during its solidification on the substrate is not more than 2°. In this way one can discard the variation of the angle during crystallization of liquid droplets and relate the values of θ found for crystallized particles to values of the contact angles of liquid droplets at the temperature of their formation.

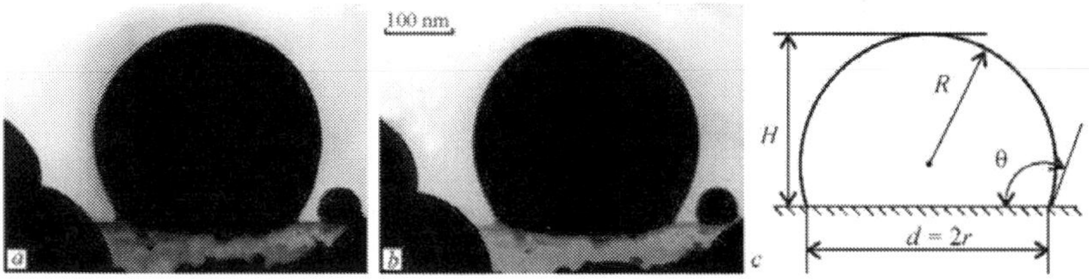

Figure 5. Electron microscope images of crystalline (a) and liquid (b) lead particles on a carbon substrate and a schematic representation of a liquid droplet on a solid substrate (c)

In the case when the gravity effect can be disregarded, the shape of small droplets is a segment of a sphere (estimations show that this is knowingly true for metals with particle size below 10^5 nm). To find the angle θ it is sufficient to measure any two of the three quantities that define droplets on the substrate: the radius of droplet surface curvature R, the diameter of its base d ($d = 2r$), and the height H (Figure 5c).

The methods suggested in [41, 42] differ in approaches to measure geometric parameters of droplets. The most frequently used cleavage and convolution methods are based on measurement of the said parameters during direct observation of droplet profiles with an optical or electron microscope [41, 42]. In this case, the contact angle is determined from the relations

$$\theta = 2\,\mathrm{arctg}\frac{2H}{d} = \arccos\left(1-\frac{H}{R}\right) = \begin{cases} \arcsin\dfrac{d}{2R}, & \theta < 90°, \\ 180° - \arcsin\dfrac{d}{2R}, & \theta > 90°. \end{cases}$$

The developed complex of methods [41, 42] makes it possible to investigate the wetting of surfaces with small droplets, with the size of the latter ranging within 3–10^5 nm.

3.1. Size effect in wetting

Wetting in the liquid–solid system is defined by the equilibrium contact angle θ, which is related to surface energies of contact phases with Young's equation

$$\cos\theta = (\sigma_u - \sigma_{ul})/\sigma_l, \tag{10}$$

where the indices u and l refer to solid (substrate) and liquid (particle) phase, respectively. One would expect that the size dependence of surface energy will cause a difference of wetting patterns in nanodispersed systems from known ones for macroscopic objects. This may change the contact angle with increasing dispersity of both liquid and solid phases. In order to provide a theoretical description of these phenomena, it is necessary to solve the problem of the equilibrium shape of the microdroplet and its contact angle, that is, obtain an equation analogous to Young's equation (10) with regard for the relationship σ(R).

Consider, following [42, 43, 45], a small droplet of liquid on a flat solid surface. The total free energy of the system F is comprised of the hydrostatic energy pV (here, pressure p may be regarded as the undetermined Lagrange multiplier that accounts the constant droplet volume V) and surface forces energy

$$F = -pV + \int_{S_l} \sigma_l dS + \int_{S_{ul}} (\sigma_{ul} - \sigma_u) dS, \tag{11}$$

where S is the interfacial area.

In accordance with existing concepts [11, 45] the surface energy σ_l is viewed as dependent on the average surface curvature C at a given point

$$\sigma_l = \sigma_l^{\infty}\left(1-\alpha C\right). \qquad (12)$$

For spherical surface ($C = 1/R$) relation (12) agrees with expression (2) known in the literature as the Tolman formula.

When finding equilibrium conditions, one should take account of the size dependence of the interfacial energy of droplet – substrate boundary σ_{ul}. It would be natural to consider this dependence as a relation not to the radius of the surface curvature R, but to the radius of the wetted perimeter r [42, 45], that is, for σ_{ul} use relationship in the form

$$\sigma_{ul} = \sigma_{ul}^{\infty}\left(1-\beta/r\right). \qquad (13)$$

Expressions (12) and (13) apply at $1/C \gg \alpha$ and $r \gg \beta$. Finding equilibrium characteristics of the droplet does not require any assumptions as to the sign and value of parameters α and β.

Due to the axiality of the problem, it can be solved using polar coordinates with their origin in the center of the circle of the wetted perimeter and the vertical axis z perpendicular to the substrate plane. The profile of the free surface of the droplet is defined by the function $z(\rho)$. Without loss of generality, one can regard $z(\rho)$ function as single-valued, that is, consider the case of $\theta < 90°$ (it can be shown that the obtained results will be valid on the whole range of angles θ, if z is chosen as an independent variable and the droplet surface is set single-valued at any θ using the function $\rho(z)$).

The equilibrium shape of the droplet is found by minimizing the functional (11), which, with regard to relations for droplet volume and areas of its boundary surfaces, is written as follows:

$$\begin{aligned} F &= 2\pi\int_0^r\left[-pz+\sigma_l\left(C\right)\left(1+z'^2\right)^{1/2}+\sigma_{ul}\left(r\right)-\sigma_u\right]\rho d\rho, \\ C &= -\frac{1}{2}\left\{z''\left(1+z'^2\right)^{-3/2}+\frac{z'}{\rho}\left(1+z'^2\right)^{-1/2}\right\}. \end{aligned} \qquad (14)$$

The summand $(\sigma_{ul}(r) - \sigma_u)\rho$ in the expression under the integral sign does not contain $z(\rho)$ and its derivatives, that is, the relation $\sigma_{ul}(r)$ defines only boundary conditions and has no effect on the shape of the droplet.

Functional variation (14) in δz gives the Euler equation, which after term-by-term integration takes on the form

http://dx.doi.org/10.5772/60900

$$-\frac{p\rho^2}{2}=\rho\left(1+z'^2\right)^{1/2}\frac{d\sigma_l}{dC}\frac{\partial C}{\partial z'}+\frac{\sigma_l\rho z'}{\left(1+z'^2\right)^{1/2}}-\frac{d}{d\rho}\left[\rho\left(1+z'^2\right)^{1/2}\frac{d\sigma_l}{dC}\frac{\partial C}{\partial z''}\right]. \tag{15}$$

The integration constant in (15) is equal to zero from the equal-zero condition of one of the non-integral summands δF at the point $\rho=0$. The complexity of equation (15) makes its general solution unlikely, which fact urges us to use specification of relation σl(C) in the form (12). Substitution in (15) of relation (12) and expressions for derivatives $d\sigma_l/dC, \partial C/\partial z'$ and $\partial C/\partial z''$ yields a nonlinear differential first-order equation

$$\left[z'\left(1+z'^2\right)^{-1/2}\right]^2+\frac{2\rho}{\alpha}\left[z'\left(1+z'^2\right)^{-1/2}\right]+\frac{p\rho^2}{\alpha\sigma_l}=0,$$

which solution by separation of variables gives equilibrium shape of the droplet surface in the form of a sphere truncated by plane $z = 0$:

$$\left(z-z_0\right)^2+\rho^2=R^2. \tag{16}$$

The sphere radius satisfies the relation

$$p=\frac{2\sigma_l^{\infty}}{R}\left(1-\frac{\alpha}{2R}\right),$$

which shows that the undetermined Lagrange multiplier p is nothing, but the Laplace pressure adjusted for the dependence σ(*R*). Integration constant $z_0=\pm\sqrt{R^2-r^2}$ has the meaning of z-coordinate of the sphere center (16) and is determined from the condition $z(r) = 0$.

The wetting angle θ can be found from the boundary condition or, since function $z(\rho)$ is defined, from the condition of minimum free energy of the droplet at a constant volume. For this purpose the surface energy of the droplet must be expressed in terms of *R* and θ

$$F_s=\pi R^2\left\{2\sigma_l\left(R\right)\left(1-\cos\theta\right)+\left[\sigma_{ul}(r)-\sigma_u\right]\sin^2\theta\right\}.$$

By setting the derivative dF_s/dR to zero we can obtain the equation, which, considering constant volume, gives the equilibrium condition of a microdroplet on a substrate

$$\cos\theta=\left(\sigma_u-\sigma_{ul}-R\frac{d\sigma_l}{dR}-\frac{r}{2}\frac{d\sigma_{ul}}{dr}\right)\Big/\left(\sigma_l+R\frac{d\sigma_l}{dR}\right). \tag{17}$$

Equation (17), naturally, is different from Young's equation (10) by presence of summands containing surface energy derivatives with respect to size.

By using expressions (12) and (13) for $\sigma_l(R)$ and $\sigma_{ul}(R)$, we can write the relation for the contact angle of the microparticle through the parameters α and β that define the size dependence of respective surface energies as follows [42, 45]:

$$\cos\theta = \cos\theta_\infty - \frac{\alpha}{R} + \frac{\beta}{2R}\frac{\sigma_{ul}^\infty}{\sigma_l^\infty}\frac{1}{\sin\theta}. \tag{18}$$

Naturally, in the extreme case at $\sigma \to \sigma^\infty$ ($\alpha/R \to 0$, $\beta/R \to 0$) all the expressions received go into known relations of the capillary theory. It should be further noted that equation (18) agrees by the type of functional dependence on size with relations received in the scope of the line tension model [46, 47].

3.1.1. Contact angle size dependencies and surface energy of the solid–liquid interface

The size effect in wetting of the flat surface of the solid substrate with small metal droplets was first found for vacuum-condensed island tin and indium on an amorphous carbon substrate [42–44]. A combination of optical and electron microscopy [41, 42] allowed to determine wetting contact angles over the range of particles of 1–10^4 nm. According to measurements using optical microscopy the contact angle in the Sn/C system for micron-sized droplets is constant and makes 151°±2°, which fact agrees with the known data for the tin-carbon system. For measurement of contact angle in islands of smaller size we applied the methods of convolution and photometric analysis of electron microscope pictures.

The results of measurement of θ for tin on carbon substrate are presented in Figure 6a, which demonstrates that for big particles (R > 30 nm) values obtained are close to respective values of micron-sized droplets. As the size of the particles decreases (R < 30 nm) one can observe decrease of the contact angle.

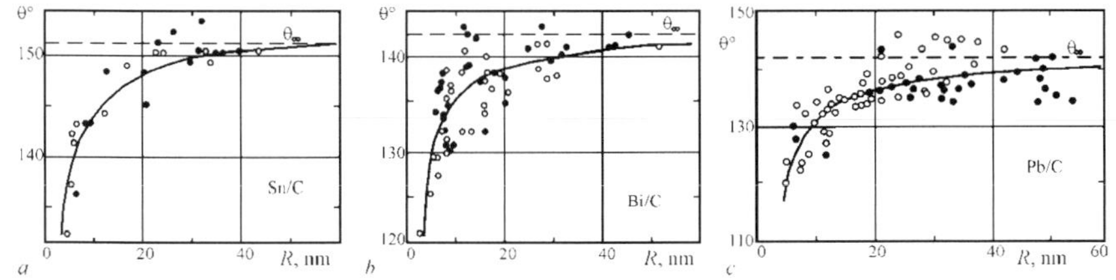

Figure 6. The plots of the contact angle against the radius of particles of tin (a), bismuth (b), and led (c) on a carbon substrate (O – data using convolution method, ● – using photometric measurement of electron microscope pictures) [9, 42–44]

The preceding research was followed later by the study of wetting in the systems of "island metal (Bi, Pb, Au) film–amorphous carbon film" and "Pb–amorphous silicon film" depending on the size of particles [9, 42–44]. For all investigated systems it was obtained that with particle sizes $R > 100$ nm and thicknesses of carbon and silicon films $t > 20$ nm, contact angles of liquid droplets in island films well agree with data of respective contact systems in bulk state. With particle sizes $R < 30$ nm the value θ decreases so that $\Delta\theta = \theta_\infty - \theta \approx (20°–25°)$ at $R = (4–5)$ nm (Figure 6). The conclusion regarding reduction of the contact angles of nanosized droplets to the extent of their spreading was obtained in [48] by modeling using the molecular dynamics method.

Influence of sample preparation conditions on the contact angles of microparticles in the gold–carbon system was discussed in detail in [50]. It showed that the amount of pressure of residual gases during the preparation of gold island films has no effect on the values of surface energies corresponding to the bulk state, though it has a slight impact on their size dependence.

The results of the study of wetting obtained [9, 42–44] and presented in Figure 6 are of separate interest, in particular, for some practical applications (e.g., the formation of ordered nano-structures by film melting [51–53]), but at the same time allow to obtain new physical information regarding properties of microparticles. Thus, using the experimental relations $\theta(R)$ and $\sigma_l(R)$ one can establish the size dependence of interfacial energy of the microparticle–substrate boundary σ_{ul}.

For analysis of results of the size effect of wetting in island films expression (18) was used, from which, using the relation $\theta(R)$ at known quantities of the parameter α and the value of the surface energy σ_u, the value of the interfacial energy of the microparticle–substrate boundary and its size dependence were found. The parameter α can be found from data on the kinetics of evaporation of small particles [8]. It may be evaluated using the relation $\alpha \approx 0.916 v_a^{1/3}$ (v_a – atomic volume) too [26]. Calculation of α using this relation yields the value of $\alpha(Pb) = 0.27$ and $\alpha(Au) = 0.23$ nm, that is, quantities close to those found experimentally in [8]. This allows to use at first approximation the preceding relation to estimate the parameter α for metals that lack experimental data for the relation $\sigma_l(R)$. The value of the surface energy of carbon film was determined and presented in work [9, 43] from the data on wetting of free films of different thickness with microdroplets of indium, tin, and lead, which makes $\sigma_u = 120 \pm 30$ mJ/m^2.

Using these data authors of work [45] found values of σ_{ul} and the parameter β, which are presented for the investigated metal-carbon systems in Table 2.

The parameters α and β are positive, which is an evidence of decreasing surface energy of microparticles and interfacial energy at the substrate boundary with a decrease of the radius. Values of α approximately correspond to the thickness of the surface layer at the liquid–vacuum boundary. The value β, which defines the width of the transition zone between the liquid particle and the substrate and depends on the nature of contacting phases, is 2–4 times as big as α.

Metal	σ_l^∞, mJ/m²	α, nm		σ_{ul}^∞, mJ/m²	β, nm	θ_∞
		Calculated	Experiment [8]			
Au	1130	0.24	0.23	955	1.0	138.4
Sn	531	0.28	-	574	0.53	152.4
Pb	450	0.29	0.27	463	0.91	140.9
Bi	376	0.30	-	407	0.5	141.0
In	559	0.27	-	566	0.55	143

Table 2. Results of wetting size effect in metal-carbon systems [42, 43, 45]

3.1.2. Wetting hysteresis in condensed microdroplets

The observed reduction of the contact angle θ with decreasing radius of the droplet is accounted for by the size dependence σ_l and σ_{lu}, due to the growing relative contribution of interface regions. However, wetting parameters are also subject to the influence of substrate elastic deformation, which was disregarded earlier when deriving equations (17) and (18). The effect of deformation on angle θ in the case when substrate is a thin film was dealt in detail in [54], for elastic half-space in [55], but because of approximation, the results presented in [55] cannot be applied to droplets with the size below 20–50 nm, that is, when wetting size effect is observed. Works [42, 56] offer solutions of the problem of determination of the value of the equilibrium wetting angle θ for the microdroplet with a radius of less than 50 nm with regard to elastic deformation of the substrate. On assumption that the force of liquid tension along the wetting perimeter is uniformly distributed along the ring of finite width, it was shown that elastic deformation effect is insufficient (1–2° for the systems considered earlier), and, hence, reduction of the contact angle is primarily defined by the size dependence of specific energies of interface surfaces. Nevertheless, for substrates with low value of Young's modulus ($E \sim 10^9$ N/m²) additional deflection of contact angle for small droplets reaches 5–6°.

At the same time, as a result of the small size of droplets and increased diffusion coefficients in nanodispersed systems [57–59], the wetting perimeter under the action of surface tension forces may experience irreversible changes, which may be registered with electron microscopy as circular traces on the substrate left after evaporated droplets. In addition, the plot of the radius of evaporating droplet against time at constant temperature, as a rule, has periodic deviations of experimental points from the continuous curve. Works dedicated to direct measurement of the wetting contact angle also demonstrate fluctuations $\Delta\theta \approx$ 10–15°, while the precision of the convolution and photometry methods makes 3–5° [41, 42]. In both cases these deviations may be accounted for by wetting hysteresis. This phenomenon consists in fixation of the wetting perimeter, which under certain conditions, for example, during evaporation, significantly changes the behavior of the liquid droplet. Work [60] examines the reasons causing this effect in microdroplets, it analyses the effect of wetting hysteresis on parameters of the droplet–substrate system.

http://dx.doi.org/10.5772/60900

A number of works were concerned with wetting hysteresis, for example [46, 61], and the commonest causes of this effect are considered to be microroughness and inhomogeneity of the substrate. However, many assumptions underlying these works, for example, the roughness height of ~ 1 μm cannot be applied to microdroplets. In several systems fixation of the wetting perimeter is achieved by partial mutual dissolution of solid and liquid phases. Nevertheless, hysteresis may be as well observed for systems with minor mutual solubility, such as Au/C. Some authors noted that at high temperatures under the action of liquid surface tension forces, the substrate may be subject to inelastic deformation. In this case the triple contact area develops a prominent welt. Comparison of different mechanisms of mass transfer at small ($\sim 10^{-8}$ m) distances implies a conclusion about the defining role of surface diffusion. As follows from estimations made in work [60] characteristic time of deformation in the Au/C systems makes about 0.1 s. Since the time of condensation of films is about 10^2 s, in the process of droplet growth the welt has enough time to form even with quite frequent jumps of the wetting perimeter, that is, the droplet creates substrate roughness itself.

The values of the contact angles corresponding to perimeter breakdown with changing volume of the droplet were received [60] from the condition of the system's minimum free energy taking into account elastic deformations of the substrate and their partial relaxation in the triple contact area trough surface diffusion. According to [60] the contribution of relaxed elastic deformation energy reaches significant values, for example, for the Au/C system the wetting hysteresis, that is, difference between advancing θ_a and receding θ_r contact angles makes about 3°.

In this way, with changes of the volume of the droplet, for example, during evaporation, its wetting perimeter will be fixed until the contact angle is reduced to the critical value θ_r. Upon this wetting the perimeter will break, and the droplet will take the position corresponding to equilibrium (Young's) value θ_0. In this position the wetted perimeter develops a new welt and the process recurs. It may be noted that for sufficiently big droplets (with their base radius considerably larger than the width of the welt) the dependence of values of the angles θ_r and θ_a on the size is small. Thus, for the Au/C system with the droplet radius growing from 20 to 1000 nm the value of these angles changes by 1° due to the increase of elastic energy contribution.

Considerable amount of Laplace pressure in very small (less than 10 nm) droplets results in elastic deformation being able to relax not only in the triple contact area, but also directly under the droplet. In this case, along with the welt along the wetted perimeter dimples may form under the droplet, which causes higher hysteresis value, with the difference $\theta_0 - \theta_r$ growing much faster than $\theta_a - \theta_0$, that is, the minimum contact angles diverge from the equilibrium value more than the maximum ones. Taking into account the decrease of the value of the contact angle, one may conclude that, for example, for the same Au/C system the contact angle of small (2–5 nm) droplets may reach 65–70° at $\theta_0 = 138°$.

3.2. Wetting in droplet–thin film–substrate systems

Among the factors that define wetting in dispersed metal–metal systems, in addition to size effect of wetting, one can single out the following: discontinuity of intermediate film and resulting heterogeneity of the substrate, mutual solubility of components in each other, formation of chemical compounds at the solid and liquid phase interface, and oxidation of the metal film. Therefore, wetting processes in ultradispersed systems are defined by a number of parameters, which are quite inseparable.

A case study research of the influence of fineness of the solid phase on the contact angle is presented in work [62], which investigated wetting of thin films of different thickness deposited on bulk substrate. It showed that in the melt (Ag, Cu, Sn, Pb)–metal film (Mo, V, Fe)–nonmetal substrate (sapphire, quartz, graphite) system the contact angle is subject to linear variation within the range of values corresponding to wetting of clean substrate (at thickness of film $t \to 0$) and wetting of film substance in compact state (at $t > t_c$). The values of critical thicknesses t_c, below which change of the contact angle is observed for the investigated systems is in the range of 20–50 nm. At the same, when carbon film covered germanium is wetted with tin the change of the contact angle is observed up to a thickness of 3 nm, while remaining unchanged further, and corresponds to the wetting of compact graphite. The authors explain the obtained results by discontinuity of carbon films at $t < 3$ nm.

Wetting in triple systems Pb/Ni/[NaCl, Si, GaAs], Sn/[C, Al, Al_2O_3]/KCl, Bi/Fe/KCl as a function of metal film thickness (2 nm $< t <$ 200 nm) was investigated in [9, 42–44, 63]. These systems substantially differ by interaction behaviors: Sn–C, Sn–Al_2O_3, Bi–Fe – complete insolubility in solid and liquid states; Sn–Al – solubility 0.5 wt. % Al in Sn; and Pb–Ni – up to 4 wt. % Ni in Pb. Test samples were prepared as follows. Variable thickness intermediate film (Al, Fe, Ni, C, Al_2O_3) was condensed on monocrystal substrates (KCl, NaCl, Si, GaAs) in a vacuum of 10^{-6}–10^{-8} mm Hg. The studied metal (Sn, Bi, Pb) was condensed on this film by the vapor–liquid mechanism without deterioration in vacuum. The substrate temperature during condensation was 653 K for Pb, 523 K for Sn, and 560 K for Bi.

In all investigated systems degree of wetting strongly depends on intermediate film thickness, though the range of thicknesses, on which change of θ occurs is different. The common feature for the analyzed systems is that the contact angle is defined at a fist approximation by heterogeneity of the wetted surface and changes within extreme limits corresponding to wetting of clean substrate ($t \to 0$) and intermediate film material in bulk state ($t > t_c$). The critical thickness t_c, at which complete screening of a bulk substrate by a thin film is observed, depends on the character of interaction of the systems' components and varies from nanometers (no interaction) to tens and hundreds of nanometers (dissolution of the film in the melt, formation of chemical compounds). The analysis of the obtained results together with data [62] allowed to classify the main types of the relations $\theta(t)$ for wetting with a melt of thin film on the surface of a bulk substrate (Figure 7):

a. **Noninteracting systems** – Figure 7a. The value t_k in such systems is defined by the microstructure of intermediate film and to some extent may depend on technological parameters of its production (substrate temperature, condensation rate, etc.). The variation of the contact angle is defined by transition from discontinuous to continuous film and the dependence of its surface energy on thickness. Sn/C/KCl (Figure 8a), Sn/Al_2O_3/KCl, Sn/C/Ge [62] may serve as examples of such systems. Since wetting angle in such systems changes for intermediate film thinner than 10 nm, effects related to size variation

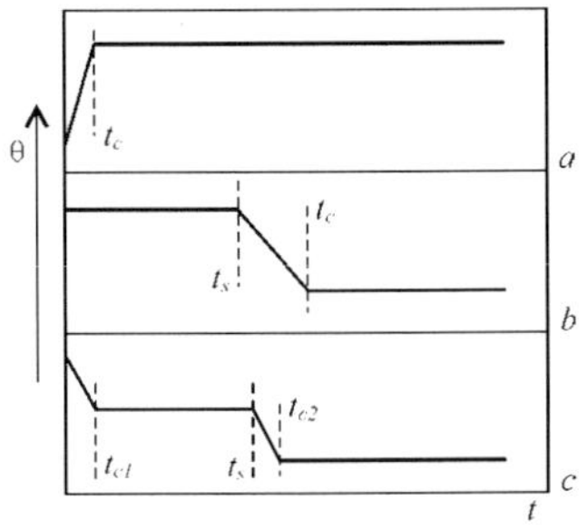

Figure 7. Main types of dependencies $\theta(t)$ for systems melt – film – substrate: a) noninteracting systems; b) systems with dissolution of film in the melt; c) systems with chemical interaction at the film–substrate interface

of film surface energy are also possible. Thus, in the Sn/C/KCl system variation of θ is observed in the range of $2 < t < 7$ nm. At the same time, electron microscope analysis of clean carbon films suggests their continuity with decrease of thickness down to 1.5–2 nm. This gives ground to assumption that apart from discontinuity of carbon film the relation $\theta(t)$ in the Sn/C/KCl system is also stipulated by their surface energy. Since with decreasing t changes not only σ_u, but also interface energies of the film–particle interface σ_{lu} and film–bulk substrate, the said relation, strictly speaking, reflects changes in adhesion tension $\sigma_u - \sigma_{lu}$ with decrease of thickness of carbon film on the surface of macroscopic KCl monocrystal.

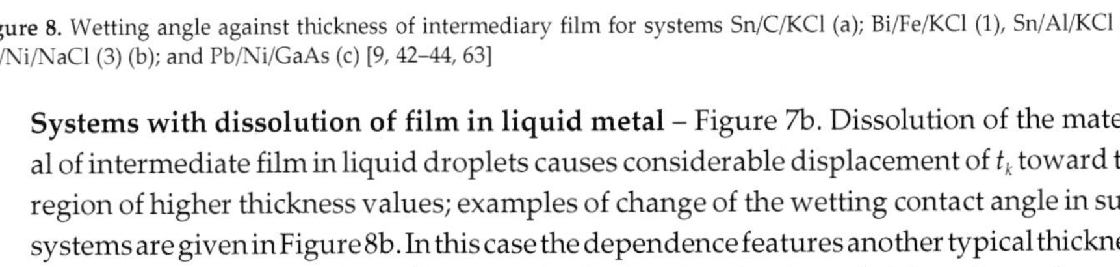

Figure 8. Wetting angle against thickness of intermediary film for systems Sn/C/KCl (a); Bi/Fe/KCl (1), Sn/Al/KCl (2), Pb/Ni/NaCl (3) (b); and Pb/Ni/GaAs (c) [9, 42–44, 63]

b. Systems with dissolution of film in liquid metal – Figure 7b. Dissolution of the material of intermediate film in liquid droplets causes considerable displacement of t_k toward the region of higher thickness values; examples of change of the wetting contact angle in such systems are given in Figure 8b. In this case the dependence features another typical thickness t_s, by which the intermediary film completely dissolves in the melt. The value t_s depends on the solubility of the film material in liquid metal at given temperature. A partial dissolution of film in liquid metal is observed within the interval of thickness $t_s < t < t_c$, which causes its discontinuity, that is, the substrate becomes heterogeneous. Since solubility of film material in the melt for the investigated systems is limited, the degree of substrate heterogeneity is a function of film thickness, which fact accounts for the observed dependence $\theta(t)$. Dependencies of this type are observed in Pb/Ni/NaCl, Bi/Fe/KCl, Sn/Al/KCl systems (Figure 8b) and for a number of systems studied in [62]: [Cu, Ag, Pb, Sn]/[Mo, V, Fe]/[quartz, sapphire, graphite];

c. **Systems with chemical interaction of film with substrate**– Figure 7c (Pb/Ni/Si и Pb/Ni/GaAs [42, 63], Figure 8c). As could be seen from Figure 8c, in the Pb/Ni/GaAs system changes in wetting occur in two stages. First, the contact angle decreases to intermediary $\theta \approx 75°$, then over the range of thickness $6 < t < 14$ nm the plot $\theta(t)$ features a plateau, which has never been observed in similar systems investigated earlier, and further the contact angle changes again until it reaches a value corresponding to the Pb/Ni system. This variation of $\theta(t)$ directly suggests the presence of at least two mechanisms of wetting variation with film thickness, one of which occurs on the interval $0 < t < 14$ nm, and the other one at $t > 14$ nm. Hence, this type of system can be divided into two subsystems, and, accordingly, is specified by the two values of critical thickness. The second subsystem (regions B, C, D in Figure 8c) belongs to type (a) or (b). In the first subsystem substrate heterogeneity on the transition segment $0 < t < t_{c1}$ (region A in Figure 8c) is caused by the growth of new phase island – a chemical combination of film with substrate (results of phase analysis of double-layer Ni–GaAs films produced in different conditions are presented in works [64, 65]), and the value t_{c1}, which corresponds to the formation of continuous film of the compound, is defined by the mechanism of interaction of intermediate layer with substrate.

It should be noted that plots presented in Figure 7 are simplest and influence other factors, for example, interaction with the residual atmosphere (Sn/Al/KCl [42]) may cause a more complex variation of the contact angle with changing thickness of the intermediary film.

3.3. Wetting of thin free films

When interpreting results of wetting in three-component systems liquid–thin film–bulk substrate, it is difficult to separate effects due to film thickness itself and the influence of bulk substrate. Therefore, it was considered expedient to investigate wetting of thin free films depending on their thickness [42–44]. The obtained results did not make it possible to find size dependence of surface energies of thin carbon films. However, these results are of interest in themselves because it is possible for highly dispersed systems, when liquid particles wet not the surface of bulk solid bodies, but that of free thin films. In this case, specific effects connected to deformation of film under the liquid droplet are observed.

A theory of half-space wetting was constructed in [55]. It suggests that the droplet deforms the region near the line of contact of three phases to form a welt. In case of thin films deformation may be significant that makes it possible to find it by experiment. Therefore, in the following text we give an outline theoretical analysis of wetting of thin free films on assumption of constant surface energies σ_l, σ_u and σ_{ul}, made in work [54], and respective experimental results of the research [42–44].

3.3.1. *Wetting of elastically deformed film with small droplets*

According to [42–44, 54], the equilibrium characteristic of the system comprised of free elastically deformed film with the thickness t and the wetting droplet (Figure 9) is found in the same way as in the problem discussed earlier – from the minimum free energy condition, with introduction to the expression of which a summand corresponding to energy of the film:

$$F = 2\pi\int_0^L \Big\{\Big[-p(z-\zeta)+\sigma_l\sqrt{1-z'^2}+(\sigma_{ul}-\sigma_u)\sqrt{1+\zeta'^2}\Big]\rho\Theta(r-\rho)+ \\ +2\sigma_u\rho\sqrt{1+\zeta'^2}+\psi(\zeta',\zeta'',u,u',\rho)\Big\}d\rho, \tag{19}$$

where L is the radius of the film fixation circle, functions $\zeta(\rho)$ and $z(\rho)$ define the radial profile of the surfaces of the film and the droplet, respectively; $\Theta(x)$ is the Heaviside step function.

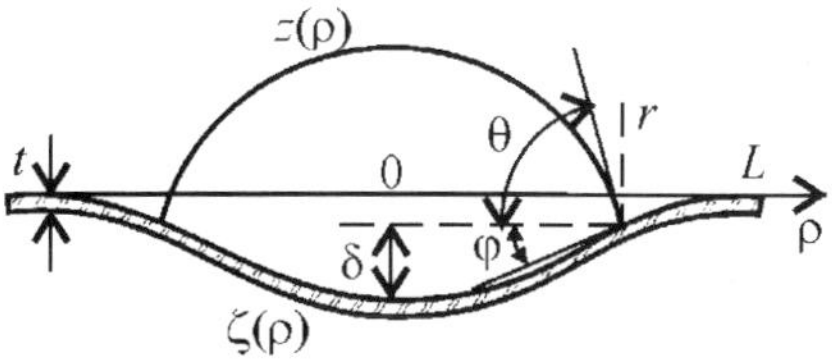

Figure 9. Schematic diagram of a liquid droplet on a thin elastic film

The function ψ is equal to the sum of contributions of elastic energies of pure bending ψ_1 and longitudinal extension of the film ψ_2, written, according to [66], with regard to axiality in the following form:

$$\psi_1 = \frac{1}{2}D\rho\left(\zeta''+\frac{2\nu}{\rho}\zeta''\zeta'+\frac{1}{\rho^2}\zeta'^2\right), \\ \psi_2 = \frac{6}{t^2}D\rho\left(u'^2+\frac{2\nu}{\rho}u'u+\frac{u^2}{\rho^2}+\frac{1}{4}\zeta'^4+u'\zeta'^2+\frac{\nu}{\rho}u\zeta'^2\right), \tag{20}$$

where ν is Poisson's ratio, u is the radial component of two-dimensional displacement vector, $D = Et^3/[12(1-\nu^2)]$ is the stiffness factor (E is Young's modulus).

The shape of the free surface of liquid is found by variation of the functional F in δz on $0 \le \rho \le r$. Upon double integration, the relevant Euler equation yields function $z(\rho)$ as a sphere (16) with the radius $R = 2\sigma_l/p$.

Variation of F in $\delta\xi$ and δu allows to obtain equations to define film deformation, which, after partial integration and substitution of the ψ_1 and ψ_2 from (20), assume the form

$$\zeta''' + \frac{1}{\rho}\zeta'' - \frac{1}{\rho^2}\zeta' - \frac{12}{t^2}\left(u' + \frac{\nu}{\rho}u + \frac{1}{2}\zeta'^2\right) - \\ -\frac{1}{D}\left[\left(\sigma_{ul} - \sigma_u\right)\Theta\left(r - \rho\right) + 2\sigma_u\right]\frac{\zeta'}{\sqrt{1+\zeta'^2}} = -\frac{pr}{2D}\Theta\left(r - \rho\right); \tag{21}$$

$$u'' + \frac{1}{\rho}u' - \frac{1}{\rho^2}u = -\zeta'\zeta'' - \frac{1-\nu}{2\rho}\zeta'^2. \tag{22}$$

The boundary conditions for equations (21) and (22) follow from non-integral summands of the variation δF going to zero. Two of them were used to draw equations (21), (22), and the rest may be written as follows: $u(0) = 0$; $\xi'(0) = 0$; $\xi(L) = 0$; $\xi'(L) = 0$; $u(L) = 0$; besides that, the point $\rho = r$ shall require continuity of the functions $\zeta(\rho)$, $\zeta'(\rho)$, $u(\rho)$, $\zeta''(\rho)$, and $u'(\rho)$.

The condition of equilibrium value of the contact angle θ may be determined by variation of functional (19) with δr. Here we obtain an expression, which is Young's equation written along the tangent line to the film surface at point $\rho = r$: $\sigma_l \cos(\theta - \phi) = \sigma_u - \sigma_{ul}$, where $\phi = \mathrm{arctg}\zeta'(r)$ – film inclination angle at the point $\rho = r$. The same Young's equation corrected for elastic surface inclination angle was obtained in [55].

Another relation connecting ϕ and θ follows from equation (21) and boundary conditions at the point $\rho = r$:

$$\sigma_l \cdot \sin\theta = \left(\sigma_{ul} - \sigma_u\right)\sin\phi + D\left[\zeta'''\left(r+0\right) - \zeta'''\left(r-0\right)\right]. \tag{23}$$

As it is seen from (21) and (23), the contact angle depends on film deformation and is determined by the jump of the third derivative $\zeta(\rho)$ on the line of three-phase contact.

Features of deformation for small and big film deflections, that is, with prevailing bending and tensile deformation, respectively, were evaluated in [54]. In the case when the maximum film deflection δ is less than its thickness, equations (21) and (22) become linear, and their solution yields the following expression

$$\frac{E}{1-\nu^2} = \frac{9\sigma_l \cdot \sin\theta_\infty}{8t^3}\left(\lim_{r\to 0}\frac{\delta}{r^3}\right)^{-1}, \quad \phi \sim \frac{4}{3}\frac{\delta}{r}, \tag{24}$$

which relates Young's module to parameters measurable by experiment. At greater film bends ($\delta > t$), approximate solutions of equations (21) and (22) can be obtained resulting in the estimation

http://dx.doi.org/10.5772/60900

$$\frac{\delta}{r} \sim \left(2\sigma_l \cdot \sin\theta_\infty / Et\right)^{1/3}. \tag{25}$$

In the case of a very thin film $t \leq 10\sigma/E$ (this can be the case for films with low elastic modulus) its shape under the droplet tends to sphere with the radius $R_{ul} = r(\sigma_u + \sigma_{ul}) / \sigma_l \sin\theta$ and remains flat outside the droplet. Smooth transition from one shape to another takes place in a narrow region with the width of the order of film thickness, and the value of the contact angle $\theta = \lim_{t \to 0} \theta(t)$ is defined only by surface phase energies in accordance with the equations

$$\begin{aligned} &\sigma_l \cos\theta_0 + \left(\sigma_u + \sigma_{ul}\right)\cos\phi = 2\sigma_u; \\ &\sigma_l \sin\theta_0 = \left(\sigma_u + \sigma_{ul}\right)\sin\phi. \end{aligned} \tag{26}$$

3.3.2. *Wetting of free carbon films with metal droplets*

As already noted, experimental research of wetting of free amorphous carbon films of different thickness with island vacuum condensates aimed at obtaining data on surface energy of films was made in [42–44]. The research used test metals (In, Sn, Pb), which are inert to carbon films and forming with carbon films contact angles of 140°–150°.

Test samples were prepared by evaporation and condensation of carbon and metal in a vacuum of 10^{-6} mm Hg on carbon films of different thickness located on copper grids with the mesh size of 60 μm. During the experiment the temperature was kept above the melting point for the relevant metal. Carbon film thickness varied on the range of 4–30 nm, and the size of liquid metal particles was 30–500 nm and was limited, on the one hand, by the need to exclude size effect due to dispersity of the liquid phase, and by strength of carbon films on the other.

Electron microscope examination of profiles of crystallized metal droplets (Figure 10) revealed substantial difference in the shape of the interphase boundary droplet–substrate for micro-particles condensed on free films and films on a solid surface. The difference is that when the film is thin enough it gets deformed by the droplet (Figure 10), while in particles condensed on a solid surface, liquid–substrate interface remained flat.

For the analyzed systems it was determined that at $t < 30$ nm the contact angle decreases with film thickness (Figure 11). At $t > 30$ nm the angle θ approaches the constant value θ_∞ that corresponds to wetting of bulk material. Analysis of profiles of droplets on free films showed that the film is deformed by the droplet, with the degree of deflection being a function of thickness and becoming a factor at $t < 10$ nm.

The obtained results were interpreted in [42–44] within the framework of wetting of elastically deformable carbon films [54], whose basic concepts were given earlier. As follows from relations (24) and (25), at t = const the relation of maximum film deflection δ to the droplet base radius r will be different for cases with prevailing deformation of bending ($\delta \sim r^3$ at $\delta < t$) and stretching ($\delta \sim r$ at $\delta > t$).

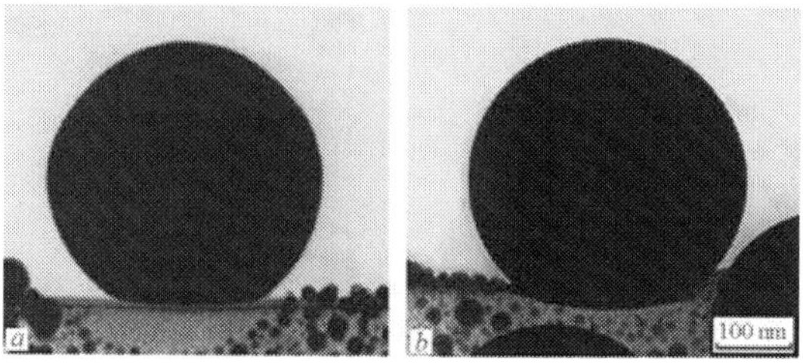

Figure 10. Micrographs of tin droplets on free carbon films with the thickness of 20 (a) и 10 (b) nm

Experimental relations δ(r) for the Sn/C system at t = 10 nm (Figure 12) corroborate this conclusion. It is apparent from the chart that the linear dependence δ(r) is observed at $\delta < t$ on the coordinates "r^3 — δ," and at $\delta > t$ – on the coordinates "r — δ."

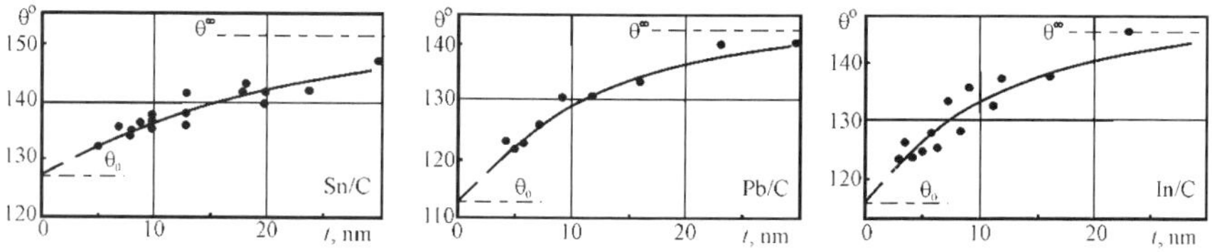

Figure 11. Dependence of the angle of wetting of free carbon films with tin (a), lead (b), and indium (c) on their thickness [42–44]

It follows from the same charts that $\lim_{r\to 0}(\delta / r^3)\,|_{\delta<t} = 7.5\cdot 10^{-6}$ nm^{-2} and $(\delta / r)\,|_{\delta>t} = 0.11$. Since Poisson's ratio is usually within the limits $1/4 < \nu < 1/2$, then from relation (24) one can estimate Young's modulus for carbon film, which turns out to be equal to $E \approx (3{,}4\text{–}4{,}2)\cdot 10^{10}$ N/m^2. With regard to this value from (25) one gets $(\delta / r)\,|_{\delta>t} \approx 0.10 \pm 0.01$, which agrees well with the experiment. Theoretical prediction of the contact angle at t = 10 nm, too, gives the value of $\theta = (140 \pm 1)°$, close to the experimental one of $\theta \approx 138°$.

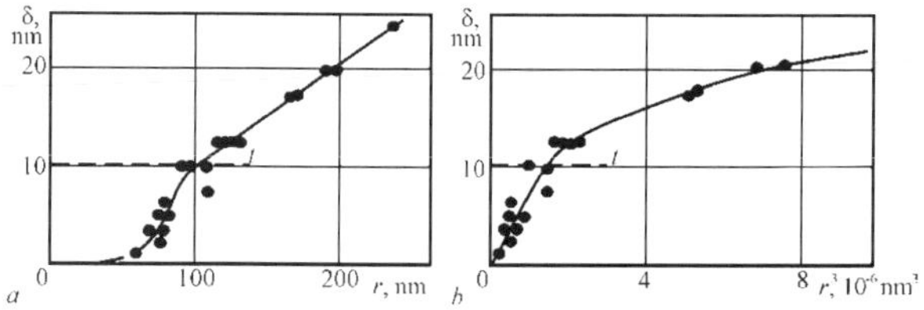

Figure 12. The relation δ(r) on the coordinates "r – δ" (a) and "r^3 – δ" (b) for tin droplets on a free carbon film with the thickness of 10 nm [42–44]

The results of theoretical consideration for very thin films were used to evaluate the surface energy of carbon films. As follows from (26) and (10), in the case when deformation energy in comparison with surface ones may be neglected, the quantity σ_u is defined by the expression

$$\sigma_u = \sigma_l \sin\theta_\infty / 4\left(\cos\theta_0 - \cos\theta_\infty\right). \tag{27}$$

The range of free films thickness on which this relation is satisfied is not attainable experimentally (t < 1 nm); however, the angle θ_0 may be found by extrapolation of the relation $\theta(t)$. The values of carbon films surface energy estimated with the help of (27) from experimental data for the In/C, Sn/C, and Pb/C system make about 120 ± 30 mJ/m^2. Since data regarding surface energy of amorphous carbon films are unavailable, works [42–44] compare obtained values of σ_u with surface energy data for different modifications of carbon. Values of σ_u for carbon found by different authors quite well agree with the results [42–44], whereas large values of σ_u (500–2500 mJ/m^2) reported in literature, as a rule, refer to high temperature and crystalline modifications of carbon. It should be noted that great values of σ_u most likely cannot be taken as characteristics of amorphous carbon film because calculation of θ_0 even at σ_u = 350 mJ/m^2 gives values (130, 140 и 135° for In, Sn, and Pb, respectively) that exceed those measured experimentally.

Variation of θ with thickness of free films as a result of theoretically predicted size dependence of their surface energy is about an order of magnitude smaller than the change in the contact angle due to deformation. Hence, the studies did not allow to trace the dependence of $\sigma(t)$ for free carbon films, though they made it possible to determine the value of surface energy for them.

3.4. Wetting in supercooled island condensates

It is known that above the melting temperature the surface energy σ_l decreases linearly with increase of temperature. However, existing knowledge and experimental data on temperature dependence of surface energy of supercooled liquids are ambiguous [67]. According to [67] at significant supercooling values, one may expect inversion of the temperature dependence $\sigma_l(T)$.

The temperature dependence of the surface energy of metals (Ga, In, Sn, Bi, Pb) [68] has been studied only in the rage of small supercoolings down to $0.1T_s$. Using containerless electrostatic levitation techniques [69–71] for a number of refractory metals supercoolings $(0.12–0.18)T_s$ were reached and it was determined that the dependence $\sigma_l(T)$ is linear ($d\sigma_l/dT<0$) in the supercooled region.

Measurement of surface energy of supercooled melts is hard to make because considerable supercoolings are normally obtained in microvolumes, and traditional methods to determine σl require large amount of melt [1, 40]. However, temperature dependence of surface energy on at $T < T_s$ can be estimated by analyzing wetting of the solid substrate with supercooled droplets.

3.4.1. Inversion of wetting temperature dependence in island films

Contact pairs being island films of tin, indium, bismuth, and copper on amorphous carbon substrates and indium on aluminum substrate were investigated in [42, 43, 72]. The test samples were prepared by condensation in a vacuum of $5 \cdot 10^{-6}$–$2 \cdot 10^{-8}$ mm Hg on the circular substrate with temperature gradient (200–900 K) set along it. As a result, condensation to equilibrium or supercooled phase with the formation of microdroplets occurred according to the condensation diagram [73–76]. The obtained samples were cooled to room temperature and then wetting contact angles on crystallized droplets condensed at different substrate temperatures were measured. Due to wetting hysteresis, which occurs even on an absolutely smooth and uniform surface because of deformation of the substrate in the region of triple contact [60], the droplet base radius remains constant during cooling. The contract angles were measured on electron microscope pictures of droplet profiles (Figure 13) and averaged for 10–20 droplets. As long as condensation takes place on a substrate with temperature gradient, the relation $\theta(T)$ can be measured in single experiment on a wide range of temperatures with arbitrarily small temperature step.

Figure 13. Micrographs of tin particles condensed in a vacuum of $5 \cdot 10^{-6}$ mm Hg at temperatures 400 K (a), 570 K (b), and 730 K (c) [72]

The results of measurement of wetting angles in the Sn/C and In/C systems [72] are presented in Figure 14a, b. The obtained dependencies are characterized by the maximum at temperatures 550 and 500 K for tin and indium, respectively. Below T_s the wetting angle gradually decreases with decrease of temperature. Decrease of θ for the investigated systems makes about 25° at maximum achieved supercoolings $\Delta T_{Sn} = 160$ K и $\Delta T_{In} = 100$ K. Better wetting is also observed above T_s at growing temperature, where for indium and tin θ decreases on the same temperature range: $550 < T < 650$ K. Over 700 K the contact angle in the Sn/C system shows behavior typical for noninteracting systems consisting in small decrease θ with the growth of temperature. Here the relation $\theta(T)$ is close to linear with the slope coefficient $d(\cos\theta)/dT \approx 0.0001$ K^{-1}.

In the Bi/C system temperature dependence of wetting, similarly to the In/C and Sn/C systems considered earlier, is nonmonotonic and is characterized by considerable decrease of the contact angle when approaching the temperature of maximum supercooling (Figure 14c). However, the maximum value of θ for bismuth is achieved at $T = 430$ K, that is, in supercooled state unlike tin and indium, for which the maximum $\theta(T)$ is above the melting temperature. The temperature range where the wetting angle decreases appears

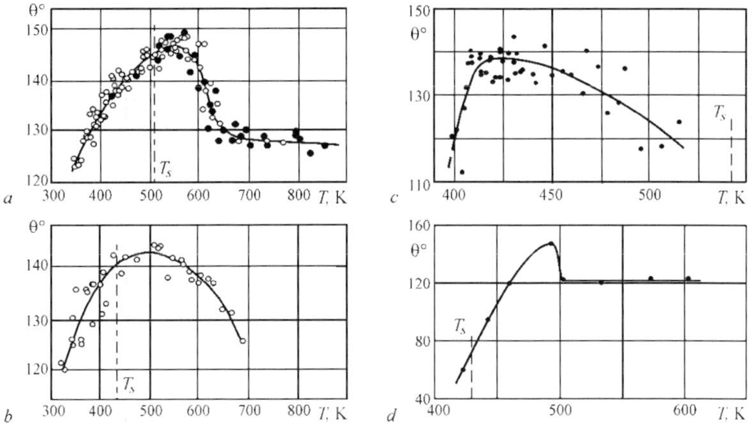

Figure 14. Temperature dependencies of wetting for island condensates of metals on different substrates. On the left – Sn/C (a) (base substrate: O – NaCl; ● – Al_2O_3; vacuum 5 10^{-6} mm Hg) and In/C (b), on the right – Bi/C (c), and In/Al (d) [42, 43, 72]

quite narrow: 400 < T < 420 K. On these interval θ decreases by 25°, this corresponds to reduction of adhesion tension by 50%.

Nonmonotonic dependence of the contact angle on temperature is also a feature of In/Al system (Figure 14d). This is similar to the relation θ(T) in the Sn/C system, but is almost completely above the melting temperature for indium. For In/Al relatively low supercoolings ($\Delta T/T_s \approx 0.05$) were received, which is generally typical for metal condensates on metal substrates [73, 74, 76]. Starting with the temperature T = 420 K (θ = 60°), the contact angle is growing with the growth of temperature, and at T = 490 K takes the maximum value of θ = 143°. It is followed by fast reduction of θ, and at T > 500 K the contact angle is constant with the value θ = 120°. Characteristically, at the melting temperature and below, in supercooled state, indium wets aluminum substrate. Transition from wetting to nonwetting, that is, the change of sign of adhesion tension, in the In/Al system is observed at T = 440 K.

For the Cu/C system wetting temperature dependence does not have any peculiarities: on the interval 1200 < T < 1300 K decrease of the contact angle is observed with growing temperature ($d(\cos\theta)/dT \approx 0.001$ K^{-1}). This, on the one hand, is similar to the behavior of θ(T) for the Bi/C system with the same values of relative supercoolings, and on the other the linear relation θ(T) is typical for contact systems with noninteracting components [1, 40].

Observed changes of the contact angle in the supercooled region, as noted in [42, 43, 72], is probably stipulated by abnormal behavior of either liquid metal surface energy or interface energy of the metal–carbon boundary. If σ_{ul} is assumed to be constant or growing with decrease of temperature, then according to Young's equation experimental data for θ(T) is an evidence of sharp increase in liquid metal surface energy. Thus, for tin at $T \leq 400$ K the variable σ_l, found under assumption of constant adhesion tension, exceeds the relevant quantity for solid metal. Hence, crystallization of tin at $T < 400$ K will be accompanied by decreasing surface energy, which is not in agreement with existing theoretical models and experimental data. In this way, the assumption regarding constant let alone growing σ_{ul} with increasing supercooling leads to a contradiction. Therefore, the data for wetting in supercooled stated most likely suggest significant decrease of interfacial energy of the supercooled droplet–substrate boundary with decreasing temperature. The relations $\sigma_{ul}(T)$, calculated using linear extrapolation of the data for surface energy temperature dependence of liquid tin and indium [68] to the supercooling region, are presented in Figure 15.

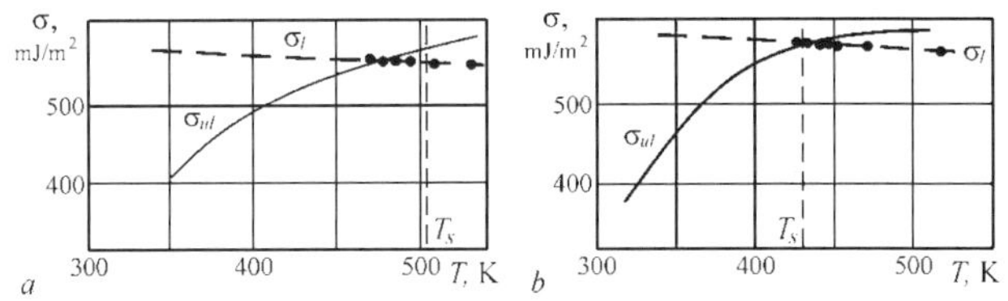

Figure 15. The plots of liquid phase surface energy [68] and interfacial energy of the droplet–substrate boundary against temperature for Sn/C (a) and In/C (b).

Among the reasons causing such a significant decrease in interfacial energy works [42, 43, 72] cite adsorption of gaseous impurities, which value increases with decrease of temperature or inversion of the surface energy of metal in supercooled state. However, considering the fact that for a number of analyzed metals (In, Sn, Bi) inversion of wetting temperature dependence occurs within approximately the same temperature range, while in the Cu/C system at high temperature it was not found at all, one should probably consider the adsorption of impurities from residual gases that grows at such temperatures crucial and causing decrease of interfacial energy on the carbon substrate boundary with increased supercooling, and hence, better wetting observed experimentally. The increase of temperature of the substrate above 500–600 K increases σ_u of carbon film due to sharp reduction of absorption of gases on its surface, which results in better wetting.

3.4.2. Influence of pressure of residual gases on wetting of carbon substrate with tin

The experimental relations $\theta(T)$ (Figure 14) may not be explained by linear change of surface energies of the contacting phases. In addition, one of the possible explanations of nonmonotonous behavior of $\theta(T)$ is the influence of adsorbed gaseous impurities. For this reason, the works [42, 43, 77] investigated temperature dependencies of contact angles for island films of tin on carbon substrates, condensed under the controlled composition of the residual atmosphere.

The samples were prepared using the technique described earlier [72] at the pressure of residual gases of 10^{-7}–10^{-9} mm Hg. A mass spectrometer was employed to control residual atmosphere, and its content was changed by leaking gas into the unit pumped to a pressure of 10^{-9} mm Hg. The wetting contact angles were measured on micrographs of particle profiles on rolled-up (Figure 16) or inclined (Figure 17) spots of carbon film (convolution and angular observation methods [41, 42]); the quantity θ for a fixed temperature was found by averaging the contact angle values for 10–20 microparticles.

Figure 16. Micrographs of crystallized tin droplets condensed in a vacuum of $2 \cdot 10^{-8}$ mm Hg on carbon substrates at temperatures 350 K (a), 410 K (b), and 500 K (c) [42, 43, 77]

The results of measurement of θ(*T*) for tin droplets obtained under different vacuum are presented in Figure 18. Comparison of experimental data on wetting in tin films prepared in a vacuum of 10^{-6} и 10^{-8} mm Hg shows that they are different, firstly, by absence of a maximum in dependence θ(*T*) (pressure 10^{-8} mm Hg) and, secondly, by the fact that with better vacuum this dependence shifts to the region of smaller values of wetting angles (for pressure of 10^{-8} mm Hg this displacement makes 20–30°).

At substrate temperatures $T > 500$ K for films prepared in a vacuum of 10^{-8} mm Hg, relation θ(*T*) goes constant, and the contact angles become approximately equal to the angles θ for films produced at $p = 10^{-5}$ mm Hg, but at temperatures above 650 K. This suggests that the maximum on the wetting temperature dependence for island condensates of tin, indium, and bismuth obtained in a vacuum of 10^{-5}–10^{-6} mm Hg [72] is defined by the influence of gaseous impurities adsorbed from residual atmosphere, which respectively modify surface energies of the contacting phases. It is typical that the relation θ(*T*) for films condensed in a vacuum of 10^{-7} mm Hg assumes intermediate position between the values obtained for samples at $p = 10^{-6}$ and 10^{-8} mm Hg.

It is worth noting that for a supercooled state of tin decrease of temperature in all cases leads to a decrease of the contact angles. At the pressure of residual gases of 10^{-8} mm Hg improvement of wetting gets quite significant and makes $\Delta\theta \approx 50°$. Variation of wetting with temperature is well illustrated by micrographs of particle profiles (Figure 16) and inclined spots of film near the temperature of maximum supercooling (Figure 17). In this case, as could be seen from the chart (Figure 18, Curve 3) and micrographs (Figs. 16, 17), transition from nonwetting ($\theta > 90°$, i.e., $\sigma_{ul} > \sigma_u$) to wetting ($\theta < 90°$, $\sigma_{ul} < \sigma_u$) is observed in the region of deep supercoolings (at $T < 350$ K). Existence of such transition, which is essentially a change of sign of the adhesion tension $\sigma_u - \sigma_{ul}$ directly corroborates the conclusion about significant decrease in interfacial energy of the droplet–substrate supercooling boundary with temperature, which is determined by adsorption of impurities from residual gases growing with decrease of temperature [42, 43, 72]. The results in work [77] give reason to assume that nonmonotonous wetting dependence in the Sn/C system is due to the interaction of methane group gases produced in the process of operation of the hetero-ion pumping system.

Figure 17. Micrographs of tin island film condensed in a vacuum of $2\cdot10^{-8}$ mm Hg on a carbon substrate at the temperature of 315 K (θ = 82°)

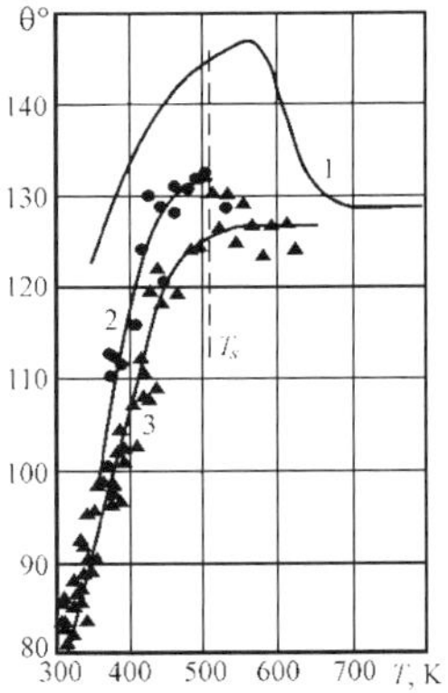

Figure 18. Temperature dependence of the contact angle of wetting of carbon substrates with tin islands prepared under different pressure of residual gases: 1 – $5\cdot10^{-6}$ mm Hg [72], 2 – $3\cdot10^{-7}$ mm Hg, and 3 – $2\cdot10^{-8}$ mm Hg. [42, 43, 77]

The analysis of the outlined results allows us to assume that the supercooled state of the metal itself is not the main cause of sharp improvement of wetting with decrease of temperature for fusible metals. This is also suggested by the fact that inversion of wetting temperature dependence is observed both above (Sn/C, In/C, In/Al) and below (Bi/C) of melting point, while for the Cu/C system there is no inversion at all. However, this conclusion may not be considered final since for the Cu/C system the relation θ(*T*) was studied at insignificant relative supercoolings ($\Delta T/T_s \approx 0.12$), and, probably, this is why wetting inversion was not detected. In this way, data available now suggest the generality of wetting inversion, though they are not sufficient to give a definite answer to the question regarding its mechanism.

3.4.3. *Size effect in wetting in supercooled droplets*

As it has been shown earlier using the contact systems (Sn/C, In/C, Bi/C, Pb/C, Au/C, Pb/Si) as a case study, wetting of amorphous neutral substrates with liquid metals is improving with decrease of the size of microdroplets [9, 42–44]. This effect is a result of decrease in surface energy of metal droplets σ_l themselves and droplet–substrate interfacial energy and has been only investigated for temperatures above the melting point of the metal. At the same time it is known that crystallization of small particles takes place at significant supercoolings [73–76], and for description of this process we must know both absolute values of the contact angles at relevant temperatures and their size dependence.

Such investigations for island tin films on amorphous carbon substrate were made in [72, 78]. The results of measurement of contact angles in the Sn/C system at temperatures 400 K and 315 K are presented in Figure 19, which shows that for supercooled droplets as well as for equilibrium ones (see Figure 6a) the contact angle decreases with decrease of droplets size. However, numerical values of the contact angles for droplets of equal size turn out to be different, and the relation θ(*R*) (*R* is the droplet radius) for supercooled droplets is displaced to the region of lower values of θ by quantity $\Delta\theta \approx 15°$ and 55° for 400 K and 315 K, respectively.

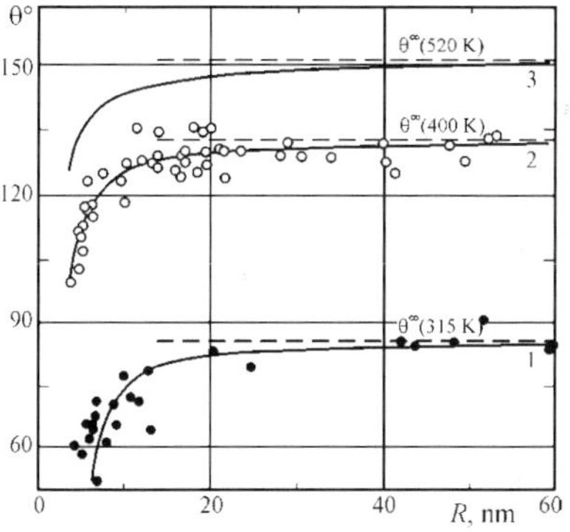

Figure 19. The plot of wetting angle against the radius of supercooled (1 – T = 315 K, 2 – T = 400 K [72, 78]) and equilibrium (3 – T = 520 K [9, 42–44]) tin microdroplets

Comparison with known results regarding the size effect of wetting at $T > T_s$ [9, 42–44] provides grounds to believe that the mechanism of the effect for metastable droplets (at $T < T_s$) is the same as for equilibrium ones, that is, it is defined by the dependence of the droplet–substrate interfacial energy on the size the droplet itself. This is also supported by the fact that obtained dependencies θ(R) are linear in the coordinates "cosθ – 1/R," which is typical for the contact systems investigated earlier [9, 42–44]. Since extrapolation of surface energies to such significant supercoolings looks unjustified, it is not possible to numerically estimate size dependencies of surface and interface energies in the contact system. At the same time, the full similarity of relations θ(R) for equilibrium and supercooled particles offers grounds to argue that they are driven by the same laws.

4. Conclusions

The experimental data and their analysis outlined herein show that the application of island vacuum condensates to study surface energy and wetting allowed to obtain a number of general results of crucial importance for physics and chemistry of surface phenomena.

Quantitative data of surface energies of solid Au, Pb, and Bi nanoparticles and parameters that define size dependence of surface energy of small particles have been obtained from the investigations of evaporation processes in condensed films.

Options for measuring surface energy in the solid phase and its temperature dependence are discussed and are based on the data regarding the decrease of melting temperature of small metal particles. Temperature dependencies of surface energy for In, Sn, Bi, Pb, Al, Au, and Pt have been calculated on the basis of the preceding consideration. The significant reduction of surface energy for all metals studied when approaching melting temperature has been shown.

The detailed investigation and theoretical description of the size effect of wetting, which consists in the decrease of the wetting angle with the decrease of particles size, have been provided. Size dependencies of wetting angle in In/C, Sn/C, Bi/C, Pb/C, Pb/Si, and Au/C have been obtained for equilibrium and supercooled (Sn/C) liquid microdroplets. On the basis of these studies, size dependencies of interfacial energy of In/C, Sn/C, Bi/C, Pb/C, Au/C couples were obtained.

Experimental data and theoretical description of the effect of thickness of free carbon film on wetting angle have been obtained for In/C, Sn/C, and Pb/C couples. These studies enabled to determine the surface energy of thin carbon films, which makes $\sigma_u = 120 \pm 30$ mJ/m^2.

Temperature dependencies of contact angle in In/C, Sn/C, Bi/C, In/Al, and Cu/C couples have been obtained for equilibrium and supercooled liquid microdroplets.

Author details

Sergei Dukarov, Aleksandr Kryshtal* and Vladimir Sukhov

*Address all correspondence to: aleksandr.p.kryshtal@univer.kharkov.ua

V.N. Karazin Kharkiv National University, Kharkiv, Ukraine

References

[1] Adamson AW, Gast AP. Physical Chemistry of Surfaces. New Delhi: Wiley India; 2012.

[2] Bormashenko EYu. Wetting of Real Surfaces. Berlin: De Gruyter; 2013.

[3] Fisher IZ. Statistical Theory of Liquids. Chicago: University of Chicago Press; 1964.

[4] Geguzin YaE, Ovcharenko NN. Surface energy and surface processes in solids. Soviet Physics Uspekhi 1962; 5(1): 129–157. doi:10.1070/PU1962v005n01ABEH003403

[5] Kumikov VK, Khokonov KhB. On the measurement of surface free energy and surface tension of solid metals. J. Appl. Phys. 1983; 54: 1346–1350. doi:10.1063/1.332209.

[6] Sambles JR. An electron microscope study of evaporating gold particles: the Kelvin equation for liquid gold and the lovering of the melting point of solid gold particles. Proc. Roy. Soc. Lond. A 1971; 324: 339–350.

[7] Sambles JR, Skinner LM, Lisgarten ND. An electron microscope study of evaporation small particles. Proc. Roy. Soc. Lond. A 1970; 318(3): 507–522.

[8] Larin VI. Determination of surface energy of small particles from their evaporation. Funct. Mater. 1998; 5(1): 104–109.

[9] Larin VI. Capillary characteristics of thin films and small particles. Funct. Mater. 1998; 5(2): 145–164.

[10] De Gennes PG. Wetting: statics and dynamics. Rev. Mod. Phys. 1985; 57: 827–863.

[11] Tolman RC. The effect of droplet size in surface tension. J. Chem. Phys. 1949; 17: 333–337.

[12] Kirhwood JG, Buff FF. The statistical mecanical theory of surface tension. J. Chem. Phys. 1949; 17(2): 338–343.

[13] Magomedov MN. Dependence of the surface energy on the size and shape of a nanocrystal. Phys. Solid State 2004; 46(5): 954–968.

[14] Lu HM, Jiang Q. Size-dependent surface energies of nanocrystals. J. Phys. Chem. B 2004; 108(18): 5617–5619.

[15] Sdobnyakov NYu, Samsonov VM, Bazulev AN, Kuplin DA, Sokolov DN. The solubility region of Ga in PbTe films prepared on Si-substrates by modified "hot wall" technique. Funct. Mater. 2011; 18(2): 181–188.

[16] Xiong S, Qi W, Cheng Y, Huang B, Wang M, Li Yejun B. Modeling size effects on the surface free energy of metallic nanoparticles and nanocavities. Phys. Chem. Chem. Phys. 2011; 13(22): 10648–10651.

[17] Chamaani A, Marzbanrad E, Rahimipour MR, Yaghmaee MS, Aghaei A, Kamachali RD, Behnamian Y. Thermodynamics and molecular dynamics investigation of a pos-

sible new critical size for surface and inner cohesive energy of Al nanoparticles. J. Nanopart. Res. 2011; 13: 6059–6067.

[18] Zadumkin SN, Khokonov KhB. Surface energy of thin metal films. Fiz. Met. Metalloved. 1962; 13: 658–663.

[19] Shebzukhova G, Arefieva LP, Khokonov KhB. Dimensional dependence of the surface energy of thin cadmium films. Bull. Russ. Acad. Sci. Phys. 2012; 76(10): 1133–1135.

[20] Shcherbakov LM, Samsonov VM, Bazulev AN. The application of thermodynamic perturbation theory to the calculation of excess free energy of small systems: 1. The study of the dimensional dependence of specific free energy of small droplets. Colloid J. 2004; 66(6): 760–765.

[21] Samsonov VM, Sdobnyakov NY, Bazulev AN. Size dependence of the surface tension and the problem of Gibbs thermodynamics extension to nanosystems. Colloids Surf. A: Physicochem. Eng. Aspects 2004; 239: 113–117.

[22] Buffat Ph, Borel JP. Size effect on the melting temperature of gold particles. Phys. Rev. A 1976; 13(6): 2287–2298.

[23] Asoro MA, Kovar D, Ferreira PJ. In-situ transmission electron microscopy observations of sublimation in silver nanoparticles. ACS Nano 2013; 7(9): 7844–7852.

[24] Nanda KK, Kruis FE, Fissan H. Evaporation of free PbS nanoparticles: evidence of the Kelvin effect. Phys. Rev. Lett. 2002; 89: 256103-1-4.

[25] Nanda KK, Maisels A, Kruis FE et al. Higher surface energy of free nanoparticles. Phys. Rev. Lett. 2003; 91: 106102.

[26] Vogelsberger W, Marx G. Zur Krummung sabhangigkeit der Oberflach enspanung kleiner Tropfchen. Z. Phys. Chem. Leipzig 1976; 257(3): 580–586.

[27] Mei QS, Lu K. Melting and superheating of crystalline solids: from bulk to nanocrystals. Progress Mater. Sci. 2007; 52(8): 1175–1262. doi:10.1016/j.pmatsci.2007.01.001

[28] Kryshtal AP, Sukhov RV, Minenkov AA. Critical thickness of contact melting in the Au/Ge layered film system. J. Alloys Compounds 2012; 512: 311– 315.

[29] Gladkikh NT, Kryshtal'AP, Sukhov RV. Contact melting in layered film systems of the eutectic type. Phys. Solid State 2010; 52(3): 633–640.

[30] Gladkikh NT, Bogatyrenko SI, Kryshtal AP, Anton R. Melting point lowering of thin metal films (Me = In, Sn, Bi, Pb) in Al/Me/Al film system. Appl. Surf. Sci. 2003; 219: 338–346.

[31] Wronski CRM. The size dependence of the melting point of small particles of tin. Brit. J. Appl. Phys. 1967; 18: 1731–1737.

[32] Berman RP, Curzon AE. the size dependence of the melting point of small particles of indium. Canad. J. Appl. Phys. 1974; 52: 923–929.

[33] Peppiat SJ. The melting of small particles. I. Lead. Proc. Roy. Soc. Lond. A 1975; 345: 387–399.

[34] Peppiat SJ. The melting of small particles. II. Bismush. Proc. Roy. Soc. Lond. A 1975; 345: 401–412.

[35] Allen GL, Bayles RA, Gile WW, Jesser WA. Small particle melting of pure metals. Thin Solid Films 1986; 144(2): 297–308.

[36] Skripov VP, Koverda VP, Skokov VN. Size effect on melting of small particles. Phys. Status solidy (a) 1981; 66: 109–118.

[37] Pawlow PJ. Uher die Ahhangigkeit des Schmeltzpunktes von der Oberflachenenergie eines festen Korpers. Z. Phys. Chem. 1909; 65: 1–35, 545–548.

[38] Lu HM, Jiang Q. Surface tension and its temperature coefficient for liquid metals. J. Phys. Chem. B 2005; 109: 15463–15468.

[39] Morokhin VA, Pastukhov BA, Khlynov BB, Furman EL. Temperature dependence of the surface energy of iron. Poverkhnost. Fizika, Khimia, Mekhanika (Surface. Physics, Chemistry, Mechanics). 1987; 9: 121–125.

[40] Jaycock MJ, Parfitt GD. Chemistry of Interfaces. New York: Halsted Press; 1986.

[41] Gladkikh NT, Chizhik SP, Larin VI, Grigoryeva LK, Dukarov SV. Methods of determination of wetting in highly dispersed systems. Phys. Chem. Mech. Surfaces 1987; 4(11): 3465–3482.

[42] Gladkikh NT, Dukarov SV, Kryshal OP, Larin VI, Sukhov VN. Kapillyarnye svojstva ostrovkovykh plenok i malykh chastits (Capillary Properties of Island Films and Small Particles). Kharkiv: V. N. Karazin Kharkiv National University; 2015.

[43] Gladkikh NT, Dukarov SV, Kryshal OP, Larin VI, Sukhov VN, Bogatyrenko SI. Poverkhnostnye yavleniya i fazovye prevrashcheniya v kondensirovannykh plenkakh (Surface Phenomena and Phase Transformations in Condensed Films). Kharkiv: V. N. Karazin Kharkiv National University; 2004.

[44] Chizhik SP, Gladkikh NT, Larin VI, Grigorieva LK, Dukarov SV. Size effects in wetting for highly dispersed systems. Phys. Chem. Mech. Surf. 1987; 4(12): 3707–3724.

[45] Gladkikh NT, Grigor'eva LK, Dukarov SV et al. Size and temperature dependences of the surface tension of ultradisperse metallic particles. Sov. Phys. Solid State 1989; 31(5): 728–733.

[46] Yuan Y, Lee TR.contact angle and wetting properties. Surf. Sci. Tech. 2013; 5: 13–34.

[47] Amirfazli A, Chatain D, Neumann AW. Drop size dependence of contact angles for liquid tin on silica surface: line tension and its correlation with solid–liquid interfacial tension. Colloids Surf. A: Physicochem. Eng. Aspects 1998; 142: 183–188.

[48] Heine DR, Grest GS, Webb EB. Surface wetting of liquid nanodroplets: droplet-size effects. Phys. Review Lett. 2005; 95(10): 107801. doi:10.1103/PhysRevLett.95.107801

[49] Eustathopoulos N, Jound J. Interfacial tension and adsorption of metallic systems. Curr. Topics Mater. Sci. 1980; 4: 281–360.

[50] Larin V, Borodin A. Dimentional effect at wetting in Au–C system. Funct. Mater. 2001; 8(2): 309–313.

[51] Kryshtal AP. Formation of island arrays by melting of Bi, Pb and Sn continuous films on Si substrate. Applied Surf. Sci. 2014; 321: 548–553.

[52] Kryshtal AP, Gladkikh NT, Sukhov RV. Features of island nanostructures formed by melting Sn, Bi and Sn–Bi thin films on C substrates. Appl. Surf. Sci. 2011; 257: 7649–7652.

[53] Dukarov SV, Petrushenko SI, Sukhov VN, Churilov IG. Effect of temperature on the pores growth in the polycrystalline films of fusible metals. Prob. Atomic Sci. Technol. 2014; 1(89): 110–114.

[54] Rozenbaum VM, Dukarov SV, Gladkikh NT. Wetting by small droplets of free elastically deformable film. Poverkhnost. Fizika, Khimia, Mekhanika (Surface, Physics, Chemistry, Mechanics). 1989; 1: 104–112.

[55] Rusanov AI. On the theory of wetting of elastically deformable solids. Kolloidnyj zhurnal (Colloid Journal) 1977; 39(4): 704–710.

[56] Borodin AA, Gladkikh NT, Dukarov SV. Size effect on wetting of elastically deformable substrate. Vestnik KhGU, Fizika (V. N. Karazin Kharkiv State University Bulletin. Series of Physics). 1998; 417(1): 110–113.

[57] Kryshtal AP, Bogatyrenko SI, Sukhov RV, Minenkov AA. The kinetics of the formation of a solid solution in an Ag-Pd polycrystalline film system. Appl. Phys. A 2014; 116(4): 1891–1896.

[58] Minenkov AA, Bogatyrenko SI, Sukhov RV, Kryshtal AP. Size dependence of the activation energy of diffusion in multilayer Cu-Ni films. Phys. Solid State 2014; 56(4): 823–826.

[59] Bogatyrenko SI, Gladkikh NT, Kryshtal'AP, Samsonik AL, Sukhov VN. Diffusion in nanodisperse layered film systems. Phys. Metals Metall. 2010; 109(3): 255–260.

[60] Borodin AA, Dukarov SV. Wetting angle hysteresis in condensed microdrops. Funct. Mater. 2002; 9(3): 498–501.

[61] Eral HB, Mannetje DJCM't, Oh JM. Contact angle hysteresis: a review of fundamentals and applications. Colloid Polym. Sci. 2013; 291(2): 247–260.

[62] Naidich YuV, Kostyuk BD, Kolesnichenko GA, Shaikevich SS. Adhesion properties and wetting by molten metal of thin metallic films applied to nonmetallic materials. Sov. Powder Metall. Met. Ceram. 1973; 12(12): 988–993.

[63] Gladkikh NT, Dukarov SV. Lead wetting of thin nickel films deposited onto GaAs. Funct. Mater. 1996; 3(1): 97–99.

[64] Dukarov SV. γ-Phase structure in Ni-Ga-As thin film system. Funct. Mater. 1997; 4(3): 355–358.

[65] Gladkikh NT, Grebennik IP, Dukarov SV. Effect of condensation conditions on phase formation in thin two-layer Ni/GaAs films. Funct. Mater. 1998; 5(1): 44–47.

[66] Landau LD, Lifshitz EM. Theory of Elasticity. Oxford/Boston: Butterworth-Heinemann; 1995.

[67] Croxton CA. Liquid State Physics – A Statistical Mechanical Introduction. New York: Cambridge University Press; 2009.

[68] Zadumkin SN, Ibragimov KhI, Ozniev DT. Study of the surface tension and density of supercooled tin, indium, bismuth, lead and gallium. Izvestiya Vysshikh Uchebnykh Zavedenij. Tsvetnaya Metallurgiya 1979; (1): 82–85.

[69] Ishikawa T, Okada JT, Paradis P-F, Watanabe Y. Thermophysical property measurements of high temperature melts using an electrostatic levitation method. Jpn. J. Appl. Phys. 2011; 50(11S): 11RD03 doi:10.1143/JJAP.50.11RD03.

[70] Okada JT, Ishikawa T, Watanabe Y, Paradis P-F. Surface tension and viscosity of molten vanadium measured with an electrostatic levitation furnace. J. Chem. Thermodynam. 2010; 42(7): 856–859. doi:10.1016/j.jct.2010.02.008.

[71] Paradis P-F, Ishikawa T, Koike N. Thermophysical properties of molten yttrium measured by non-contact techniques. Microgravity Sci. Technol. 2009; 21(1–2): 113–118. doi: 10.1007/s12217-008-9074-8.

[72] Dukarov SV. Size and temperature effects on wetting in supercooled vacuum condensates. Thin Solid Films 1998; 323(1–2): 136–140.

[73] Gladkikh NT, Dukarov SV, Sukhov VN. Investigation of supercooling during metal crystallization under conditions close to weightlessness using island vacuum condensates. Z. fur Metallkunde 1996; 87(3): 233–239.

[74] Gladkikh NT, Dukarov SV, Sukhov VN. Overcooling on metal crystallization in island-like vacuum deposited films. Fizika Metallov i Metallovedenie 1994; 78(3): 87–93.

[75] Gladkikh NT, Dukarov SV, Sukhov VN, Churilov IG. Condensation mechanism of AgCl and NaCl island films on a nickel substrate. Funct. Mater. 2011; 18(4): 529–533.

[76] Gladkikh NT, Dukarov SV, Kryshal OP, Larin VI. Size effect upon solidification of small bismuth particles. Phys. Metals Metallurg. 1998; 85(5): 536–541.

[77] Dukarov SV, Gladkikh NT, Borodin SA. Temperature dependence of wetting in supercooled tin condensates at a carbon substrate. Fizicheskaya inzheneriya poverkhnosti (Physical Surface Engineering) 2003; 1(1): 89–93.

[78] Dukarov SV. Size effect on wetting in a supercooled island films of Sn. Adgeziya rasplavov i payka materialov (Adhesion of Melts and Brazing of Materials) 2010; 43: 3–11.

Chapter 8

Wettability of Nanostructured Surfaces

L. Duta, A.C. Popescu, I. Zgura, N. Preda and
I.N. Mihailescu

Additional information is available at the end of the chapter

http://dx.doi.org/10.5772/60808

Abstract

There are many studies in literature concerning contact angle measurements on different materials/substrates. It is documented that textiles can be coated with multifunctional materials in form of thin films or nanoparticles to acquire characteristics that can improve the protection and comfort of the wearer. The capacity of oxide nanostructures to inhibit fungal development and neutralize bacteria is a direct consequence of their wetting behavior [1–6]. Moreover, the radical modification of wetting behavior of nanostructures from hydrophilic to hydrophobic when changing the pulsed laser deposition (PLD) ambient will be thoroughly discussed.

When an implant is introduced inside the human body, its surface is first wetted by the physiological fluids. This further controls the proteins adsorption followed by the attachment of cells to the implant surface. Hence, surface wettability is considered an important criterion that dictates biocompatibility of the implant and could stand for a decisive factor for its long-term stability inside the human body.

In Section 1 of this chapter, the reader is briefly introduced to wetting phenomenon, and correlations between well-known Young, Cassie, and Wenzel approaches are made. Next, one of the most spread techniques to measure the wettability of surface, the contact angle measurement, is thoroughly explained and relevant examples are given.

Section 2 begins with a summarized table about previous works on synthesis of hydrophobic or hydrophilic nanostructures with a special focus on ZnO, SiO_x, TiO_2, and DLC materials. A short presentation of the advantages of their synthesis by PLD, sol-gel, thermal evaporation, solution based on chemical approaches, sputtering, and plasma enhanced chemical vapor deposition will be introduced.

Sections 3 includes a brief literature overview on results regarding synthesis by aforementioned techniques of different oxides (ZnO, TiO_2, and SiO_x) and DLC nanostructures onto textile (polyester, polyamide, cotton/polyester, and poly(lactic acid)) or metallic substrates for medical purposes.

The chapter ends with conclusions and references, which include books and review articles relevant to the topics.

Keywords: Wettability, contact angle measurements, oxide and diamond like carbon nanostructures, textile functionalization, medical applications, pulsed laser deposition, sol-gel, combined radio frequency plasma enhanced chemical vapor deposition, magnetron sputtering

1. Introduction

Wetting is the ability of liquids to keep in contact with solid surfaces. It is a direct result of intermolecular interactions, which occur when two media (liquid and solid) are brought together. Wettability studies usually involve the measurements of contact angle (CA), which indicates the degree of wetting when a solid and liquid interact. A low CA (<90°) corresponds to high wettability, and the fluid will spread over a large area of the surface. A high CA (>90°) corresponds to low wettability, and the fluid will minimize contact with the surface and form a compact liquid droplet. CA>150° indicates minimal contact between the liquid droplet and the surface and corresponds to a superhydrophobic behavior.

In the case of a liquid droplet on an ideal solid surface (which is flat, rigid, perfectly smooth, chemically homogeneous, and has zero CA hysteresis), which forms a CA (θ), the general formula of the well-known Young's equation (which assumes a perfectly flat and rigid surface) [7] that describes the balance (Figure 1a) between the surface tension of the liquid/vapor γ_{SV} and that of the liquid/vapor γ_{LV} and the interfacial tension of the solid/liquid γ_{LS} is as follows:

$$\cos\theta = \left(\gamma_{SV} - \gamma_{LS}\right) / \gamma_{LV} \tag{1}$$

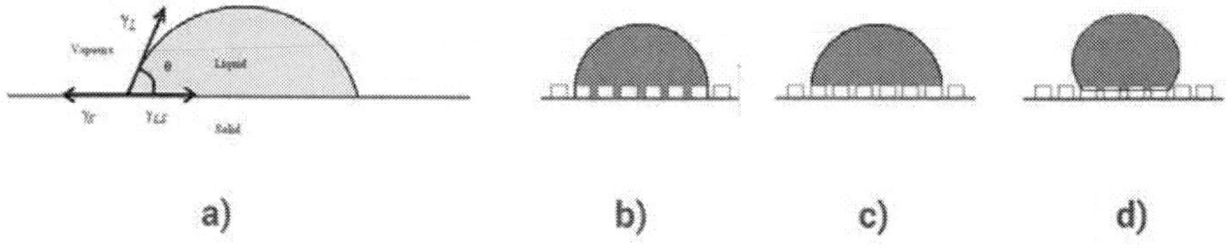

Figure 1. Wetting behavior of solid substrates: (a) Young, (b) Wenzel, (c) Cassie, and (d) intermediate state between Wenzel and Cassie regimes.

In reality, only a few solid surfaces are actually flat. The surface roughness is therefore one important parameter that should be taken into consideration when assessing the wetting behavior of a surface [8, 9]. This influence can prove significant for static and dynamic wetting.

Besides Young's theory, the Wenzel (Figure 1b) and Cassie (Figure 1c) regimes [9, 10], or an intermediate state between these two models (Figure 1d) [11], are generally used to correlate the roughness of the surface with the apparent CA of a liquid.

Several authors modeled the effect of surface roughness over CA [12–14]. The basic idea was to account for roughness through r, which is the ratio of the actual to projected area. Thus, $A_{LS}=r \cdot A_{LSapp}$ and $A_{SV}=r \cdot A_{SVapp}$ where A_{LS} and A_{SV} are the liquid-solid and solid-vapor areas, A_{LSapp} and A_{SVapp} are the liquid-solid and solid-vapor areas. In this case, Eq. (1) becomes

$$\cos\theta_{rough} = r\cos\theta_{true} \tag{2}$$

Due to surface roughness, for CA <90°, the apparent CA will decrease, while for CA>90°, its value will increase. This corresponds to the Wenzel case, as illustrated in Figure 1b, where the liquid completely fills the air pockets of the rough surface at the contact site. If the CA is large and the surface rough, the liquid may trap air. As a consequence, a composite surface effect will appear, as illustrated in Figure 1c.

In the Cassie model [15], it is presumed that in the grooves the air is trapped under the liquid droplet. This determines the appearance of a composite surface (Figure 1c). The chemical heterogeneity of a rough surface can be related, in this model, to the apparent CA, θ_{app}, through the following formula:

$$\cos\theta_{app} = f_S\cos\theta_S + f_V\cos\theta_V \tag{3}$$

with f_S and f_V as the area fractions of the solid and vapor on the surface, respectively. Since $f_S + f_V = 1$, $\theta_S = \theta$, and $\theta_V = 180°$, Eq. (3) can be written as follows:

$$\cos\theta_{app} = -1 + f_S\left(\cos\theta_{true} + 1\right) \tag{4}$$

where θ_{true} is the CA on a smooth surface [15].

From Eq. (4), it follows that in case of a true value of CA >90°, the surface roughness will determine an increase of CA's apparent value. In contradiction to Wenzel regime, in the Cassie mode, the CA value will increase even for CA values <90º due to the air pockets trapped under the liquid droplet [9]. It was shown that, when applying a physical deformation to a spherical droplet, a variation from the Cassie to Wenzel state can be achieved [16, 17] (Figure 1d). In some cases, a transition between these two modes can also occur [18–20]. Therefore, the droplet will fill the air pockets of the rough substrate resulting in a decrease of the apparent value of CA. In these cases, for the estimation of CA values, Eqs. (2) and (4) can be used before and after the transition, respectively. Taking into account these results, the following equation can be written [11]:

$$\cos\theta_{th} = (f_S - 1)/(r - f_S) \tag{5}$$

where θ_{th} represents a threshold value between Wenzel and Cassie states.

1.1. CA measurements of nanostructured surfaces

Immediately after an implant is introduced inside the human body, the first events that occur are the wetting of the material by the physiological fluids, followed by attachment of cells to the implant surface [21]. In order to evaluate the wetting behavior of a system, plenty of quantitative (CA, imbibition, and forced displacement, and electrical resistivity wettability) and qualitative (imbibition rates, microscope examination, flotation, glass slide, relative permeability curves, permeability/saturation relationships, capillary pressure curves, capillarimetric method, displacement capillary pressure, reservoir logs, nuclear magnetic resonance, and dye adsorption) methods have been developed [22]. Among these, CA measurement is probably the most adopted technique to investigate the average wettability of a surface [23]. Moreover, this type of investigation has been extensively applied to assess the wetting behavior of different nanostructured surfaces, used for various medical applications. Some relevant literature examples limited to oxides (ZnO, TiO_2, SiO_x) and diamond-like carbon (DLC) structures are briefly summarized in Table 1.

Material	Envisaged application	Cited reference
ZnO	Self-cleaning coatings and antifogging materials	[24]
	Antibacterial properties	[25, 26]
	Environmental sensing	[27]
	Micro/nanodevices	[28]
SiO_x	Antibacterial properties	[29]
	Cells migration on artificial surfaces	[30]
	Bioactive properties	[31]
	Superoleophobic surfaces	[32]
TiO_2	Antibacterial properties	[25]
	Microbial fuel cells and bioremediations	[33]
	Blood-contacting biomaterials	[34]
DLC	Temporary blood-contacting medical devices (cardiovascular and interventional devices, artificial organs, pacemakers)	[35, 36]
	Femoral head and the acetabulum hip joint components	[37]
	Resistance to corrosion	[38]
	Antifogging	[39]
	Resistance to microbial adhesion	[40]

Table 1. CA measurements of hydrophobic/hydrophilic ZnO, SiO_x, TiO_2, and DLC nanostructures used for medical applications.

CA can be classified into static or dynamic. Static CA is measured when liquid droplet is standing alone on the surface, without needle insertion, and the solid/liquid/air boundary is not moving. These measurements are used in quality control and research and product development. One can measure the dynamic CA when the solid/liquid/air boundary is moving. In this way, advancing and receding CA are measured. CA hysteresis, which represents the difference between these two angles, comes from surface chemical and topographical heterogeneities, solution impurities absorbing on the surface, or swelling, rearrangement or alteration of the surface by the solvent [41, 42].

The hydrophobic behavior of a surface is generally assessed by the apparent water CA, in static measurements. Moreover, when evaluating a surface repellency, one should take into consideration the sliding-down (which is evaluated by measuring the sliding angle, α, at which a liquid droplet begins to slide down an inclined plate) and rolling-off behaviors of liquid droplets [9]. Due to the CA hysteresis [43, 44], the liquid droplets do not slide off easily on a surface presenting a high value of static CA. Eq. (6) [43, 45] quantitatively describes the relationship between the hysteresis and the sliding angle:

$$mg(\sin\alpha)/w = \gamma_{LV}(\cos\theta_R - \cos\theta_A) \tag{6}$$

where θ_A and θ_R are the advancing and receding CAs, respectively (Figure 2), g is the gravitational force, m is the mass, and w is the width of the droplet.

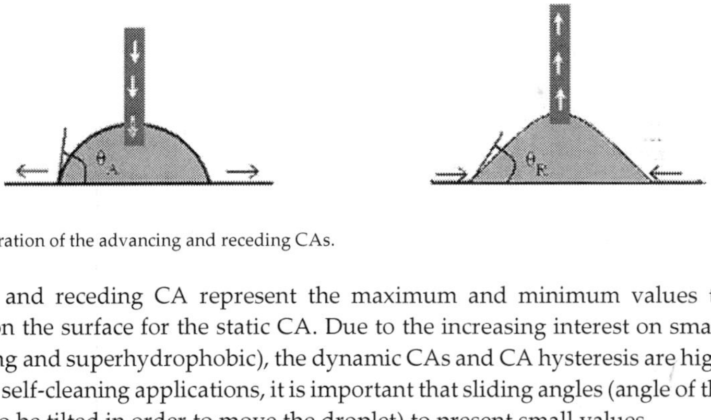

Figure 2. Illustration of the advancing and receding CAs.

Advancing and receding CA represent the maximum and minimum values that can be measured on the surface for the static CA. Due to the increasing interest on smart materials (self-cleaning and superhydrophobic), the dynamic CAs and CA hysteresis are highly applied [46, 47]. For self-cleaning applications, it is important that sliding angles (angle of the substrate which has to be tilted in order to move the droplet) to present small values.

From Eq. (6), it can be inferred that a lower droplet mass and smaller difference between the advancing and receding CAs will result in a smaller angle α. It is worthy to note that the surface roughness has a strong effect on the CA hysteresis [43].

Zisman observed for the first time that $\cos\theta$ increases linearly as the surface tension of the liquid (γ_{LV}) decreases [48, 49]. He investigated the wettability of solids by determining the critical surface tension using CA. This method is used to determine the so-called critical surface free energy (γ_c), that differs from the solid surface free energy, γ_S. According to his method, the value of γ_C of a solid is equal to the value of γ_L of a liquid, which is in contact with the solid and for which the CA is zero. The value for γ_c is determined from empirical investigations, consisting of the CA measurements for the studied solid and the liquids of a homologous series of organic compounds like n-alkanes. The values are plotted with the y-axis corresponding to the cosine values of the CA (θ) and the x-axis relating to the γ_L values for the studied liquids. The values of $\cos\theta$ for the liquids of a series of n-alkanes form approximately a straight line. Extrapolation of this line to the point of $\cos\theta = 1$ yields the value of γ_C equal to γ_L.

Despite the fact that γ_C is not the solid surface free energy, the critical surface tension has been shown to correlate with the known surface chemistry of several solids.

The Zisman method has been widely used to assess the critical surface tension γ_C of different organic films or low-energy solids deposited on high-energy solids (e.g., metals, glass [50, 51]). In this approach, by using series of homologous nonpolar liquids (e.g., n-alkanes), one can obtain the total solid surface energy of a nonpolar solid and the dispersion component (γ_S^d) of the total surface energy of a polar solid. We note that, when using polar liquids on polar and nonpolar solids, one can obtain the deviation from rectilinear relation. Also by using polar liquids, the determination of any component of the solid free energy it is not possible.

2. Alternative deposition techniques employed for the synthesis of hydrophobic/hydrophilic nanostructured surfaces (thin films or nanoparticles)

Many methods were employed to synthesize hydrophobic or hydrophilic nanostructures (thin films, TFs, and nanoparticles, NPs), and some literature examples limited to ZnO, SiO_x, TiO_2, and DLC are summarized in Table 2.

Among these methods, pulsed laser deposition (PLD), sol-gel (SG), thermal evaporation (TE), solution based on chemical approaches, sputtering, and plasma enhanced chemical vapor deposition (PECVD) will be briefly described hereinafter. They are easy to use, low cost, and yield high throughput of micro- and nanostructures.

Material	Structure type	Deposition technique	Cited reference
ZnO	TFs	SG	[52]
	TFs	Metal-organic vapor deposition	[53]
	NPs	Microwave plasma	[54]
	TFs	Magnetron sputtering, MS	[55]
	TFs	Electrodeposition	[56]

Material	Structure type	Deposition technique	Cited reference
	TFs	Atomic layer deposition	[57]
	TFs	Electrochemical anodization	[58]
	TFs	PLD	[59]
SiO_x	TFs	Sputtering	[60]
	TFs	Ion beam evaporation, IBE	[61]
	TFs	PECVD	[62–64]
	NPs	SG	[65, 66]
TiO_2	NPs	IBE	[67]
	TFs	PECVD	[68, 69]
	TFs	Sputtering	[70, 71]
	NPs	(Dip-)pad-dry-cure	[72, 73]
	NPs	Impregnating TiO_2 particles in a resin and this composite deposited into the yarn-array	[74]
	NPs	SG	[75, 76]
	TFs	PLD	[77, 78]
DLC	TFs	PECVD	[79–81]
	TFs	MS	[82, 83]
	TFs	Combined PECVD and MS	[84]
	TFs	Filtered pulsed arc discharge	[85, 86]
	TFs	Ion beam deposition	[87]
	TFs	PLD	[88, 89]

Table 2. Different deposition techniques used for the synthesis of hydrophobic/hydrophilic ZnO, SiO_x, TiO_2, and DLC nanostructures.

In the field of TFs growth, PLD has proven to be among the most versatile methods [59], with features superior to conventional deposition techniques (fast processing, reliability and low production cost). In this technique, high power laser energies are used. They are focused onto a target in order to evaporate its surface under vacuum or different gas ambient atmospheres. The vaporized material consisting of ions, atoms, or molecules is subsequently deposited onto a generally parallel substrate. Repeated laser pulses will result in the deposition of the TFs in form of a coating on the substrate.

One important advantage of PLD method is the stoichiometric transfer of different materials from the targets in the deposited films [59, 90, 91]. This represents a direct consequence of the high ablation rate that allows all elements to evaporate simultaneously [92]. This technique ensures an excellent adherence of the deposited structure to substrates, the high accuracy control of the growth rate (10^{-2}–10^{-1} Å/pulse), the absence of contamination, the simplified growth of materials and combinations of materials of technological interest [93], and the good control of the final crystalline state of the coatings [59, 94].

The SG process is a synthesis route consisting in the preparation of a sol and successive gelation and solvent removal. This technique represents one of the simplest approaches to produce TFs. It presents many advantages in comparison with traditional deposition techniques, such as low working temperature, possibility to cover large surfaces, and high purity of the working conditions.

Compared to the physical route where harsh conditions such as high temperature or special equipment are usually required and consequently generating high costs, the solution based on chemical approaches [95–97] presents several advantages, including the simplicity in operation, low fabrication costs, low process temperatures (below 90 °C), and ambient pressure processing.

Thermal vacuum deposition or TE method is used to fabricate TFs under a high vacuum environment. In this method, an electron beam (e-beam) or resistive heating is usually used to evaporate the desired material inside the vacuum chamber, which then adheres to a substrate positioned above it.

The uniformity, high quality, and adherence of the deposited materials on large areas; the high deposition rate; and the versatility of sputtering techniques have made them attractive for the production of TFs [98–101]. In plasma sputtering deposition, plasma is used as the source of ions. These ions bombard a solid material, commonly known as the cathode or the target, with a typical kinetic energy of several hundreds electron volts. The ion bombardment produces the emission and acceleration of the secondary electrons, which play an important role in maintaining the plasma around the cathode [102]. The ionizing energetic electrons are confined close to the cathode allowing operation at high plasma densities and low pressures.

3. Synthesis of hydrophobic/hydrophilic oxides and DLC nanostructures onto textiles and metallic medical substrates

A brief literature review on results regarding oxide (ZnO, TiO_2, SiO_x) and DLC nanostructures synthesized by PLD, SG, TE, solution based on chemical approaches, sputtering, and PECVD onto textile or metallic substrates will be presented hereinafter. Polyester, cotton/polyester, and poly(lactic acid) woven fabrics can be coated with multifunctional oxide materials in form of TFs or NPs to get properties that increase the protection and comfort of the wearer. When covering the surface with NPs, a new roughness is added leading thus to a dual-size surface roughness. Therefore, the study of wettability properties is a tool to test the surface functionalization [103]. It is well known that wetting of a surface by a liquid is affected by surface roughness [104]. In the case of textile materials, the roughness is related to the geometry which is very complex [105]. Due to the fiber topography, the construction of the yarn, and the construction of the fabric, polymer, natural, and synthetic fibers might be made from porous materials that can absorb water from the environment. Fabrics have thus pronounced texture, porosity, and also (oriented) in-plane capillarity along the threads [103]. CAs on textile substrates can be useful quantities for comparative measurements in order to characterize the effects of surface modification, especially if the textile is distinctly hydrophobic [105]. Titanium

(Ti) stands for the most used metallic material for medical applications due to its unique properties such as biocompatibility, excellent mechanical properties in bulk, relative to the low mass density, and high corrosion and ductility resistance [106].

3.1. ZnO

ZnO is an *n*-type metal oxide semiconductor having a wide band gap, high electron mobility, and thermal conductivity. It mainly crystallizes in the wurtzite phase, being intrinsically polar, and thus exhibiting interesting piezoelectric properties. In addition, in the form of TFs or NPs, ZnO possesses promising antibacterial and antifungal, photocatalytic, electrical, electronic, and optical properties [107–115]. Recently, combinations ZnO-organic were tested for various applications requiring antimicrobial properties [116, 117]. Also, ZnO has probably the richest family of structures' morphology including rods, prisms, wires, whiskers, or tubes [95–97, 118–123]. Moreover, morphology influences other properties such as wettability, another significant characteristic of ZnO covered surfaces bringing great advantages in a wide variety of applications in industry and daily life [124–127]. For example, wettability is critical in cosmetics and textile fields where ZnO can be used due to its biocompatibility property.

Hydrophobins are a class of small-size cysteine-rich proteins synthesized by filamentous fungi [128]. They form ~5–10 nm thick self-assembled monolayers [129] on different substrates, changing their surface wetting properties. Namely, hydrophobic surfaces can be turned to hydrophilic, while hydrophilic materials become hydrophobic [130] after immersion in an aqueous solution of hydrophobin. Textile materials can be finished with various functionalization agents, such as chitosan microcomposites [131] or nanocomposites [132, 133], medicinal herbs [134], nisin [135], polyhexamethylene biguanide [136], or PMMA nanocomposites [137], in order to obtain new surface properties like antimicrobial, hydrophobicity, resistance to laundering, or protection against decoloration. Due to exceptional surface properties and to the tuning opportunities, their use is envisaged in cosmetic industry, polymer emulsion synthesis, and biosensing [138].

3.1.1. ZnO nanostructures synthesized by PLD onto cotton/polyester textiles

Yang et al. [139] and Papadopoulou et al. [140] demonstrated that the structures synthesized by PLD can be controlled in terms of wetting behavior. Therefore, ZnO structures showed a hydrophilic behavior after exposure to UV and were converted to hydrophobic after thermal treatment or storage in complete darkness. In this respect, a one-step PLD procedure to obtain either hydrophobic or hydrophilic ZnO structures (TFs or NPs), without any complementary post-deposition treatments of the surface, was recently proposed [141]. Depending on the number of applied laser pulses, well-separated NPs (for 10 pulses) or compact TFs (for 100 pulses) were synthesized. By varying the ambient gas nature and pressure inside reaction chamber, hydrophilic or hydrophobic surfaces were obtained. The expected properties of the textiles coated with ZnO were evaluated at room temperature (RT) by static CA measurements.

The TFs deposited on textiles (Figure 3) in a flux of 13 Pa oxygen were highly transparent and had a hydrophilic behavior (Figure 3a), while those obtained in vacuum were opaque and showed a hydrophobic behavior (Figure 3b).

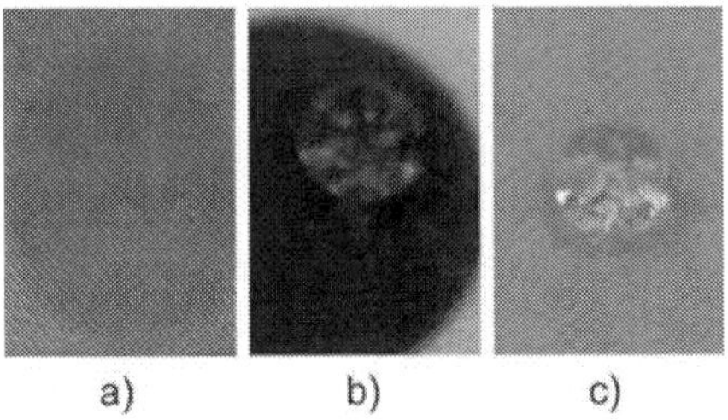

Figure 3. Textiles coated with ZnO nanostructures: (a) hydrophilic TF deposited in 13 Pa oxygen, (b) hydrophobic TF deposited in vacuum, and (c) hydrophobic NPs deposited in vacuum.

A CA of 157° (Figure 4) was measured, which qualified these films as superhydrophobic.

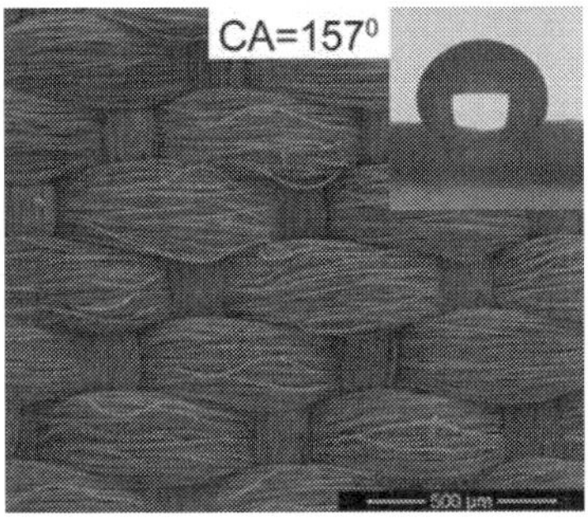

Figure 4. SEM micrograph of the superhydrophobic textile coated with ZnO TF in vacuum. Inset: water droplet in static mode with the CA of 157°.

In the case of NP samples, eye examination confirmed a hydrophilic behavior for the structures deposited in the oxygen flux and a hydrophobic one after deposition in vacuum (Figure 3c).

The macroscopic and microscopic observations have revealed a smoother surface in case of TFs deposited in vacuum characterized by a six times smaller RMS and negative values for surface skewness (S_{sk}) and kurtosis (S_{ku}) (Table 3).

Amplitude parameters	Sample type/scanning area			
	TFs oxygen/10×10 μm²	TFs oxygen/2×2 μm²	TFs vacuum/10×10 μm²	TFs vacuum/2×2 μm²
RMS (nm)	36.817	36.793	6.578	5.796
S_{sk}	0.404	0.421	–0.113	–0.0731
S_{ku}	0.0274	0.24	–0.375	–0.357

Table 3. Amplitude parameters for ZnO TFs deposited in 13 Pa oxygen flux and vacuum. Reproduced from Popescu et al. [141].

Figure 5 shows two-dimensional AFM images of the TFs deposited in 13 Pa oxygen flux and vacuum. The grains (of ~140 nm) visualized by AFM (Figure 5b, d) were in fact consisting of very small crystallites (of ≤10 nm), as proved by the XRD patterns.

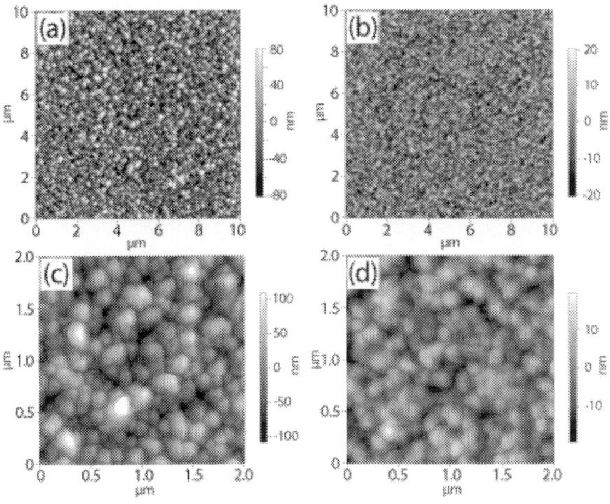

Figure 5. Two-dimensional AFM topography images of the TFs deposited in (a, c) a 13-Pa oxygen flux and (b, d) vacuum at different scales: (a, b) (10 × 10) µm^2 and (c, d) (2 × 2) µm^2. Reproduced from Popescu et al. [141].

In order to account for the significant difference observed in the wetting behavior of the TFs and NPs deposited in a flux of oxygen and in vacuum, a model was proposed for surface wetting. The numerous gaps between crystallites are filled with air acting as a support "buffer" for the water droplet, in contact to the surface in a few small nanometric sites only. Conversely, the TFs deposited in an oxygen flux (Figure 5a, c) consist of larger crystallites and a few intergranular pores only. Thus, the air "buffer" is rarefied, so the contact between the water droplet and the ZnO surface is extending over a larger area (Figure 6). The droplet weight prevails over the counter pressure exerted by the ZnO surface and eventually collapses under its own weight. Figure 6 shows schematically the water droplet in contact with ZnO structures synthesized in vacuum (Figure 6a) and oxygen flux (Figure 6b).

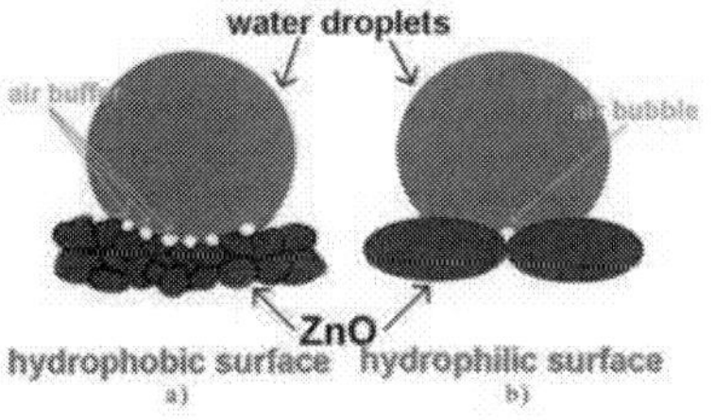

Figure 6. Schematic of the water droplet in contact with ZnO surface deposited in (a) vacuum and (b) 13 Pa oxygen flux. Reproduced from Popescu et al. [141].

The NP depositions in vacuum consist of a large number of small crystallites, which include a huge amount of vapor pockets. Their action cumulates with the effect of the air, which is present in the space between NPs to more efficiently support the droplet weight. This model is in accordance with other studies on hydrophobic plant leaf surfaces [142]. Accordingly, the largest contact area between the water droplet and the leaf surface corresponds to flat and microstructured surfaces but is generated in case of nanostructures as an effect of vapor pockets entrapment.

The electric charging of the surface should be considered when explaining the affinity or repellency to water of ZnO structures. XRD investigations demonstrated that the ambience in the interaction chamber also showed the combinations between Zn and O atoms in the crystalline lattice [141]. In case of structures deposited in vacuum, there is a mix in each crystalline plane of positive and negative charges. One should note that the water droplet is neutral from the electrical point of view. Accordingly, the deposited structures do not interact electrically with the water droplet. Oppositely, the structures deposited in an oxygen flux have only one type of atoms per plane that induce a positive (Zn) or negative (O) charging of surface [141]. The synthesized structures interact electrically with the droplet to reach the neutral status, thus attracting the water bubble toward the ZnO surface causing supplementary stress that contributes to the collapse of the bubble.

In a parallel study, the capacity of these oxide nanostructures to completely inhibit fungal development and neutralize bacteria was found to be a direct consequence of their wetting behavior [1-6].

The intercalation of a hydrophobin nanolayer between substrate and ZnO film, which can boost the oxide efficiency against microorganisms with a higher natural resistance, was recently studied and an explanation of the observed phenomena was proposed [143]. In case of ZnO TFs deposited on bare textiles, the adhesion is governed by physical mechanisms only (as e.g., mechanical or dispersion adhesion [144]), while in case of a buffer layer of hydrophobin interposed between textile and ZnO, chemical bonding occurs, the fastening between the ZnO and the textile substrate becoming much stronger. When used alone, the hydrophobin had no effect on both *Candida albicans* colonies and six strains of filamentous fungi. In case of simple finishing with ZnO, the reduction rate was of 50% and 70% of the colonies in 24 h (Figure 7a, b).

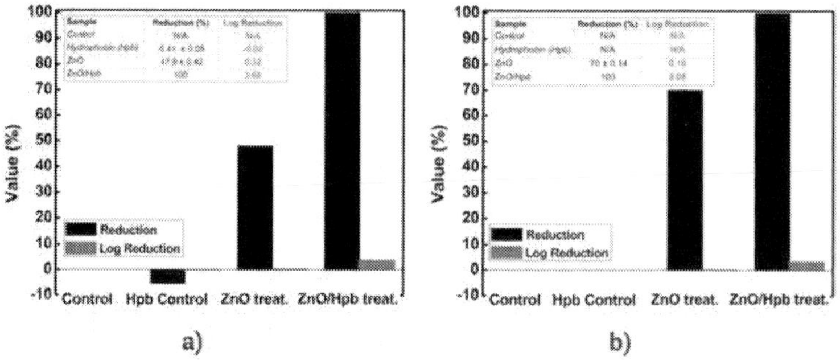

Figure 7. Percentage and logarithmic reduction of (a) *C. albicans* population and (b) mold mix inoculum after 24 h cultivation on untreated and ZnO treated textiles. Reproduced from Popescu et al. [143].

In order to improve ZnO efficiency against resistant fungi, the oxygen concentration on films' surface was increased by covering the textile fibers with hydrophobin and then adding an upper layer of ZnO. As an effect, the orientation and shape of ZnO crystallites were changed, the (001) film texturing becoming more pronounced and nanocrystallites elongated, with more polar planes (001) parallel to the surface (Figure 8a). Depending on the orientation of the *c*-axis, these planes may contain oxygen atoms only (Figure 8b). The ZnO film deposited on hydrophobin proved in this case 100% efficient in reducing colonies of both *C. albicans* and a mold mix of filamentous fungi (Figure 7a, b). This significant enhancement was attributed to the higher texturing of the oxide film when growing on hydrophobin interlayer, resulting in an increased presence of oxygen species on surface.

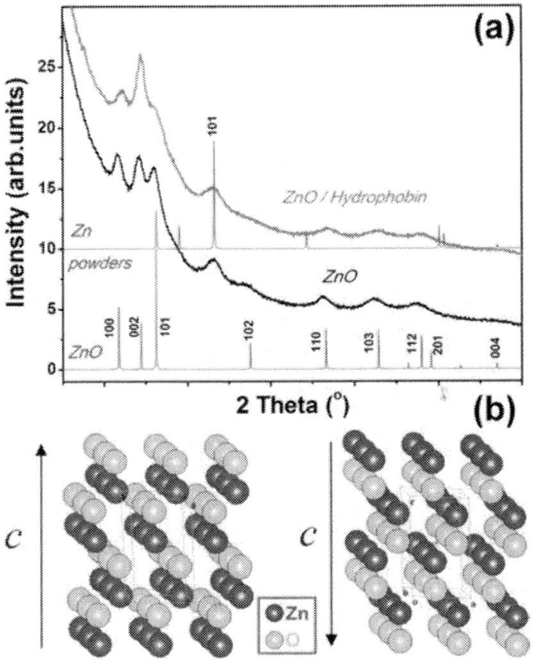

Figure 8. XRD patterns of ZnO TFs (a); the orientation of the (001) ZnO crystallites grown on hydrophobin, resulting in outer termination either in O or in Zn atoms only (b). Reproduced from Popescu et al. [143].

ZnO is recognized to possess antibacterial and antifungal properties. Nair et al. [109] assessed the microbiological activity of ZnO against a mold mix of microbes and associated the high reduction ratio to the generation of surface oxygen species. Sawai et al. [110] and Premanathan et al. [111] suggested that these oxide species form in wet media hydroxyl radicals and hydrogen peroxide. As known, the hydroxyl radical is the most reactive one, able to interact with almost every type of molecule of the living cells of bacteria and fungi, causing irreversible damage to cellular components and eventual apoptosis. Applerot et al. [112] advanced a mechanism for the reactive oxygen species formation on ZnO surface. The oxygen atoms present on surface interact with water molecules, forming OH^- radicals. A chain reaction occurs, resulting in exponential multiplication of these radicals on surface.

We note that no negative side effects of hydrophobins when in contact with human tissue were reported [145], and to the benefit of biomedical applications, they were able to form, in specific cases, resistant monolayers with antimicrobial activity [146]. Moreover, the proposed antimicrobial finishing procedure of fabrics with a conjunction of a thin layer of hydrophobin and a ZnO layer can find applications in the medical field, where solutions are constantly required for elimination of microbial contamination, thus reducing the risks of infections during surgery.

3.1.2. ZnO nanostructures synthesized by solution based on chemical approaches onto solid (glass) substrates

In the synthesis process of ZnO nanostructures using solution based on chemical approaches, a zinc salt and a basic compound are brought together. The involved chemical reactions can be described as follows:

i. Using a weak base ($(CH_2)_6N_4$)

$$
\begin{aligned}
&Zn(NO_3)_2 \rightarrow Zn^{2+} + 2NO_3^- \\
&(CH_2)_6N_4 + 6H_2O \rightarrow 6HCHO + 4NH_3 \\
&NH_3 + H_2O \rightarrow NH_4^+ + HO^- \\
&Zn^{2+} + 3NH_4^+ \rightarrow \left[Zn(NH_3)_4\right]^{2+} \\
&\left[Zn(NH_3)_4\right]^{2+} + HO^- \rightarrow Zn(OH)_2 + 4NH_3 \\
&Zn(OH)_2 \rightarrow ZnO\downarrow \ + H_2O
\end{aligned}
$$

ii. Using a strong base (NaOH)

$$
\begin{aligned}
&Zn(NO_3)_2 \rightarrow Zn^{2+} + 2NO_3^- \\
&Zn^{2+} + 2HO^- \rightarrow Zn(OH)_2 \\
&Zn(OH)_2 + 2HO^- \rightarrow \left[Zn(OH)_4\right]^{2-} \\
&\left[Zn(OH)_4\right]^{2-} \rightarrow ZnO\downarrow \ + 2HO^- + H_2O
\end{aligned}
$$

iii. Using a reducing agent ($(CH_3)_2NHBH_3$) [147]

$$Zn(NO_3)_2 \rightarrow Zn^{2+} + 2NO_3^-$$
$$(\mathrm{CH_3})_2\mathrm{NHBH_3} + 2\mathrm{H_2O} \rightarrow \mathrm{HBO_2} + (\mathrm{CH_3})_2\mathrm{NH_2}^+ + 5\mathrm{H}^+ + 6\mathrm{e}^-$$
$$\mathrm{NO_3}^- + \mathrm{H_2O} + 2\mathrm{e}^- \rightarrow \mathrm{NO_2}^- + 2\mathrm{HO}^-$$
$$\mathrm{Zn}^{2+} + 2\mathrm{HO}^- \rightarrow \mathrm{Zn(OH)_2}$$
$$\mathrm{Zn(OH)_2} \rightarrow \mathrm{ZnO}\downarrow + \mathrm{H_2O}$$

The ZnO structures were examined by XRD (Figure 9). The diffraction peaks observed at 2θ = (31.8°, 34.5°, 36.3°, 47.5°, 56.6°, 63.0°, 66.4°, 68.0°, and 69.1°) are characteristic to ZnO hexagonal wurtzite phase (JCPDS file no. 36-1451), with corresponding Miller indexes at (100), (002), (101), (102), (110), (103), (200), (112), and (201). The strong and sharp diffraction patterns suggest that the as-obtained structures are well crystallized.

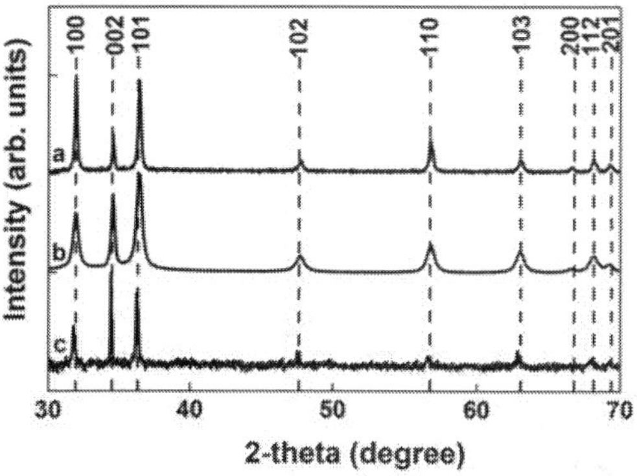

Figure 9. XRD patterns of ZnO samples synthesized in the presence of (a) $(CH_2)_6N_4$, (b) NaOH, and (c) $(CH_3)_2NHBH_3$.

SEM images of the samples (Figure 10) revealed the following morphologies for the ZnO micro/nanostructured TFs: rods (4.5 μm in length and 330 nm in diameter; Figure 10a, b), flowers (1–2 μm in dimension; Figure 10c, d), and hexagonal prisms (400 nm in length and 200 nm in diameter; Figure 10e, f). Insets to Figure 10 show the influence of the ZnO surface morphology on wetting behavior. The corresponding CA values of the ZnO samples are 164.8° (rods), 94.3° (flowers), and 79.4° (prisms). An explanation for the different values of CA can be related to the numerous gaps between the ZnO structures filled with air. For this reason, the film containing a higher volume of air trapped between the ZnO structures at the solid/water interface has a superhydrophobic behavior. The CA results were confirmed by AFM measurements (Figure 11). The RMS values were as follows: 390 nm (rods), 120 nm (flowers), and 50 nm (prisms).

Due to their morphology, the ZnO structures present different degrees of compactness, trapping more or less air in-between. In this way, the CA value is linked to the RMS value of the sample.

Figure 10. SEM images, at different magnification, of the ZnO samples synthesized in the presence of (a, b) $(CH_2)_6N_4$, (c, d) NaOH, and (e, f) $(CH_3)_2NHBH_3$. Insets: optical photographs of the water droplets shape on the ZnO surfaces with the corresponding CA values.

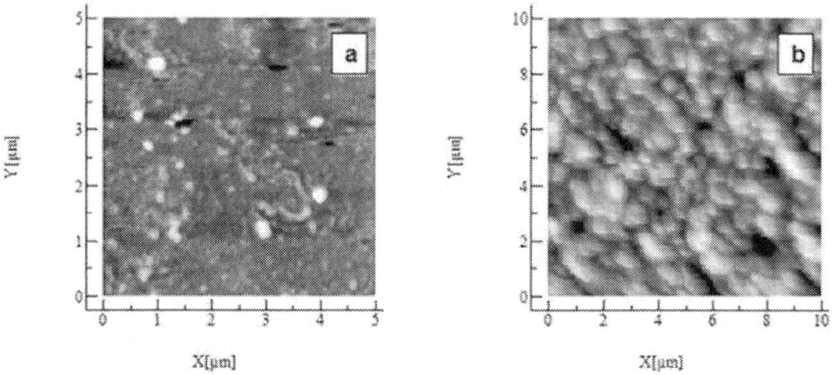

Figure 11. AFM images of the ZnO samples synthesized in the presence of (a) NaOH and (b) $(CH_3)_2NHBH_3$.

3.2. TiO_2

Titanium dioxide (TiO_2) is a transition metal oxide with UV absorbing properties with many technological applications [148, 149]. High photocatalytic efficiency, great stability, and low cost of production are in favor of TiO_2's photocatalytic properties [149]. In addition to bulk applications, TiO_2 TFs were obtained for UV blocking, antibacterial or/and photocatalytic properties [149].

3.2.1. TiO_2 structures synthesized by SG and sputtering onto textile substrates

Some properties of the substrates used in the experiments are summarized in Table 4.

Sample code	Textile 2D element/thread	Nature of fibers	Color	Thickness (mm)	Density (g/cm³)
P2	Knitted/interlock/Nm 70/1			0.82	0.25
P3	Knitted/interlock/Nm 50/1	Polyester		0.89	0.26
P28	Fabric		White	0.46	0.47
P30	Fabric			0.52	0.41
PLA	Nonwoven	Poly(lactic acid)		0.64	0.31

Table 4. Characteristics of different textiles functionalized with TiO_2.

XRD and SEM investigations [103] indicated, for both deposition techniques, that TiO_2 NPs were amorphous. Sputtered layers consisted of aggregates randomly distributed on substrate, while the SG layers showed a uniform distribution of NPs, with a mosaic-like structure. SEM images (Figure 12) suggest the formation of NPs, which are not singularly distinguishable. The sputtered layers consist of NP aggregates (in coalescence) with less than 20 nm diameter, randomly scattered on substrate. In the case of the SG layer, there are bridge-aggregated NPs leading both to a mosaic-like structure and to cracks and interfiber bonds [150].

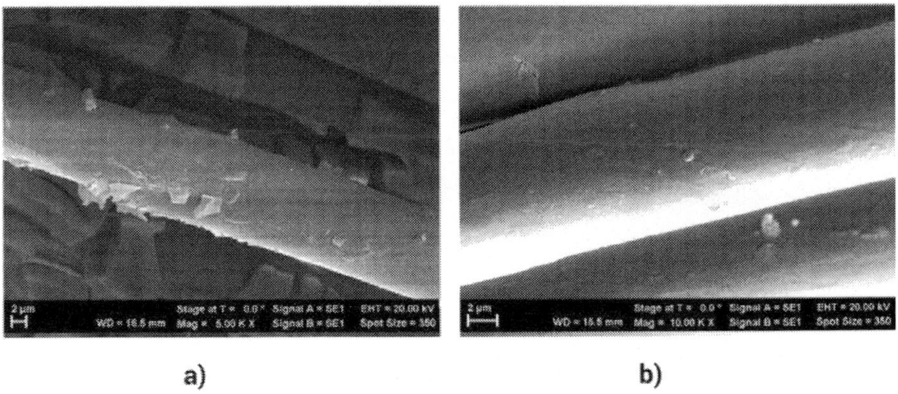

Figure 12. SEM images of TiO_2 samples deposited on P2 substrates by (a) SG and (b) sputtering.

A highly polar liquid–water was recommended [151] as testing liquid in CA measurements, for estimating the wettability of polar solids as polyester materials. The water repellency was thus regarded as indicating the performances of the coated layers and was evaluated by measuring static (equilibrium) CAs at RT [152]. In order to have a general idea of the samples' wetting behavior, different measurement points on each sample were thus considered (Figure 13).

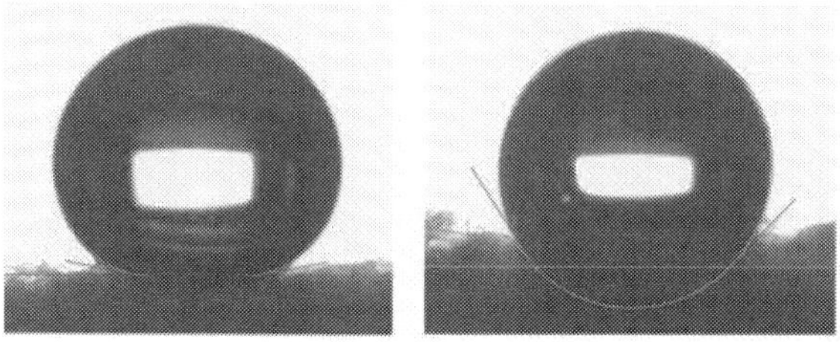

Figure 13. Water droplets on TiO_2 deposited on P2 samples by (a) SG and (b) sputtering.

The mean CA values of the raw or coated samples are summarized in Table 5.

Sample code	CA (°)	$f = \frac{1+\cos\theta_c}{1+\cos\theta_0}$ *
P2	136.9	0.244
P3	138.1	0.231
P28	152.1	0.105
P30	124.8	0.389
PLA	129.6	0.328
TiO_2 SG/P2	169.3	0.017
TiO_2 SG/P3	169.7	0.016
TiO_2 SG/P28	152.7	0.110
TiO_2 SG/P30	158.9	0.067
TiO_2 SG/PLA	140.6	0.225
TiO_2 SP4/P2	133.8	0.209
TiO_2 SP4/P3	166.0	0.020
TiO_2 SP4/P28	Hydrophilic	0.681
TiO_2 SP4/P30	155.8	0.059
TiO_2 SP4/PLA	150.3	0.089

* θ_0 = 84° (for raw), 89.4° (for SG), and 62.1° (for sputtered) samples.

Table 5. Water CA values measured onto different investigated surfaces.

From Table 5, one can see that the CAs increase more by fabric modification (after Titania deposition). CAs were influenced by air, water droplet, and surface of fabric, which formed a nanorough substrate. One sample (P28) is hydrophilic meaning that water passes through it; this sample has voids large enough, and the margins become hydrophilic by deposition of hydrophilic particles. The behavior might be approximated by the Cassie–Baxter equation:

$$\cos\theta_c = f\cos\theta_0 - (1 - f) \tag{7}$$

Here, θ_C is the composite CA formed on the treated fabric and θ_0 is the CA formed on untreated fabric [153]. The parameter f represents the fraction of the surface in contact with the water droplet. Knowing the corresponding CAs, its values can be calculated for each raw-treated pair of samples, using the following equation:

$$f = (1 + \cos\theta_c) / (1 + \cos\theta_0) \tag{8}$$

These values are summarized in Table 5. However, it seems that Cassie–Baxter equation (or Wenzel equation) should be applied to superhydrophobic surfaces with caution [154]. TiO_2 can be used to obtain hydrophobic surfaces by producing artificial roughness via micro structuring [155].

3.3. SiO_x

Silicon oxide was deposited onto polymeric substrates as a viable alternative to metallic depositions used for packing materials due to their transparency, recyclability, microwave use, and impressive barrier properties [156], to produce textiles with hydrophobic properties [157]. In addition, silica NPs immobilized on textiles can lead to flame retardant properties [158]. SG-deposited layer can be compared to the one deposited in vacuum at low angle because in both cases the deposit is awaited (super) hydrophilization evidence since SiO_2 brings its OH groups which print to the media hydrophilic properties. However, due to the columnary nanostructured relief of deposited layer, it was expected that the roughness of the textile surface would be increased. In addition, information was acquired on vacuum deposition at small angle [159–162]. Thus, we preferred this technique for a SiO_x deposition onto textile materials [163].

3.3.1. SiO_x structures synthesized by thermal evaporation at small angles onto polyester (P), polyamide (PA), poly(lactic acid) (PLA), and natural cellulosic hemp (H) substrates

The differences between the investigated textiles [163] are summarized in Table 6.

Sample code	Textile 2D element/thread	Nature of the fibers	Color	CA (°)	
				Raw textile	SiO_x/textile
P1	Knitted/interlock/Nm 70/1			136.9	139.2
P2	Knitted/interlock/Nm 50/1			138.1	128.9
P3	Knitted/glat/Nm 50/1			158.2	154.9
P4	Fabric/Nm 70/2 + Nm 40/2	Polyester		136.9	139.2
P27	Fabric		White	Hydrophilic	Hydrophilic
P28	Fabric			152.1	105.3
P30	Fabric			124.8	75.5
PA	Knitted	Polyamide		165.1	97.6
PLA	Nonwoven	Poly(lactic acid)		129.6	Hydrophilic
H	Fabric	Hemp	Beige	126.9	135.0

Table 6. Different functionalized textiles and their corresponding static CA values.

The XRD diffractograms pointed to an amorphous phase of the SiO_x deposited layers [163].

SEM morphologies of SiO_x particles synthesized on fabrics are presented in Figure 14. The raw material images showed defects like kink bands, dislocations, nodes, and slip planes, which are common characteristics of hemp materials [164]. SEM images of synthesized samples showed that SiO_x particles were grown on the fiber surfaces in a continuous and noncolumnar layer (Figure 14). Apparently, each individual fiber of samples looks uniformly covered by an amorphous layer [163].

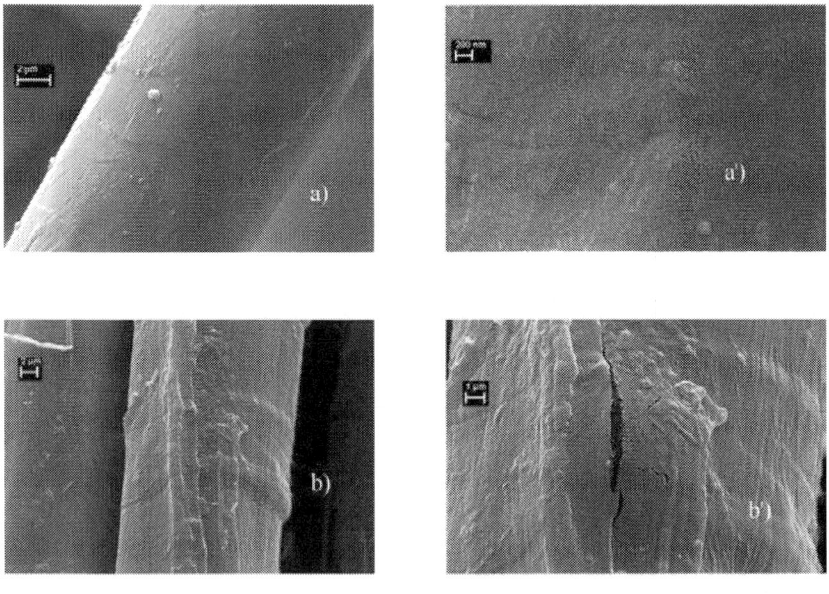

Figure 14. SEM images of SiO_x layers deposited on (a, a′) P2 and (b, b′) H substrates, at two different magnifications.

The wettability properties were evaluated by measuring static (equilibrium) CAs. The measurements were carried out at RT [152]. The images were processed using specific programs to fit the profile with the Young–Laplace equation in order to obtain the value of static CA.

In Figure 15, the image of the water droplet onto the deposited P2 sample and the corresponding CA is represented. The measurements were performed for a direction parallel to the privileged one of the knitted matter (vertically advance geometry). When following a direction perpendicular to the privileged one, the measurements evidence differences of few degrees only.

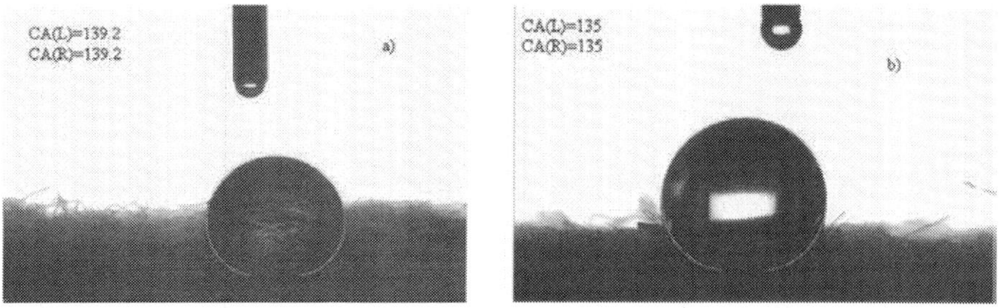

Figure 15. Water droplets on SiO_x TFs deposited on (a) P2 and (b) H samples. Reproduced from Frunza et al. [163].

From Figure 15, one can observe that the investigated surfaces are not flat, smooth, or homogeneous. These characteristics of the samples make it difficult to apply a specific model algorithm. Moreover, wetting of fabric surfaces is complicated by the heterogeneity, the diffusion of liquid into the fiber, and the capillary action of the fiber assembly. Under these conditions, the experimentally measured CA is an apparent one and can differ considerably from the actual value [163].

The mean CA values of the raw and deposited samples are summarized in Table 6. One can observe that the CA generally decreases after surface functionalization with SiO_x, in a range of few up to several tens of degrees. As expected, the presence of SiO_x NPs onto the fiber surface confers to the textiles a hydrophilic behavior (see Table 6).

3.3.2. *Surface free energy of SiO_2 (quartz) inferred from CA measurements*

Starting from known values of the dispersive and polar parts of the probe liquids' surface tension and obtained values of the CAs, the dispersive and polar parts of the surface tension of the solid (fused quartz) were estimated either by minimization of the equation system using the least square method or by solving the equations taken for combinations of two probe liquids [165].

Eq. (9) is a relation between the dispersive and polar parts of the solid substrate's surface tension and the same quantities of the surface tension of the wetting liquid and the corresponding CA:

$$\gamma_L(1+\cos\theta) = 2\sqrt{\gamma_S^d}\sqrt{\gamma_L^d} + 2\sqrt{\gamma_S^p}\sqrt{\gamma_L^p} \tag{9}$$

The values γ_S^d and γ_S^p were obtained by averaging the dispersive and polar components of γ_S resulted from solving Eq. (9) for all pairs of liquids that have the condition number of system matrix low enough (as defined in [166]).

CAs of water on fused silica can vary in a large interval. This behavior is in agreement with the one described in the literature, for example, with a 20º to 80º range obtained on quartz dehydroxylated by heating, slightly contaminated, or deliberately methylated [167]. The values we found can be interpreted in terms of the dependence of water CAs on sample purity; the presence of amorphous materials, chemicals, heating, and other pretreatments; and contamination by adsorption of substances from laboratory ambient. All these factors could have an influence over the increasing values of the CA. The obtained values were supposed to depend on the amounts of silanol groups and physically adsorbed water molecules on the quartz/silica surface. The investigation of "cleaner" surfaces obtained by a thermal treatment removing the hydroxyl groups at temperatures of the beginning and ending of the dehydroxylation process [168] was carried out.

The components of surface free energy of fused silica were determined by CA measurements of several liquids (see Table 7).

Treatment temperature (°C)			**CA for different liquids (°)**		
Water	Glycerol	NP5	Ethylene glycol	Dimethyl sulfoxide	
240	5.3	14.2	24.9	8.3	0
1000	33.6	14.8	24	0	6.5

Table 7. Values of CA (°) of different liquids on fused quartz treated at two different temperatures.

The fused silica plate samples were heated in atmosphere in order to remove water adsorbed on surface and most of the silanol groups. Measurements of CA on solid substrate were performed by analysis of the profile images of symmetric static liquid drops using the Drop Shape Analysis System (model DSA 100, from Krüss) [141, 152]. The samples were placed on a stage, under the tip of liquid-dispensing disposable blunt-end stainless steel needle with an outer diameter of 0.5 mm. The fixed needle was attached to a syringe pump, which was controlled by the computer for drop delivery. The volume of the drops was of ~ 2–3 μl. The CAs were determined by fitting the shape of the sessile drop with a smooth curve and then calculating the slope of the tangent to the drop at the liquid–solid–vapor interface. Low CAs ($\theta < 30$º) were determined by fitting the shape of the sessile drop with a circle, whereas larger CAs were estimated by fitting the drop shape with a polynomial equation of second degree or a circle equation. The camera was positioned to observe the droplet under an angle of about 2°–3° in respect to the sample surface supporting the droplet. The tests were carried out at RT.

CAs were obtained with an uncertainty of ±2° due to combined effects of drop asymmetry, surface heterogeneity, and variation in drop position on the plate.

Representative images of the observed water droplets on plates are given in Figure 16.

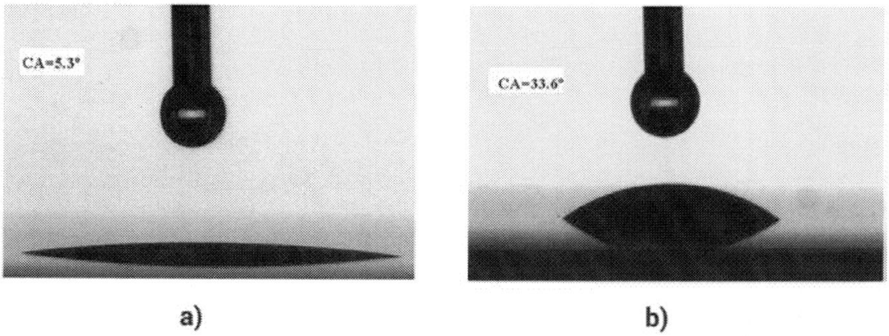

Figure 16. Water droplets on the SiO_x plates treated at (a) 240°C and (b) 1000°C, and the corresponding CAs.

Based on literature values [169–171] of the polar and dispersion parts of the liquid surface tension (see Table 8) and using the methods of geometric or harmonic mean for the interaction term, the calculation of the two components for fused silica (see Table 9) gave some differences, but their sum did not differ much. Moreover, our tests seem to indicate the method of harmonic mean as better than that one of Owens–Wendt [165].

Sample liquid	Dipole moment (D)	γ_L (mN/m)	γ_L^d (mN/m)	γ_L^p (mN/m)
Water	1.85	72.8	21.8	51.0
Glycerol	2.56	63.3	33.6	29.7
Ethylene glycol	2.69	48.0	33.8	14.2
Dimethyl sulfoxide	4.11	44.0	36.0	8.0
NP5	1.70	38.1	34.1	4.0

Table 8. Physical properties of different liquids used as samples.

Treatment temperature (°C)	Owens-Wendt/least squares (mN/m)		Owens-Wendt/ Average (mN/m)		Wu/ least squares (mN/m)	
	γ_s^d	γ_s^p	γ_s^d	γ_s^p	γ_s^d	γ_s^p
240	11.96	61.16	12.36	60.32	23.70	46.67
1000	15.96	46.90	15.96	49.56	25.29	37.48

Table 9. The surface tension components (dispersive γ_s^d and polar γ_s^p parts) of fused silica obtained by different calculation methods.

The polar part of the surface free energy of fused silica thermally treated is higher than the dispersive part as resulting from both methods (geometrical and harmonic mean). This might be an indication that, at the measurement moment, the plate surface was not (totally) covered by water vapors from environment.

In agreement with the decreasing number of silanol groups by the thermal treatment, the polar part of the surface tension shows a decreasing trend when increasing the pretreatment temperature.

The indirect method of CA measurements applied for the set of liquids chosen to have complementary interactions with quartz surface, allowed for obtaining values for the components of the surface free energy.

3.4. Effects of proteins from blood plasma on the hydrophobicity of DLC films

The amorphous phase of sp^3 bonded C atoms is known as DLC [82, 172, 173]. Beside high wear resistance coatings for metallic parts, DLC also proved useful in coating implants due to specific surface properties (low surface energy values and chemical inertness) that prevent blood coagulation and favor osteoblasts adhesion [90, 172]. In the biomedical field, the main necessity for DLC coatings comes from vascular prostheses. In the case of interaction with blood, it seems that DLC quality has a major influence upon clotting time. During the blood flow through these tubes, the erythrocytes and thrombocytes (platelets) aggregate in certain spots and may eventually block the blood passage. To compensate for this general weakness of vascular prostheses, DLC films can bind albumin molecules from the sanguine plasma forming a passive layer that makes the surfaces less adhesive for blood platelets [174].

The blood compatibility with carbon-based films is extremely complex and for the moment there is no relation found between hemocompatibility and surface properties such as surface energy, atomic bond structure of carbon, or composition of material. Contradictory data have been reported regarding the behavior of the material in terms of blood clotting, the adherence of platelets, or protein adsorption to surfaces. The relationship between the sp^3 bonds content of DLC and its antithrombogenicity properties is still not well understood. In vitro [82] and in vivo [88, 175] studies indicate that better results can be obtained for a higher sp^3 content.

Kwok et al. [176] pointed out that a higher surface energy of phosphorous doped a-C:H films is associated with a low adsorption of proteins, among them the albumin being the preferential one. Similar findings in terms of protein adsorption were presented by Ma et al. [177], who reported a higher albumin to fibrinogen adsorption ratios on surfaces with higher surface energy.

Jones et al. [178] explored platelet attachment on Ti, TiN, TiC, and DLC surfaces and reported that the more hydrophilic surfaces present a greater platelet spreading and fibrinogen adsorption. They suggested that the better hemocompatibility of DLC surface is linked to its low surface energy and thus high hydrophobicity. Okpalugo et al. [179] also noted that improved blood compatibility can be obtained when surface energy is lowered in silicon doped a-C:H films.

Recently, the correlation between activated partial thromboplastin time (aPTT) and surface energy of DLC structures with different sp^3/sp^2 bonds ratio was studied. Attention was paid to the investigation of protein adsorption and platelets adherence to the surface, both acting as crucial factors for material hemocompatibility [84].

3.4.1. *Types of bonds in the films*

XPS analysis, indicating the C 1s core level variation, was used in order to assess the amount of sp^2 and sp^3 bonded C in three types of samples (D20, D60, and D100; see Table 10).

Sample	Component (%)			sp^3/sp^2 ratio
	C=O/–COO	sp^2-C	sp^3-C	
D20	5.7±0.3	35.9±1.2	58.4±2.0	~1.6
D60	4.3±0.2	17.3±0.6	78.4±2.4	~4.5
D100	2.8±0.15	9.7±0.5	87.5±3.4	~9.0

Table 10. XPS peak separation data for the C 1s line of DLC films. Reproduced from Popa et al. [84].

From the XPS analysis, the amount of sp^3-bonded C and sp^2-C, as the ratio between the integral intensities of each component, could be extracted. The XPS spectra exhibited a very complex shape pointing to the existence of different chemical states for C 1s (Figure 17).

Three components were needed in order to assure a good fit, associated with the sp^3-C (286 eV) and sp^2-C (284.3 eV) contributions, as well as to C–O, C=O, and/or O–C=O bonds (287.5–289.9 eV) owing most probably to the contamination of the sample surface [180–182]. The deconvolution studies of the C 1s spectra generally reveal two main distinct peaks assignable to sp^2- and sp^3-C hybridization [182]. The peak placed at a higher binding energy (BE) is assigned to sp^3-bonded carbon (C–C and C–H), and that at lower BE corresponds to the sp^2 hybridization state of carbon. From the analysis of the main components of C 1s core level spectra, one could assume that the amount of sp^2 bonded C decreases from 36% in D20 sample to about 10% in D100. When the methane dilution is increased (D60 and D100), the sp^3-C concentration strongly increases (to ~78 and 87%, respectively).

The increase in the sp^3 content with the augmentation of the methane concentration has been confirmed both by Raman and XPS. A significant sp^3 content augmentation from sample D20 to D100 was measured. This could be the effect of the initial sp^3 hybridization of carbon in the methane molecule. Bugaev et al. [183] also reported that high-quality DLC films can be obtained from pure methane, their results pointing that most probably methyl mechanism is favoring diamond-like bonds formation. It is known that CH_3 are the most abundant species in pure methane discharges, while carbon dimer C2 is the most abundant in methane highly diluted in argon discharges [184, 185].

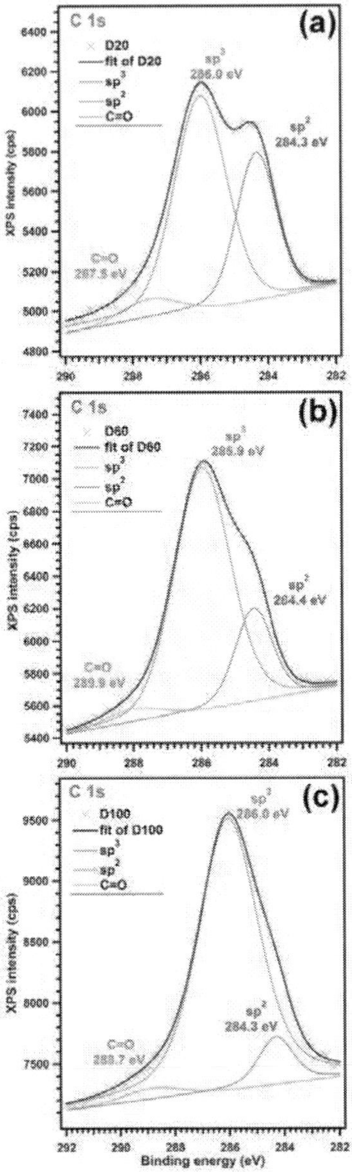

Figure 17. High-resolution XPS spectra for C 1s core level photoelectron after sputter cleaning: samples (a) D20, (b) D60 and (c) D100. Reproduced from Popa et al. [84].

3.4.2. Surface energy

Using deionized water and formamide as standard solvents, solid surface energy calculations based on CA measurements were performed. The measurements of the prepared DLC structures were carried out using the goniometric method, the two solvents being dropped onto the surface and the CA estimated. The drop size and the drip distance were kept constant in all cases. The CA values were determined by the evaluation of the tangent angle of a sessile liquid drop on the DLC solid surface. The surface energy was calculated using the Owens–Wendt approximation [186, 187].

The surface energy values recorded for DLC/Ti structures were lower than those of the bare medical grade Ti and PMMA control substrates (see Table 11).

Sample	Deposition atmosphere composition	Surface energy (mJ/m^2)
Bare Ti	N/A	37.85 ± 0.94
PMMA	N/A	36.35 ± 0.78
D20	20% CH_4 + 80% Ar	32.09 ± 0.73
D60	60% CH_4 + 40% Ar	30.50 ± 0.70
D100	100% Ar	28.71 ± 0.34

Table 11. Surface energy values recorded for the DLC TFs, and for the Ti and PMMA controls. Reproduced from Popa et al. [84].

One notices a decrease of the surface energy with the increase of methane dilution in the reactor chamber (Table 11). The two tailed *t*-testing showed statistically significant differences ($p < 0.05$) between the surface energy values recorded for all samples. An important decrease (with ~25%) of the surface energy was obtained when applying the DLC coating: from 37.85 ± 0.94 mJ/m^2 for the bare Ti substrate down to 28.7 ± 0.34 mJ/m^2 for the D100 structure.

3.4.3. DLC films interaction with blood

Platelets were obtained by centrifugation of whole blood and their adherence to the DLC films surface was investigated by Western blot method. The detailed procedures for platelets isolation and for the Western blot technique are described in Ref. [84].

The obtained signal is proportional to the amount of beta-actin, a structural protein present in all cells and, therefore, to the number of platelets adhered on the sample surface at the moment of lysis. As visible from Figure 18, there was almost the same number of platelets present on the surface of bare titanium and D20 samples.

The number of platelets adhered on D60 and D100 was significantly lower. The DLC coatings ensure conditions for a weaker platelet–surface interaction, which in vivo can conduct to a lower platelet activation and subsequently a prolonged time of coagulation. One can assert that this effect derives from the fact that all cells have a negatively charged cellular membrane,

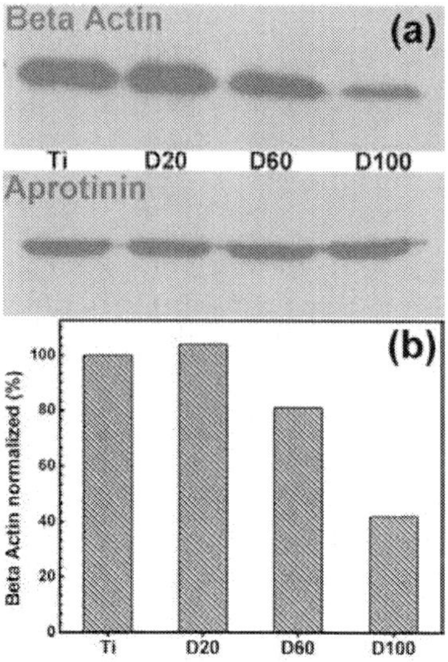

Figure 18. (a) Western blot analysis of beta-actin and aprotinin present in platelets adhered on DLC and bare Ti samples; (b) optical density histograms of normalized quantity of beta-actin present in platelets adhered on DLC and bare Ti samples. Reproduced from Popa et al. [84].

which tends to interact/adhere to positively charged surfaces (hydrophilic surfaces) rather than to hydrophobic ones.

The polyvinylidene fluoride membranes were also probed with aprotinin (a protease inhibitor with proteic structure and a mass of ~6 kDa), which was present in the same concentration in all samples, since it was added to the lysis buffer formulation. This is an internal quality control which ensures that all steps of the technique are properly done.

The results of protein adsorption on the DLC surfaces are shown in Figure 19.

Figure 19a shows that serum albumin was adsorbed in greater quantities on all DLC surfaces than on the bare titanium surface [albumin molecular weight (MW) ~66,483 Da]. Other proteins (G immunoglobulins) presented a roughly similar pattern (G immunoglobulins MW ~134,350 Da). Another important peak is that of 28,900 Da, which can be assigned to the factor XIIa light chain and is more prominent on the titanium sample (Figure 19b). The factor XII, the activator of surface contact coagulation cascade, could not be identified because it had a mass similar to that of albumin (factor XII MW ~67,792 Da).

Albumin is a protein that has hydrophobic moieties, being a blood carrier for many hydrophobic molecules. Since our DLC surfaces tend to be more hydrophobic, it is expected to find more adsorbed albumin than on titanium as confirmed by mass spectroscopy spectra. The vast majority of proteins in blood are glycosylated, which makes them more hydrophilic and more susceptible to polar interactions. The quantity of albumin adsorbed on the surface shields the surface of the sample, making it difficult for the different proteins and coagulation factors to

http://dx.doi.org/10.5772/60808

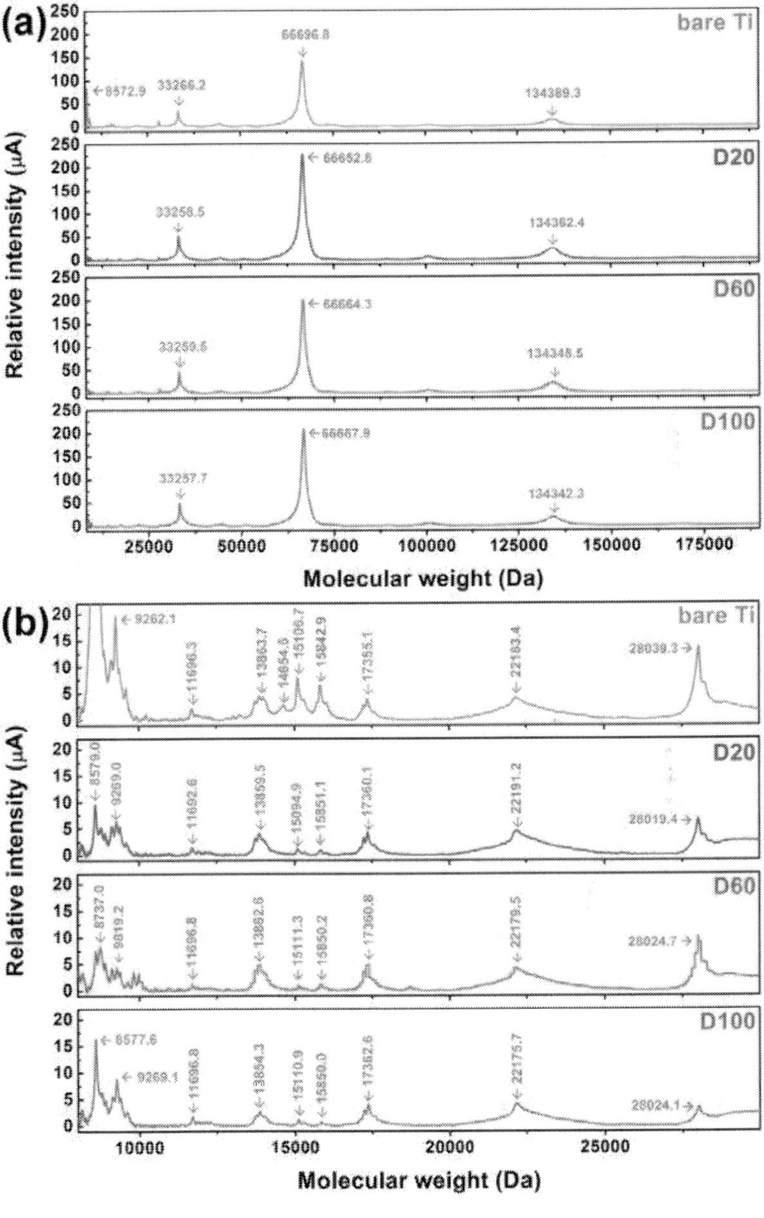

Figure 19. (a) SELDI-ToF complete spectra of proteins adsorbed on DLC and bare Ti samples from fresh blood plasma; (b) SELDI-ToF detailed spectra in MW range 8000–30,000 Da. Reproduced from Popa et al. [84].

reach the sample and activate the coagulation cascade (Figure 19b). These findings are in line with Liu et al. [188], who showed that the albumin adsorption on DLC inactivates the surface for blood clotting. One can state that the coagulation time for each material is in line with surface energy data, with the platelet–surface adherence properties and protein adsorption profiles, and so advocates for a cause–effect relationship between these factors.

4. Conclusions

Wettability of solid substrates represents an important phenomenon for many natural systems and can play a key role in a wide range of applications such as coatings, tunable surfaces, design of hydrophobic/superhydrophobic, or hydrophilic surfaces. It is well known that the wettability of a solid surface is governed by both surface structure and chemistry. After a brief introduction on wettability of nanostructures and the possibility to investigate it by contact angle (CA) measurements, this chapter focused on hydrophobic and hydrophilic structures (oxide and DLC TFs or NPs) synthesized by various deposition techniques (PLD, SG, TE, solution based on chemical approaches, sputtering, and PECVD).

The possibility of tuning the wetting behavior of textile materials by their functionalization with oxide TFs or NPs was reviewed. Depending on the deposition ambience, the TFs can change their behavior from hydrophilic when obtained in an oxygen flux to superhydrophobic when deposited in vacuum. The hydrophobicity was found consistent with the organization of the deposits in vacuum consisting of nanometric crystallites. The subsequent treatment with a TF of a fusion hydrophobin, deposited by soaking in solution, and a ZnO TF finishing in vacuum boosted the antifungal efficiency of the structure by 100%. This significant enhancement was attributed to the higher texturing of the oxide film when growing on hydrophobin interlayer, resulting in an increased presence of oxygen species on surface. In complementary studies, fabrics functionalized with oxide layers showed improved UV protective performances. These results might offer guidance for laser manufacturing in one technological step of stable superhydrophobic and antifungal textile surfaces, used for everyday garments and medical clothing.

ZnO structures can present different degrees of compactness, and as a consequence, they can trap more or less air. This result can be explained by the Cassie–Baxter model. Due to the morphology of the deposited ZnO structure, which is made of a large number of small prisms, the roughness presents high values. The apparent CA is therefore enhanced as compared to the one measured on a similar smooth surface. When the space between the ZnO structures is large enough, the water droplet can penetrate, and an explanation of the phenomenon can be based on the Wenzel model. There exists also the possibility to obtain a transition between these two regimes, and the apparent CA could be different than the one inferred for a smooth surface.

CA measurements confirmed that the presence of SiO_x particles on fiber surfaces can change the wetting behavior of the structure. Since it brings OH groups to the surface, the deposition of SiO_x is therefore expected to provide hydrophilic properties to the textiles.

Although bulk polyester is hydrophobic, water droplets can be sucked into the fibers due to high porosity (void areas) of the material. The void areas were drastically reduced by the addition of TiO_2 particles. They decrease the voids and concomitantly increase the sample hydrophobicity. Under these complex conditions, one cannot use the traditional equations like Cassie–Baxter or Wenzel to model the wettability behavior of the heterogeneous and rough samples.

Protein adsorption using fresh blood plasma from healthy patients was also studied. In the case of DLC films with the highest sp^3 content, albumin was preferentially adsorbed (due to the affinity between the surface and the hydrophobic moieties of the protein), thus shielding the surface and preventing the immobilization of coagulation factors.

The results reviewed in this chapter are devoted to improve the understanding of the wettability of nanostructured surfaces. Understanding the importance of surface wettability and succeeding to control this phenomenon at nanometric scale will hopefully facilitate the fabrication of devices with improved characteristics for top applications, especially in nanotechnology.

Acknowledgements

LD and INM acknowledge project no. 7-083/2014 (CARLA). IZ acknowledges the financial support of the Romanian Ministry of Education and Research under the Project IDEI 281/2011. ACP acknowledges the funding of this research by the Romanian National Authority for scientific research through PNII-RU-TE-2012-3-0379 (TE 16/2013).

Author details

L. Duta[1], A.C. Popescu[1], I. Zgura[2], N. Preda[2] and I.N. Mihailescu[1*]

*Address all correspondence to: ion.mihailescu@inflpr.ro

1 National Institute for Lasers, Plasma, and Radiation Physics, Magurele, Romania

2 National Institute of Materials Physics, Magurele, Romania

References

[1] Duta L., Popescu A.C., Dorcioman G., Mihailescu I.N., Stan G.E., Zgura I., Enculescu I., Dumitrescu I. ZnO thin films deposited on textile material substrates for biomedical applications. In: Vaseashta A., Braman E., Susmann P. (eds.), Technological inno-

vations in sensing and detection of chemical, biological, radiological, nuclear threats and ecological terrorism. Netherlands: Springer; 2012.

[2] Yamamoto O. Influence of particle size on the antibacterial activity of zinc oxide. International Journal of Inorganic Materials. 2001;3(7) 643–646.

[3] Rajendran R., Balakumar C., Mohammed Ahammed H.A., Jayakumar S., Vaideki K., Rajesh E.M. Use of zinc oxide nano particles for production of antimicrobial textiles. International Journal of Engineering, Science and Technology. 2010;2(1) 202–208.

[4] Brayner R., Ferrari-Iliou R., Brivois N., Djediat S., Benedetti M.F., Fievet F. Toxicological impact studies based on *Escherichia coli* bacteria in ultrafine ZnO nanoparticles colloidal medium. Nano Letters. 2006;6(4) 866–870.

[5] Yamamoto O., Komatsu M., Sawai J., Nakagawa Z.E. Effect of lattice constant of zinc oxide on antibacterial characteristics. Journal of Materials Science: Materials in Medicine. 2004;15(8) 847–851.

[6] Huang Z., Zheng X., Yan D., Yin G., Liao X., Kang Y., Yao Y., Huang D., Hao B. Toxicological effect of ZnO nanoparticles based on bacteria. Langmuir 2008;24(8) 4140–4144.

[7] Holysz L., Chibowski E., Terpilowski K. Influence of ambient humidity on the apparent surface free energy of poly(methyl methacrylate) (PMMA). In: Mittal K. (ed.), Contact angle, wettability and adhesion. VSP/Brill: Leiden; 2008. p95–113.

[8] Kubiak K.J., Wilson M.C.T., Mathia T.G., Carval Ph. Wettability versus roughness of engineering surfaces. Wear 2011;271(3–4) 523–528.

[9] Xinjian F, Lei J. Design and creation of superwetting/antiwetting surfaces. Advanced Materials 2006;18(23) 3063–3078.

[10] Whyman G., Bormashenko E., Stein T. The rigorous derivation of Young, Cassie–Baxter and Wenzel equations and the analysis of the contact angle hysteresis phenomenon. Chemical Physics Letters. 2008;450(4–6) 355–359.

[11] Bico J., Tordeux C., Quéré D. Wetting of textured surfaces. Colloid and Surfaces A: Physicochemical and Engineering Aspects. 2002;206(1–3) 41–46.

[12] Rosales-Leal J.I., Rodríguez-Valverde M.A., Mazzaglia G., Ramón-Torregrosa P.J., Díaz-Rodríguez L., García-Martínez O., Vallecillo-Capilla M., Ruiz C., Cabrerizo-Vílchez M.A. Effect of roughness, wettbility and morphology of engineered titanium surfaces on osteoblast-like cell adhesion. Colloids and Surfaces A: Physicochemical and Engineering aspects 2010;365(1–3) 222–229.

[13] Tuteja A., Choi W., Ma M., Mabry J.M., Mazzella S.A., Rutledge G.C., McKinley G.H., Cohen R.H.. Designing superoleophobic surfaces. Science 2007;318(5856) 1618–1622.

[14] Marmur A. Solid-surface characterization by wetting. Annual Review of Materials Research 2009;39 473–489.

[15] Cassie A.B.D., Baxter S. Transactions of the Faraday Society 1944;40() 546–551.

[16] Lafuma A., Quéré D. Superhydrophobic states. Nature Materials 2003;2(7) 457–460.

[17] Ishino C., Okumura K., Quéré D. Wetting transitions on rough surfaces. Europhysics Letters 2004;68(3) 419–425.

[18] Marmur A. The lotus effect: superhydrophobicity and metastability. Langmuir 2004;20(9) 3517–3519.

[19] Dupuis A., Yeomans J.M. Modeling droplets on superhydrophobic surfaces: equilibrium states and transitions. Langmuir 2005;21(6) 2624–2629.

[20] Patankar N.A. On the modeling of hydrophobic contact angles on rough surfaces. Langmuir 2003;19(4) 1249–1253.

[21] Sameer R.P., Zheng C., Wei H., Narendra B.D. Wetting effects on in vitro bioactivity and in vitro biocompatibility of laser micro-textured Ca-P coating. Biofabrication 2010;2(2) 025001.

[22] Adamson A.W., editor. Physical Chemistry of Surfaces. New York: Wiley; 1990.

[23] Tammar S., Meiron A., Sam Saguy I. Contact angle measurement on rough surfaces. Journal of Colloid and Interface Science 2004;274(2) 637–644.

[24] Deepak P.S., Dinesh K.M., Ashish S., Ujjwal M.J., Andrzej H. Study of the wettability of ZnO nanofilms. International Nano Letters 2011;1(2) 117–122.

[25] Yin-Yu C., Chih-Ho L., Jui-Ting H., Chih-Hsin T., Wan-Chuen L., Heng-Li H. Antibacterial properties and human gingival fibroblast cell compatibility of TiO_2/Ag compound coatings and ZnO films on titanium-based material. Clinical Oral Investigations 2012;16(1) 95–100.

[26] Helena T., Patrycja S-WoŸ, Marta F., Iwona K. Deposition of Zinc Oxide on the materials used in medicine. Preliminary results. Fibres and Textiles in Eastern Europe 2014;3(105) 126–132.

[27] Kaushalkumar B., Duncan R., Radhakrishna P., Pat P. LED-controlled tuning of ZnO nanowires' wettability for biosensing applications. Nano Reviews 2015;6() 26711.

[28] Mao-Gang G., Xiao-Liang X., Zhou Y., Yan-Song L., Ling L. Superhydrophobic surfaces via controlling the morphology of ZnO micro/nano complex structure. Chinese Physics B 2010;19(5) 056701.

[29] Amirhosein B., Ramin K., Mohammad E.Y. Fabrication of superhydrophobic and antibacterial surface on cotton fabric by doped silica-based sols with nanoparticles of copper. Nanoscale Research Letters 2011;15,6(1) 594.

[30] Yang S., Guixue W., Xianliang H., Qin Z., Jiang W., Chaojun T., Qingsong Y., Xiaoheng L. Surface wettability of plasma SiO_x:H nanocoating-induced endothelial

cells'migration and the associated FAKRho GTPases signalling pathways. Journal of the Royal Society Interface 2012;9(67) 313–327.

[31] Antonia T., Jose IV-P., de la Orden E., Yubero F., Gonzalez-Caballero J.L., González-Elipe A.R., Vilches J., Salido M. Osteoconductive potential of barrier nanoSiO_2 PLGA membranes functionalized by plasma enhanced chemical vapour deposition. Bio-Medical Research International 2014;2014() 253590.

[32] Brown P.S., Bhushan B. Mechanically durable, superoleophobic coatings prepared by layer-by-layer technique for anti-smudge and oil-water separation. Scientific Reports 2014;5() 8701.

[33] Anitha V.C., Lee J.H., Lee J., Banerjee A.N., Joo S.W., Min B.K. Biofilm formation on a TiO_2 nanotube with controlled pore diameter and surface wettability. Nanotechnology 2015;26(6) 065102.

[34] Hyde G.K., Stewart S.M., Scarel G., Parsons G.N., Shih C.C., Shih C.M., Shing-Jong L., Yea-Yang S., Monteiro-Riviere N.A., Narayan R.J. Atomic layer deposition of titanium dioxide on cellulose acetate for enhanced hemostasis. Biotechnology Journal 2011;6(2) 213–223.

[35] Terumitsu H., So N., Aki K., Myoung-Woon M., Yousuke K., Atsushi H., Kwang-Ryeol L., Koki T., Takuji Y., Tetsuya S. Hydrophobicity and non-thrombogenicity of nanoscale dual rough surface coated with fluorine-incorporated diamond-like carbon films: biomimetic surface for blood-contacting medical devices. Diamond and Related Materials 2013;38 14–18.

[36] Terumitsu H., Atsushi S., Tetsuya S., Yoshiaki M., Toshiya S., Satoshi Y., Aki K., Nobuyuki S., Mutsumi H., Kanako K., Hirokuni Y., Sachio K. Fluorinated diamond-like carbon as antithrombogenic coating for blood-contacting devices. Journal of Biomedical Materials Research Part A 2006;76A(1), 86–94.

[37] Salgueiredo E., Vila M., Silva M.A., Lopes M.A., Santos J.D., Costa F.M., Silva R.F., Gomes P.S., Fernandes M.H. Biocompatibility evaluation of DLC-coated Si3N4 substrates for biomedical applications. Diamond and Related Materials 2008;17(4–5) 878–881.

[38] Shanhong W., Liping W., Qunji X. Super-hydrophilic properties of TiO_2–DLC nanocomposite films fabricated by the simple electrochemical process. Applied Surface Science 2011;257(23) 10000–10004.

[39] Leonard R.L., Terekhov A.Y., Thompson C., Erck R.A., Johnson J.A. Antifog coating for bronchoscope lens. Surface Engineering 2012;28(6) 468–472.

[40] Jones D.S., Garvin C.P., Dowling D., Donnelly K., Gorman S.P. Examination of surface properties and in vitro biological performance of amorphous diamond-like carbon-coated polyurethane. Journal of Biomedical Materials Research Part B: Applied Biomaterials 2006;78(2) 230–236.

[41] Gao I., McCarthy T.J. Contact angle hysteresis explained. Langmuir 2006;22(14) 6234–6237.

[42] www.attension.com (last accessed 14 April 2015).

[43] Miwa M., Nakajima A., Fujishima A., Hashimoto K., Watanabe T. Effects of the surface roughness on sliding angles of water droplets on superhydrophobic surfaces. Langmuir 2000;16(13) 5754–5760.

[44] McHale G., Shirtcliffe N.J., Newton M.I. Contact angle hysteresis on super-hydrophobic surfaces. Langmuir 2004;20(23) 10146–10149.

[45] Furmidge C.G.L. Studies at phase interfaces. I. The sliding of liquid drops on solid surfaces and a theory for spray retention. Journal of Colloid and Interface Science 1962;17(4) 309–324.

[46] Feng L., Li S., Li Y., Li H., Zhang L., Zhai J., Song Y., Liu B., Jiang L., Zhu D. Superhydrophobic surfaces: from natural to artificial. Advanced Materials 2002;14(24) 1857–1860.

[47] Blossey R. Self-cleaning surfaces - virtual realities. Nature Materials 2003;2 301–306.

[48] Zisman W.A. Relation of the equilibrium contact angle to liquid and solid constitution. In: Fowkes F.M. (ed.), Advances in Chemistry. American Chemical Society. 1964; p1–51.

[49] Fox H.W., Zisman A.W. The spreading of liquids on low-energy surfaces. II. Modified tetrafluoroethylene polymers. Journal of Colloid and Interface Science 1952;7(2) 109–121.

[50] Hejda F., Solar P., Kousal J. Surface free energy determination by contact angle measurements—a comparison of various approaches. In: Safrankova J, Pavlu J. (eds.), WDS'10 Proceedings of Contributed Papers: Part III—Physics, 1–4 June 2010, Prague. Matfyzpress.

[51] Sharma P.K., Rao K.H. Analysis of different approaches for evaluation of surface energy of microbial cells by contact angle goniometry. Advances in Colloid and Interface Science 2002;98(3) 341–463.

[52] Patra S., Sarkar S., Bera S.K., Ghosh R., Paul G.K. Hydrophobic self-cleaning surfaces of ZnO thin films synthesized by sol-gel technique. Journal of Physics D: Applied Physics. 2009;42(7) 075301.

[53] Zhang Z., Chen H., Zhong J., Saraf G., Lu Y. Fast and reversible wettability transitions on ZnO nanostructures. Journal of Electronic Materials. 2007;36(8) 895–899.

[54] Irzh A., Genish I., Klein L., Solovyov L.A., Gedanken A. Synthesis of ZnO and Zn nanoparticles in microwave plasma and their deposition on glass slides. Langmuir 2010;26(8) 5976–5984.

[55] Han J., Gao W. Surface wettability of nanostructured zinc oxide films. Journal of Electronic Materials. 2009;38(4) 601–608.

[56] Patra S., Sarkar S., Bera S.K., Paul G.K., Ghosh R. Influence of surface topography and chemical structure on wettability of electrodeposited ZnO thin films. Journal of Applied Physics. 2010;108(8) 083507.

[57] Kekkonen V., Hakola A., Kajava T., Sahramo E., Malm J., Karppinen M., Ras R.H.A. Self-erasing and rewritable wettability patterns on ZnO thin films. Applied Physics Letters. 2010;97(4) 044102.

[58] He S., Zheng M., Yao L., Yuan X., Li M., Ma L., Shen W. Preparation and properties of ZnO nanostructures by electrochemical anodization method. Applied Surface Science. 2010;256(8) 2557–2562.

[59] Eason R., editor. Pulsed laser deposition of thin films: applications-lead growth of functional materials. Wiley-Blackwell; 2007.

[60] Singh A., Davis E.A. The a-SiO_x:H_y thin film system I. Structural study by IR spectroscopy. Journal of Non-Crystalline Solids 1990;122(3) 223–232.

[61] Leplan H., Robic J.Y., Pauleau Y. Kinetics of residual stress evolution in evaporated silicon dioxide films exposed to room air. Journal of Applied Physics. 1996;79(9) 6926–6932.

[62] da Silva Sobrinho A.S., Czeremuszkin G., Latreche M., Wertheimer M.R. Defect-permeation correlation for ultrathin transparent barrier coatings on polymers. Journal of Vacuum Science and Technology A. 2000;18(1) 149–158.

[63] Madocks J., Rewhinkle J., Barton L. Packaging barrier films deposited on PET by PECVD using a new high density plasma source. Materials Science and Engineering: B. 2005;119(3) 268–273.

[64] Leterrier Y. Durability of nanosized oxygen-barrier coatings on polymers. Progress in Materials Science. 2003;48(1) 1–55.

[65] Anastasescu C., Anastasescu M., Teodorescu V.S., Gartner M., Zaharescu M. SiO_2 nanospheres and tubes obtained by sol-gel method. Journal of Non-Crystalline Solids. 2010;356(44–49) 2634–2640.

[66] Bae G.Y., Min B.G., Jeong Y.G., Lee S.C., Jang J.H., Koo G.H. Superhydrophobicity of cotton fabrics treated with silica nanoparticles and water-repellent agent. Journal of Colloid and Interface Science 2009;337(1) 170–175.

[67] Yang C., Fan H., Xi Y., Chen J., Li Z. Effects of depositing temperatures on structure and optical properties of TiO_2 film deposited by ion beam assisted electron beam evaporation. Applied Surface Science. 2008;254(9) 2685–2689.

[68] da Silva Sobrinho A.S., Czeremuszkin G., Latreche M., Wertheimer M.R. Defect-permeation correlation for ultrathin transparent barrier coatings on polymers. Journal of

Vacuum Science and Technology. A, Vacuum, Surfaces, and Films. 2000;18(1) 149–157.

[69] Sonnenfeld A., von Rohr P.R., Hauert R. UV Absorptance of titanium dioxide thin films by plasma enhanced deposition from mixtures of oxygen and titanium-tetrakis-isopropoxide. Plasma Chemistry and Plasma Processing. 2006;26(3) 319–334.

[70] Carneiro J.O., Teixeira V., Nascimento J.H.O., Neves J., Tavares P.B. Photocatalytic activity and UV-protection of TiO_2 nanocoatings on poly(lactic acid) fibres deposited by pulsed magnetron sputtering. Journal of Nanoscience and Nanotechnology 2011;11(10) 8979–8985.

[71] Xu Y., Xu W., Huang F., Wei Q.F. Preparation and photocatalytic activity of TiO_2-deposited fabrics. International Journal of Photoenergy. 2012;852675.

[72] Ojstrsek A., Kleinschek K.S., Fakin D. Characterization of nano-sized TiO_2 suspensions for functional modification of polyester fabric. Surface and Coatings Technology. 2013;226 68–74.

[73] Pisitsak P., Samootsoot A., Chokpanich N. Investigation of the self-cleaning properties of cotton fabrics finished with nano-TiO_2 and nano-TiO_2 mixed with fumed silica. KKU Research Journal. 2013;18(2) 200–211.

[74] Watanabe Y., Kobayashi T., Kirihara S., Miyamoto Y., Sakoda K. Cotton-yarn/TiO_2 dispersed resin photonic crystals with straight and wavy structures. European Physical Journal B. 2004;39(3) 295–300.

[75] Asadi M., Montazer M. Multi-functional polyester hollow fiber nonwoven fabric with using nano clay/nano TiO_2/polysiloxane composites. Journal of Inorganic and Organometallic Polymers and Materials. 2013;23(6) 1358–1367.

[76] Macwan D., Dave P.N., Chaturvedi S. A review on nano-TiO_2 sol-gel type syntheses and its applications. Journal of Materials Science. 2011;46(11) 3669–3686.

[77] Ning L., Hongyi L., Hong W., Tao C., Jinshu W., Lei C. Studies on the TiO_2 modified microchannels for microfluidic applications. Materials Letters. 2012;89 247–250.

[78] Fusi M., Maccallini E., Caruso T., Casari C.S., Li Bassi A., Bottani C.E., Rudolf P., Prince K.C., Agostino R.G. Surface electronic and structural properties of nanostructured titanium oxide grown by pulsed laser deposition. Surface Science. 2011;605(3–4) 333–340.

[79] Petersen M., Bandorf R., Brauer G., Klages C.P. Diamond-like carbon films as piezoresistors in highly sensitive force sensors. Diamond and Related Materials. 2012;26 50–54.

[80] Oliveira E.C., Cruz S.A., Aguiar P.H.L. Effect of PECVD deposition parameters on the DLC/PLC composition of a-C:H thin films. Journal of the Brazilian Chemical Society. 2012;23(9) 1657–1662.

[81] Vaghri E., Khalaj Z., Ghoranneviss M., Borghei M. Characterization of diamond-like carbon films synthesized by DC-plasma enhanced chemical vapor deposition. Journal of Fusion Energy. 2011;30(5) 447–452.

[82] Logothetidis S. Haemocompatibility of carbon based thin films. Diamond and Related Materials. 2007;16(10) 1847–1857.

[83] Flege S., Hatada R., Ensinger W., Baba K. Properties of hydrogenated DLC films as prepared by a combined method of plasma source ion implantation and unbalanced magnetron sputtering. Journal of Materials Research. 2012;27(5) 845–849.

[84] Popa A.C., Stan G.E., Husanu M.A., Pasuk I., Popescu I.D., Popescu A.C., Mihailescu I.N. Multi-layer haemocompatible diamond-like carbon coatings obtained by combined radio frequency plasma enhanced chemical vapor deposition and magnetron sputtering. Journal of Materials Science: Materials in Medicine. 2013;24(12) 2695–2707.

[85] Myllymaa S., Kaivosoja E., Myllymaa K., Sillat T., Korhonen H., Lappalainen R., Konttinen Y.T. Adhesion, spreading and osteogenic differentiation of mesenchymal stem cells cultured on micropatterned amorphous diamond, titanium, tantalum and chromium coatings on silicon. Journal of Materials Science: Materials in Medicine. 2010;21(1) 329–341.

[86] Soininen A., Levon J., Katsikogianni M., Myllymaa K., Lappalainen R., Konttinen Y.T., Kinnari T.J., Tiainen V.M., Missirlis Y. In vitro adhesion of staphylococci to diamond-like carbon polymer hybrids under dynamic flow conditions. Journal of Materials Science: Materials in Medicine. 2011;22(3) 629–636.

[87] Srinivasan S., Tang Y., Li Y.S., Yang Q., Hirose A. Ion beam deposition of DLC and nitrogen doped DLC thin films for enhanced haemocompatibility on PTFE. Applied Surface Science. 2012;258(20) 8094–8099.

[88] Jelinek M., Smetana K., Kocourek T., Dvorankovac B., Zemek J., Remsa J., Luxbacher T. Biocompatibility and sp3/sp2 ratio of laser created DLC films. Materials Science and Engineering: B. 2010;169(1–3) 89–93.

[89] Pısarık P., Jelınek M., Smetana K. Jr., Dvorankova B., Kocourek T., Zemek J., Chvostova D. Study of optical properties and biocompatibility of DLC films characterized by sp3 bonds. Applied Physics A. 2013;112(1) 143–148.

[90] Nelea V., Jelinek M., Mihailescu I.N. Biomaterials: new issues and breakthroughs for biomedical applications. In: Eason R. (ed.), Pulsed Laser Deposition of Thin Films: Applications-Lead Growth of Functional Materials. Wiley & Sons; 2007. p421–p459.

[91] Boyd I.W. Thin film growth by pulsed laser deposition. Ceramics International 1996;22() 429–434.

[92] Bao Q., Chen C., Wang D., Ji Q., Lei T. Pulsed laser deposition and its current research status in preparing hydroxyapatite thin films. Applied Surface Science 2005;252(5) 1538–1544.

[93] Mihailescu I.N., Gyorgy E. Pulsed laser deposition: an overview. In: Asakura T. (ed.), International Trends in Optics and Photonics. Heidelberg: Springer; 1999. p201–214.

[94] Liste S., Gonzalez P., Serra J., Borrajo J.P., Chiussi S., Leon B., Perez Amor M., Garcıa Lopez J., Ferrer F.J., Morilla Y., Respaldiza M.A. Study of the stoichiometry transfer in pulsed laser deposition of bioactive silica based glasses. Thin Solid Films 2004;453–454 219–223.

[95] Baviskar P.K., Nikam P.R., Gargote S.S., Ennaoui A., Sankapal B.R. Controlled synthesis of ZnO nanostructures with assorted morphologies via simple solution chemistry. Journal of Alloys and Compounds. 2013;551 233–242.

[96] Xu S., Wang Z.L. One-dimensional ZnO nanostructures: solution growth and functional properties. Nano Research. 2011;4(11) 1013–1098.

[97] Pauporte T. Design of solution-grown ZnO nanostructures. In: Wang Z.M. (ed.), Toward Functional Nanomaterials, Lecture Notes in Nanoscale Science and Technology. LLC: Springer Science+Business Media; 2009. p77–125.

[98] Nelea V., Morosanu C., Iliescu M., Mihailescu I.N. Hydroxyapatite thin films grown by pulsed laser deposition and radio-frequency magnetron sputtering: comparative study. Applied Surface Science 2004;228(1–4) 346–356.

[99] Nelea V., Morosanu C., Iliescu M., Mihailescu I.N. Microstructure and mechanical properties of hydroxyapatite thin films grown by RF magnetron sputtering. Surface and Coatings Technology. 2003;173(2–3) 315–322.

[100] Stan G.E. Adherent functional graded hydroxylapatite coatings produced by sputtering deposition techniques. Journal of Optoelectronics and Advanced Materials. 2009;11(8) 1132–1138.

[101] Schneider J.M., Rohde S., Sproul W.D., Matthews A. Recent developments in plasma assisted physical vapour deposition. Journal of Physics D: Applied Physics 2000;33(18) R173.

[102] Berg S., Nyberg T. Fundamental understanding and modeling of reactive sputtering processes. Thin Solid Films. 2005;476(2) 215–230.

[103] Zgura I., Frunza S., Frunza L., Enculescu M., Florica C., Cotorobai V.F., Ganea C.P. Polyester fabrics covered with amorphous titanium dioxide layers: combining wettability measurements and photoinduced hydrophilicity to assess their surface properties. Romanian Reports in Physics. 2016;68(1).

[104] Miller J.D., Veeramasuneni S., Drelich J., Yalamanchili M.R., Yamauchi G. Effect of roughness as determined by atomic force microscopy on the wetting properties of PTFE thin films. Polymer Engineering and Science. 1996;36(14) 1849–1855.

[105] Bahners T.J. The do's and don'ts of wettability characterization in textiles. Journal of Adhesion Science and Technology. 2011;25(16) 2005–2021.

[106] Oldani C., Dominguez A. Titanium as a biomaterial for implants. In: Fokter S. (ed.). Recent Advances in Arthroplasty. Rijeka: InTech; 2012. p149–p162.

[107] Reddy K.M., Feris K., Bell J., Wingett D.G., Hanley C., Punnoose A. Selective toxicity of zinc oxide nanoparticles to prokaryotic and eukaryotic systems. Applied Physics Letters. 2007;90(21) 2139021- 2139023.

[108] Jones N., Ray B., Ranjit K.T., Manna A.C. Antibacterial activity of ZnO nanoparticle suspensions on a broad spectrum of microorganisms. FEMS Microbiology Letters. 2008;279(1) 71–76.

[109] Nair M.G., Nirmala M., Rekha K., Anukaliani A. Structural, optical, photo catalytic and antibacterial activity of ZnO and Co doped ZnO nanoparticles. Materials Letters. 2011;65(12) 1797–1800.

[110] Sawai J., Shoji S., Igarashi H., Hashimoto A., Kokugan T., Shimizu M., Kojima H. Hydrogen peroxide as an antibacterial factor in zinc oxide powder slurry. Journal of Fermentation and Bioengineering. 1998;86(5) 521–522.

[111] Premanathan M., Karthikeyan K., Jeyasubramanian K., Manivannan G. Selective toxicity of ZnO nanoparticles toward Gram-positive bacteria and cancer cells by apoptosis through lipid peroxidation. Nanomedicine: Nanotechnology, Biology, and Medicine. 2011;7(2) 184–192.

[112] Applerot G., Lipovsky A., Dror R., Perkas N., Nitzan Y., Lubart R., Gedanken A. Enhanced antibacterial activity of nanocrystalline ZnO due to increased ROS-mediated cell injury. Advanced Functional Matterials. 2009;19(6) 842–852.

[113] Carminna O., Andrea L., Marco F., Valentina C. Wetting behavior of hierarchical oxide nanostructures: TiO_2 nanotubes from anodic oxidation decorated with ZnO nanostructures. Journal of the Electrochemical Society. 2014;161(10) D484-D488.

[114] Djuriisic A.B., Ng A.M.C., Chen X.Y. ZnO nanostructures for optoelectronics: material properties and device applications. Progress in Quantum Electronics. 2010;34(4) 191–259.

[115] Ozgur U., Alivov Ya.I., Liu C., Teke A., Reshchikov M.A., Dogan S., Avrutin V., Cho S.-J., Morkoc H. A comprehensive review of ZnO materials and devices. Journal of Applied Physics. 2005;98(4) 041301.

[116] Sontakke T.K., Jagtap R.N., Singh A., Kothari D.C. Nano ZnO grafted on MAA/BA/MMA copolymer: an additive for hygienic coating. Progress in Organic Coatings 2012;74(3) 582–588.

[117] Hochmannova L., Vytrasova. Photocatalytic and antimicrobial effects of interior paints. Progress in Organic Coatings. 2010;67(1) 1–5.

[118] Arya S.K., Saba S., Ramirez-Vick J.E., Gupta V., Bhansali S., Singh S.P. Recent advances in ZnO nanostructures and thin films for biosensor applications: review. Analytica Chimica Acta. 2012;737 1–21.

[119] Loh L., Dunn S. Recent progress in ZnO-based nanostructured ceramics in solar cell applications. Journal of Nanoscience and Nanotechnology 2012;12(11) 8215–8230.

[120] Zhang Y., Yan X., Yang Y., Huang Y., Liao Q., Qi J. Scanning probe study on the piezotronic effect in ZnO nanomaterials and nanodevices. Advanced Materials. 2012;24(34) 4647–4655.

[121] Singh D.P. Synthesis and growth of ZnO nanowires. Science of Advanced Materials. 2010;2 245–272.

[122] Wang Z.L. ZnO nanowire and nanobelt platform for nanotechnology. Materials Science and Engineering: R 2009;64(3–4) 33–71.

[123] Baruah S., Dutta J. Hydrothermal growth of ZnO nanostructures. Science and Technology of Advanced Materials. 2009;10(1) 013001.

[124] Wood T.J., Hurst G.A., Schofield W.C.E., Thompson R.L., Oswald G., Evans J.S.O., Sharples G.J., Pearson C., Petty M.C., Badyal J.P.S. Electroless deposition of multifunctional zinc oxide surfaces displaying photoconductive, superhydrophobic, photowetting, and antibacterial properties. Journal of Materials Chemistry. 2012;22(9) 3859–3867.

[125] Lee M., Kwak G., Yong K. Wettability control of ZnO nanoparticles for universal applications. ACS Applied Materials and Interfaces. 2011;3(9) 3350–3356.

[126] Lee S. Developing UV-protective textiles based on electrospun zinc oxide nanocomposite fibers. Fibers and Polymers. 2009;10(3) 295–301.

[127] Zhang J., Huang W., Han Y. Wettability of zinc oxide surfaces with controllable structures. Langmuir 2006;22(7) 2946–2950.

[128] Maheswari R., editor. Fungi: Experimental Methods in Biology. Boca Raton: CRC press; 2011.

[129] Kisko K., Szilvay G.R., Vuorimaa E., Lemmetyinen H., Linder M.B., Torkkeli M., Serimaa R. Self-assembled films of hydrophobin proteins HFBI and HFBII studied in situ at the air/water interface. Langmuir 2009;25(3) 1612–1619.

[130] Cox P.W., Hooley P. Hydrophobins: new prospects for biotechnology. Fungal Biology Reviews. 2009;23(1–2) 40–47.

[131] Liu J., Liu C., Liu Y., Chen M., Hu Y., Yang Z. Study on the grafting of chitosan–gelatin microcapsules onto cotton fabrics and its antibacterial effect. Colloids and Surfaces B: Biointerfaces. 2013;109 103–108.

[132] Perelshtein I., Ruderman E., Perkas N., Tzanov T., Beddow J., Joyce E., Mason T.J., Blanes M., Molla K., Patlolla A., Frenkele A.I., Gedanken A. Chitosan and chitosan-ZnO-based complex nanoparticles: formation, characterization, and antibacterial activity. Journal of Materials Chemistry B 2013;1(14) 1968–1976.

[133] Zemljic L.F., Sauperl O., Kreze T., Strnad S. Characterization of regenerated cellulose fibers antimicrobial functionalized by chitosan. Textile Research Journal 2013;83(2) 185–196.

[134] Nithyakalyani D., Ramachandran T., Rajendran R., Mahalakshmi M. Assessment of antibacterial activity of herbal finished surface modified polypropylene nonwoven fabric against bacterial pathogens of wound. Journal of Applied Polymer Science. 2013; 129(2) 672–681.

[135] Kerkeni A., Behary N., Dhulster P., Chihib N.E., Perwuelz A. Study on the effect of plasma treatment of woven polyester fabrics with respect to nisin adsorption and antibacterial activity. Journal of Applied Polymer Science. 2013;129(2) 866–873.

[136] Gao Y., Yu X., Pierlot A.P., Denning R.J., Cranston R. A simultaneous antimicrobial and shrink resistance treatment of wool woven fabrics using the polymeric biocide polyhexamethylene biguanide. Journal of Materials Science. 2011;46(9) 3020–3026.

[137] Gowri V.S., Almeida L., de Amorim M.T.P., Pacheco N.C., Souto A.P., Esteves M.F., Sanghi S.K. Functional finishing of polyamide fabrics using ZnO-PMMA nanocomposites. Journal of Materials Science. 2010;45(9) 2427–2435.

[138] Subkowski T., Karos M., Subkowski T. Industrial performance proteins: hydrophobin-learning from nature. Journal of Biotechnology. 2007;131(2) S212-S213.

[139] Yang J., Zhang Z., Men X., Xu X., Zhu X. Reversible superhydrophobicity to superhydrophilicity switching of a carbon nanotube film via alternation of UV irradiation and dark storage. Langmuir. 2010;26(12) 10198–10202.

[140] Papadopoulou E.L., Zorba V., Pagkozidis A., Barberoglou M., Stratakis E., Fotakis C. Reversible wettability of ZnO nanostructured thin films prepared by pulsed laser deposition. Thin Solid Films. 2009;518(4) 1267–1270.

[141] Popescu AC, Dorcioman G, Duta L, Mihailescu IN, Stan GE, Pasuk I, Zgura I, Beica T, Enculescu I, Ianculescu A, Dumitrescu I. Radical modification of the wetting behavior of textiles coated with ZnO thin films and nanoparticles when changing the ambient pressure in the pulsed laser deposition process. Journal of Applied Physics 2011;110(6) 064321.

[142] Bhushan B., Jung Y.C. Natural and biomimetic artificial surfaces for superhydrophobicity, self-cleaning, low adhesion, and drag reduction. Progress in Materials Science. 2011;56(1) 1–108.

[143] Popescu A.C., Stan G.E., Duta L., Dorcioman G., Iordache O., Dumitrescu I., Pasuk I., Mihailescu I.N. Influence of a hydrophobin underlayer on the structuring and antimicrobial properties of ZnO films. Journal of Materials Science. 2013;48(23) 8329–8336.

[144] Schaaf P. (ed.), Laser Processing of Materials. Berlin: Springer; 2011.

[145] Janssen M.I., van Leeuwen M.B.M., Scholtmeijer K., van Kooten T.G., Dijkhuizen L., Wosten H.A.B. Coating with genetic engineered hydrophobin promotes growth of fibroblasts on a hydrophobic solid. Biomaterials. 2002;23(24) 4847–4854.

[146] Boeuf S., Throm T., Gutt B., Strunk T., Hoffmann M., Seebach E., Muhlberg L., Brocher J., Gotterbarm T., Wenzel W., Fischer R., Richter W. Engineering hydrophobin DewA to generate surfaces that enhance adhesion of human but not bacterial cells. Acta Biomaterialia. 2012;8(3) 1037–1047.

[147] Preda N, Enculescu M, Zgura I, Socol M, Matei E, Vasilache V, Enculescu I. Superhydrophobic properties of cotton fabrics functionalized with ZnO by electroless deposition. Materials Chemistry and Physics 2013;138(1) 253–261.

[148] Carp O., Huisman C.L., Reller A. Photoinduced reactivity of titanium dioxide. Progress in Solid State Chemistry. 2004;32(1–2) 33–177.

[149] Hashimoto K., Irie H., Fujishima A. TiO_2 photocatalysis: a historical overview and future prospects. Japanese Journal of Applied Physics. 2005;44(12) 8269–8285.

[150] I. Zgura, S. Frunza, M. Enculescu, C. Florica, F. Cotorobai. Deposition of titanium dioxide layers upon polyester textile materials: checking the adherence by ultra-sonication. Romanian Journal of Physics. Accepted June 2015.

[151] Voinea M., Vladuta C., Bogatu C., Duta A. Surface properties of copper based cermet materials. Materials Science and Engineering: B. 2008;152(1–3) 76–80.

[152] Zgura I, Beica T, Mitrofan IL, Mateias CG, Parvu D, Patrascu I. Assessment of the impression materials by investigation of the hydrophilicity. Digest Journal of Nanomaterials and Biostructures 2010;5(3) 749–755.

[153] Ashraf M., Campagne C., Perwuelz A., Champagne P., Leriche A., Courtois C. Development of superhydrophilic and superhydrophobic polyester fabric by growing zinc oxide nanorods. Journal of Colloid and Interface Science. 2013;394 545–553.

[154] Erbil H.Y., Cansoy C.E. Range of applicability of the Wenzel and Cassie–Baxter equations for superhydrophobic surfaces. Langmuir. 2009;25(24) 14135–14145.

[155] Myint M.T.Z., Kitsomboonloha R., Baruah S., Dutta J. Superhydrophobic surfaces using selected zinc oxide microrod growth on ink-jetted patterns. Journal of Colloid and Interface Science. 2011;354(2) 810–815.

[156] Howells D.G., Henry B.M., Madocks J., Assender H.E. High quality plasma enhanced chemical vapour deposited silicon oxide gas barrier coatings on polyester films. Thin Solid Films. 2008;516(10) 3081–3088.

[157] Textor T., Mahltig B. A sol-gel based surface treatment for preparation of water repellent antistatic textiles. Applied Surface Science. 2010;256(6) 1668–1674.

[158] Alongi J., Tata J., Frache A. Hydrotalcite and nanometric silica as finishing additives to enhance the thermal stability and flame retardancy of cotton. Cellulose 2011;18(1) 179–190.

[159] Beica T., Frunza S., Zgura I., Frunza L., Cotarlan C., Negrila C., Vlaicu A.M., Zaharia C.N. Nanostructured gold layers. I. Deposition by Vacuum evaporation at small angles of incidence. Journal of Optoelectronics and Advanced Materials. 2010;12(2) 347–353.

[160] Zgura I., Beica T., Frunza S., Rasoga O., Galca A., Frunza L., Moldovan A., Dinescu M., Zaharia C. Nanostructured gold layers. II. Gold deposition onto polystyrene substrate. Journal of Optoelectronics and Advanced Materials. 2010;12(2) 354–359.

[161] Zgura I., Beica T., Frunza S., Frunza L., Cotirlan-Simioniuc C., Ungureanu F., Gheorghe N., Rasoga O., Velula T., Zaharia C. Nanostructured gold layers. III. Functionalization of gold layers obliquely deposited onto polystyrene substrate. Journal of Optoelectronics and Advanced Materials. 2010;12(8) 1729–1738.

[162] Zgura I., Beica T., Frunza S., Frunza L., Ganea P., Ungureanu F., Negrila C., Nuta A., Sorescu A.-A., Bunea I., Zaharia C.N. Nanostructured SiOx layers as substrates for liquid crystal local ordering. I. Functionalization to bind bovine serum albumin. Optoelectronics and Advanced Materials—Rapid Communications. 2011;5(3) 318–323.

[163] Frunza L., Zgura I., Enculescu M., Frunza S., Ganea C.P., Rasoga O., Cotorobai F., Dorogan A. Polyester fabrics coated with SiOx nanoparticles by vacuum deposition at small angle. Structural characterization and wetting properties. Journal of Optoelectronics and Advanced Materials. 2014;16(1–2) 176–181.

[164] Dai D., Fan M. Characteristic and performance of elementary hemp fibre. Materials Sciences and Applications. 2010;1(6) 336–342.

[165] Zgura I., Moldovan R., Negrila C.C., Frunza S., Cotorobai V.F., Frunza L. Surface free energy of smooth and dehydroxylated fused quartz from contact angle measurements using some particular organics as probe liquids. Journal of Optoelectronics and Advanced Materials 2013;15(7–8) 627–634.

[166] Della Volpe C., Siboni S. In: Mittal K.L. (ed.). Acid–Base Interactions: Relevance to Adhesion Science and Technology. Utrecht: VSP BV; 2000. p55–p90.

[167] Pashley R.M., Kitchener J.A.J. Surface forces in adsorbed multilayers of water on quartz. Colloid Interface Science 1979;71(3) 491–500.

[168] Lamb R.N., Furlong D.N. Controlled wettability of quartz surfaces. Journal of the Chemical Society, Faraday Transactions 1982;78(1) 61–73.

[169] Beica T., Moldovan R., Zgura I., Frunza S., Poterasu M. Mechanism of homeotropic alignment of a doped liquid crystal. Journal of Optoelectronics and Advanced Materials 2006;8(4) 1512–1515.

[170] Kuo H.-F., Lien D.-H., Hsu W.K. Nondipolar liquid migration on carbon nanotube films at low field. Applied Physics Letters 2006;89(4) 044109.

[171] Carré A. Polar interactions at liquid/polymer interfaces. Journal of Adhesion Science and Technology. 2007;21(10) 961–981.

[172] Roy R.K., Lee K.R. Biomedical applications of diamond-like carbon coatings: a review. Journal of Biomedical Materials Research Part B: Applied Biomaterials. 2007;83(1) 72–84.

[173] Ostrikov K., Cvelbar U., Murphy A.B. Plasma nanoscience: setting directions, tackling grand challenges. Journal of Physics D: Applied Physics. 2011;44(17) 174001.

[174] Fedel M., Motta A., Maniglio D., Migliaresi C. Surface properties and blood compatibility of commercially available diamond-like carbon coatings for cardiovascular devices. Journal of Biomedical Materials Research Part B: Applied Biomaterials. 2009;90B(1) 338–349.

[175] Kocourek T., Jelınek M., Vorlıcek V., Zemek J., Janca T., Zızkova V., Podlaha J., Popov C. DLC coating of textile blood vessels using PLD. Applied Physics A. 2008;93(3) 627–632.

[176] Kwok S.C.H., Wang J., Chu P.K. Surface energy, wettability, and blood compatibility phosphorus doped diamond-like carbon films. Diamond and Related Materials. 2005;14(1) 78–85.

[177] Ma W.J., Ruys A.J., Mason R.S., Martin P.H., Bendavid A., Liu Z.W., Ionescu M., Zreiqat H. DLC coatings: effects of physical and chemical properties on biological response. Biomaterials. 2007;28(9) 1620–1628.

[178] Jones M.I., McColl I.R., Grant D.M., Parker K.G., Parker T.L. Protein adsorption and platelet attachment and activation on TiN, TiC and DLC coatings on titanium for cardiovascular applications. Journal of Biomedical Materials Research. 2000;52(2) 413–421.

[179] Okpalugo T.I.T., Ogwu A.A., Maguire P.D., McLaughlin J.A.D. Platelet adhesion on silicon modified hydrogenated amorphous carbon films. Biomaterials. 2004;25(2) 239–245.

[180] Yan X.B., Xu T., Yang S.R., Liu H.W., Xue Q.J. Characterization of hydrogenated diamond-like carbon films electrochemically deposited on a silicon substrate. Journal of Physics D: Applied Physics. 2004;37(17) 2416–2424.

[181] Okpalugo T.I.T., Papakonstantinou P., Murphy H., McLaughlin J., Brown N.M.D. High resolution XPS characterization of chemical functionalized MWCNTs and SWCNTs. Carbon. 2005;43(1) 153–161.

[182] Liao J.X., Liu W.M., Xu T., Xue Q.J. Characteristics of carbon films prepared by plasma-based ion implantation. Carbon. 2004;42(2) 387–393.

[183] Bugaev S.P., Korotaev A.D., Oskomov K.V., Sochugov N.S. Properties of diamond-like films obtained in a barrier discharge at atmospheric pressure. Technical Physics. 1997;42(8) 945–949.

[184] Riccardi C., Barni R., Fontanesi M., Tosi P. Gaseous precursors of diamond-like carbon films in CH4/Ar plasmas. Chemical Physics Letters. 2000;329(1–2) 66–70.

[185] Valentini L., Kenny J.M., Mariotto G., Tosi P. Deposition of hydrogenated amorphous carbon films from CH4/Ar plasmas: Ar dilution effects. Journal of Materials Science. 2001;36(21) 5295–5300.

[186] Owens D.K., Wendt R.C. Estimation of the surface free energy of polymers. Journal of Applied Polymer Science. 1969;13(8) 1741–1747.

[187] Roy R.K., Choi H.W., Yi J.W., Moon M.W., Lee K.R., Han D.K., Shin J.H., Kamijo A., Hasebe T. Hemocompatibility of surface-modified, silicon-incorporated, diamond-like carbon films. Acta Biomaterialia. 2009;5(1) 249–256.

[188] Liu Y., Li Z., He Z., Chen D., Pan S. Structure and blood compatibility of tetrahedral amorphous hydrogenated carbon formed by a magnetic-field-filter plasma stream. Surface and Coatings Technology. 2007;201(15) 6851–6856.

Chapter 9

Wetting Behavior of Dental Implants

In-Hye Kim, Tae-Yup Kwon and Kyo-Han Kim

Additional information is available at the end of the chapter

http://dx.doi.org/10.5772/61098

Abstract

Titanium (Ti) and titanium alloys are widely used in biomedical devices and components, because of their desirable properties, such as relatively low modulus, good fatigue strength, formability, machinability, corrosion resistance, and biocompatibility. However, Ti and its alloys cannot meet all of the clinical requirements. Therefore, surface modification of Ti has been often performed to improve the biological, chemical, and mechanical properties. Various modifications of surface properties have been investigated to predictably improve the osseointegration of Ti implants. The rate and quality of osseointegration in Ti implants are related to their surface properties. A multiplicity of implant surface forms exist engineered with mechanical features that physically interlock the implant with bone. Various strategies have been utilized to improve bone integration of Ti-based implants. For example, surface grit blasting, acid-etching and anodization methods enhance cell growth, improving implant fixation through increases in interlocking surface area and alterations of oxide thickness. On the other hand, surface composition and hydrophilicity are parameters that may play a role in implant-tissue interaction and osseointegration. Highly hydrophilic surfaces seem more desirable than hydrophobic ones in view of their interactions with biological fluids, cells and tissues. Several recent studies have shown that the surface energy of biomaterials strongly has influence the initial cell attachment and spreading of osteoblastic cells on the biomaterial surfaces. Hallab et al. said that surface energy might be a more important determinant of cell adhesion and proliferation, and might be more useful than surface roughness for generating cell adhesion and cell. It may have the influence on protein adsorption and the structural rearrangement of the proteins on the material. Therefore, understanding the relationship between surface energy and cell adhesion on different biomaterials will facilitate the design of optimized implant material surfaces and subsequently the cell attachment. Surface energy is an important parameter of the material surface. It is affected by several surface characteristics, such as chemical composition, surface charge, and microstructural topography. Many papers reported that surface energy is one of important surface characteristics parameter of modified titanium surfaces. Given the importance of surface wettability of dental implants surfaces in the achievement of osseointegration, the surface free energy values for a given material, obtained by various methods and with use of different measuring liquids, are not consistent. Thus, we provided a review article of the surface modification on titanium surface and the surface wettability. The relationship between CAs and surface preparations was determined in this review.

Keywords: surface free energy, contact angle, dental implant

1. Introduction

Titanium (Ti) and titanium alloys are widely used in biomedical devices and components because of their desirable properties, such as relatively low modulus, good fatigue strength, formability, machinability, corrosion resistance, and biocompatibility. [1] However, Ti and its alloys cannot meet all of the clinical requirements. Therefore, the surface modification of Ti has been often performed to improve the biological, chemical, and mechanical properties. [2] Various modifications of surface properties have been investigated to significantly improve the osseointegration of Ti implants. [3] The rate and the quality of osseointegration in Ti implants are related to their surface properties. A multiplicity of implant surface forms exists, which are engineered with mechanical features that physically interlock the implant with bone. Various strategies have been implemented to improve bone integration of Ti-based implants. [4, 6] For example, surface grit blasting, acid etching, and anodization methods enhance cell growth, improving implant fixation and thereby increasing interlocking surface area and altering oxide thickness.

On the other hand, surface composition and hydrophilicity are parameters that play an important role in implant–tissue interaction and osseointegration. [7] Radiofrequency glow discharge has been implemented to increase surface energy and to enhance cell binding. Highly hydrophilic surfaces seem more desirable than hydrophobic ones in view of their interactions with biological fluids, cells, and tissues. [8] Recent studies have shown that the surface energy of biomaterials strongly has influence on the initial cell attachment and spreading of osteoblastic cells on the biomaterial surfaces. [9, 10] Hallab et al. [11] suggested that surface energy may be a more important determinant of cell adhesion and proliferation and may be more useful than surface roughness for generating cells. It may have the influence on protein adsorption and the structural rearrangement of the proteins on the material. Therefore, understanding the relationship between surface energy and cell response on different biomaterials will facilitate the design of optimized implant surfaces and subsequently enhance cell responses. [12]

http://dx.doi.org/10.5772/61098

Surface energy is an important parameter of the material surface. [12] It is affected by several surface characteristics, such as chemical composition, surface charge, and microstructural topography. [13, 14] It has been reported that surface energy is one of important surface characteristic of modified titanium surfaces. [7, 15, 17] The relationship among surface factors, including surface roughness, surface energy, contact angle (CA) values, and cell adhesion to biomaterial surfaces, is presented in Figure 1. Each relationship was supported by a number of studies, which are referenced in the diagram. The understanding of surface factors, cell adhesion, and their relationships is mandatory for better understanding of the bone-implant interface.

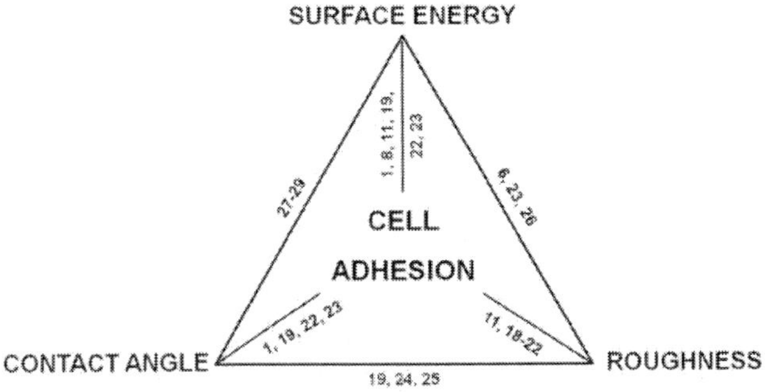

Figure 1. The relationship among contact angle, surface preparation, surface energy, roughness, and cell adhesion.

Given the importance of surface wettability of dental implants in the achievement of osseointegration, the surface free energy (SFE) values for a given material obtained by various methods using different measuring liquids are not consistent. Thus, the current review article deals with the relationship between CAs and surface preparations.

2. Surface Free Energy (SFE) and Contact Angle (CA)

Surface free energy (SFE) is defined as the work required for increasing the surface area of a substance per unit area. SFE is induced from the unfulfilled bonding potential of molecules at a surface,. These are different molecules within a material, which have less energy because they are affected by interactions with like molecules in all directions. Moreover, SFE is dictated by the surface roughness, topography, and composition of the implant and is crucial in determining which proteins are absorbed onto the surface. Surface energy, which is intimately related to wettability, [18] is a useful quantity that has often correlated strongly with biological interaction. Thus, it is usually reported that biomaterial surfaces with moderate hydrophilicity improves cell growth and higher biocompatibility. [19] This points out to the existence of a range of optimal surface energies. [20]

SFE can be determined by measuring the contact angle formed by a range of liquids on a given surface, using several diverse approaches. [21] The most useful methods for characterizing wettability on solid surfaces are static CA measurements. CA measurements are quantifiable, readily acquired using relatively low-cost instruments and simple procedures, amenable for use in environments from academic research laboratories to industrial manufacturing facilities, and an extremely powerful method for characterizing surfaces. A drop of liquids in contact with a surface will display a contact angle, traditionally measured through the liquid. Thus, examination of wetting behavior draws the conclusion that a liquid usually shows a wide range of angles on a measured solid surface. [22] The physical surface properties and the surface energy can be quantified by the wettability and by the CA of liquids with the surface. [23] The values of the CA indicate whether the surface is hydrophilic or hydrophobic. [24] Several authors have suggested that CA measurements give values ranging from 0° (hydrophilic) to 140° (hydrophobic) for titanium implant surfaces. [25, 26]

3. Calculations of surface free energy

The modified form of the Young equation is as follows:

$$\gamma_s = \gamma_{sl} + \gamma_l \cos\Theta \tag{1}$$

where Θ is the contact angle and γ_s, γ_{sl}, and γ_l are the surface tensions of the solid, solid–liquid, and liquid surfaces, respectively. The quantities γ_l and Θ, appearing in Eq. 1, can be measured. However, the quantity γ_{sl} remains unknown. The effect of adsorption of the measuring liquid on the surface of a solid should also be taken into account. Therefore, some additional assumptions concerning the relations between γ_s, γ_l, and γ_{sl} need to be made in order to solve Eq. 1.

The idea of the partition of the surface free energy into singular components includes the assumption that the quantity γ_{sl} is determined by dissimilar interfacial interactions that rely on the properties of both the measuring liquid and the γ_{sl} of the solid. Fowkes [27] assumed that the SFE of a solid (and of a liquid) is a sum of independent components, associated with specific interactions:

$$\gamma_s = \gamma_s^{d} + \gamma_s^{p} + \gamma_s^{h} + \gamma_s^{i} + \gamma_s^{ab} + \gamma_s^{o} \tag{2}$$

where γ_s^{p}, γ_s^{d}, γ_s^{ab}, γ_s^{h}, and γ_s^{i} are the polar, dispersion, acid–base components, hydrogen (related to hydrogen bonds), and inductions, respectively, while γ_s^{o} refers to all interactions. Moreover, the dispersion component of the surface free energy is related with the London force interactions, resulting from the electron dipole instability according to the theory. These interactions occur normally between neighboring atoms and molecules. The forces depend on the kind of similarly attracting elements of the matter but are independent of other types of interactions. The remaining van der Waals interactions have been regarded by Fowkes [27] as a division of the generation interactions. He investigated a solid or liquid in which the dispersion interactions appear. Considering such systems, he found out the surface free energy corresponding to the interface of a solid and liquid as follows:

$$\gamma_{sl} = \gamma_s + \gamma_l - 2\left(\gamma_s^{d}\gamma_l^{d}\right)^{0.5} \tag{3}$$

Eq. 3 is limited to the interfacial London interactions.

Owens and Wendt [28] changed the Fowkes idea while assuming that the sum of all the components occurring on the right-hand side of Eq. 2, except γ_s^d, can be considered as associated with the polar interaction (γ_s^p). Thus, they suggested the following equation:

$$\gamma_{sl} = \gamma_s + \gamma_l - 2\left(\gamma_s^{d}\gamma_l^{d}\right)^{0.5} - 2\left(\gamma_s^{p}\gamma_l^{p}\right)^{0.5} \tag{4}$$

Because the polar interaction definition by Fowkes [27] differs from that by Owens and Wendt, the meanings of γ_s^p and γ_l^p in Eq. 2 are different than those in Eq. 4.

The latest idea of the partition of surface free energy of solids and liquids into components is that presented by van Oss, Chaudhury, and Good. [29] The authors separated γ_s into two components, one containing the long-range interactions (London, Debye, and Keesom) called the Lifshitz–van der Waals component (γ^{LW}) and the other that includes the short-range interactions called the acid–base component (γ^{AB}). The latter component associated with the acid–base interactions is equal $2(\gamma^+\gamma^-)^{0.5}$, where γ^+ and γ^- mean the acidic and basic components, respectively. Consequently, the following relationship was created:

$$\gamma_{sl} = \left\{\left(\gamma_s^{LW}\right)^{0.5} - \left(\gamma_l^{LW}\right)^{0.5}\right\}2 + 2\left\{\left(\gamma_s^{+}\right)^{0.5} - \left(\gamma_l^{+}\right)^{0.5}\right\}\cdot\left\{\left(\gamma_s^{-}\right)^{0.5} - \left(\gamma_l^{-}\right)^{0.5}\right\} \tag{5}$$

4. Owens–Wendt method

In the Owens–Wendt method, [30] they made the assumptions similar to those in the Fowkes method. The two methods, being identical in the mathematical aspect, differ slightly in the way of calculating the surface free energy. The combination of Eq. 1 with Eq. 4 leads to the following relationship:

$$\left(\gamma_s^{d}\gamma_l^{d}\right)^{0.5} + \left(\gamma_s^{p}\gamma_l^{p}\right)^{0.5} = \gamma_l\left(1 + \cos\Theta\right) \tag{6}$$

5. van Oss–Chaudhury–Good method

The component γ^{ab} is equal $2(\gamma^{+}\gamma^{-})^{0.5}$ and combining Eq. 1 with Eq. 5, van Oss, Chaudhury, and Good obtained the following relationship: [30]

$$\left(\gamma_s^{LW}\gamma_l^{LW}\right)^{0.5}+\left(\gamma_s^{+}\gamma_l^{-}\right)^{0.5}+\left(\gamma_s^{-}\gamma_l^{+}\right)^{0.5}=0.5\left(1+\cos\Theta\right) \tag{7}$$

Since three unknowns, γ_s^{LW}, γ_s^{+}, and γ_s^{-}, appear in Eq. 7, the solution of a system of three independent linear equations is needed to establish these quantities. When three different liquids are used to measure the contact angle of a material, such a system is obtained. Moreover, two bipolar and one non polar liquid should form the set of the three measuring liquids. The values of the coefficients appearing in such a scheme of equations have been given somewhere else. The key of the scheme of three equations shown, as used in the van Oss–Chaudhury–Good method, cannot always be appropriate and undoubtedly interpreted. This follows from the presumed conditions and limitations, related with both the selected measuring liquids and the methods of determination of the surface free energy components such as γ_l^{LW}, γ_l^{+}, and γ_l^{-}.

6. Methods based on determination of the contact angle hysteresis

This approach is one of the latest methods for calculating the SFE of polymeric materials. [31, 32] It consists of the measurements of both the advancing CA (Θ_a) and the receding one (Θ_r) by using the same measuring liquid of a known value of r_l. The surface free energy of a tested solid can be calculated from the following equation:

$$r_s=r_l\left(\cos\Theta_r-\cos\Theta_a\right)\left\{\left(1+\cos\Theta_a\right)2/\left[\left(1+\cos\Theta_r\right)2-\left(1+\cos\Theta_a\right)2\right]\right\} \tag{8}$$

Unlike the approaches presented above, Eq. 8 takes into account adsorption at the interface. The contact angle appearing in Eq. 1 is the advancing contact angle. Thus, this equation transforms into the following one:

$$r_s=r_{sl}+r_l\cos\Theta_a \tag{9}$$

The SFE of a solid (r_{sf}), which considers adsorption occurring during the measurement of Θ_r, can be expressed by the following relationship:

$$r_{sf}=r_{sl}+r_l\cos\Theta_r \tag{10}$$

http://dx.doi.org/10.5772/61098

The following relation is valid:

$$r_{sf} = r_{sl} + \pi \tag{11}$$

where π is the equilibrium pressure of the surfaces of the measuring liquid. The adhesion can be determined from the following equation:

$$W_{sl} = r_s + r_l - r_{sl} \tag{12}$$

in which Θ_a or Θ_r is used, depending on the kind of the interfacial system.

When applying the Young and Dupre equations, the parameter Φ defined by Girifalco and Good as well as making suitable substitutions. While finding this relationship, its authors neither asked nor confirmed the basics of the knowledge in this area. Finding new relations are unquestionable contribution of the authors of this method.

The determination of the polymeric material surface energy with the use of Eq. 8 needs the measurements of Θ_a and Θ_r and the information of r_l of the measuring liquid. Nonetheless, as the authors of the method highlighted, the calculated values of the surface free energy rely on the tested measuring liquid. Therefore, they verify the results of studies regarding other methods for calculating the surface free energy of polymeric materials. [33, 34]

7. Analysis methods of wetting behavior of different dental implant surfaces

There are a number of techniques to measure the contact angle of a liquid on a substrate, including optical reflectometry, contrast interferometry, capillary rise technique, Wilhelmy plate tensiometry, and various goniometric methods (Table 1).

Surface modifications	Conditions of measurement				Ref.
	Time	Liquid	Method	Surface energy calculation	
Calcium phosphate coating	Drop diameter from 5 to 10 mm	Water, glycerol	Static drop	Owens and Wendt	[35]
Nitric acid etching	Drop diameter from 5 to 10 mm	Water, glycerol	Static drop	Owens and Wendt	[35]
Acid etching	10 mm/min	Water, fibronectin solutions	Wilhelmy plate	–	[6]

Surface modifications	Conditions of measurement				Ref.
	Time	Liquid	Method	Surface energy calculation	
Thermal oxidation	N/A	Benzylethanol, diiodmethane, formamide, and water	Static drop	Owens and Wendt	[3, 6]
Physical vapor deposition	N/A	Water, Dulbecco's modified Eagle's medium	Dynamic contact angle	–	[37]
Acid, sandblasted, and anodized treatment	N/A	Water, NaCl, DMSO, and human blood	Static drop	–	[38]
Acid, blasted, and blasted + etched	1–2 s	Water	Sessile drop	Wenzel law	[39]
None	N/A	Water, formamide, and diiodomethane	Captive air bubble	van Oss–Good	[40]
Plasma spray	10 s–20 min	Water, diiodomethane	Sessile drop	Owens and Wendt	[5]
Plasma immersion ion implantation	N/A	Glycol, glycerol, water, formamide, methylene iodide, and tricresyl phosphate	Static drop	Owens and Wendt	[41]
RGDS-coated anodized Ti	1 μL/s	Water	Sessile drop	N/A	[42]

Table 1. Contact angle and surface free energy calculation analysis of various treated dental implants on previous studies.

7.1. Static drop method

The most commonly employed technique for measuring the contact angle of drops on liquid repellent surfaces is the sessile drop method coupled with digital image analysis. A liquid drop of a volume (calculated) is silently dropped on the substrate and a camera captured the boundary of the drop. Many imaging analysis algorithms can be utilized to estimate the contact angle from the drop outline, such as rough spherical cap calculations [43] or direct fitting to arithmetical keys of the Young Laplace equation. [44]

7.2. Advancing and receding angle

The advancing contact angles of water and other liquids (diiodomethane, formamide, etc.) were measured after settling 6 μL droplets on the surface. Then after sucking of 2 μL from the droplet into the syringe, the receding contact angle was measured. On the other hand, a drop on a tilted plate is shown schematically, in which the front angle is close to the advancing angle and the rear angle is close to the receding angle on the drop before descending. When the hysteresis is small, the droplet is close to a spherical cap. Moreover, the contact angle passes from the advancing value to the receding one along the contact line. Because these angles are considered to be close to each other, it just is written that the upper half of the droplet makes the angle Θ_r, while the lower half meets the angle Θ_a.

7.3. Captive air bubble method

Although most studies addressing (super)hydrophobic behaviors have so far dealt with the wetting of low surface energy and textured substrates in air environment, the captive air bubble method, the so-called two liquid phase method, is a totally novel system and configuration involving the wetting of highly hydrophilic, textured metallic materials in liquid alkane medium.

7.4. Wilhelmy plate method

The dynamic contact angle measurements are performed on the basis of the Wilhelmy plate technique. The force acting on plates immerse (wetting) and emerge (dewetting) in a liquid is recorded by means of an electrobalance. The hysteresis force loops are used to calculate advancing and receding contact angles (CAs) during immersion and emersion according to the following equation:

$$\cos\Theta = F / L_{\gamma} \tag{13}$$

where Θ is the advancing or receding CA, F is the wetting force, L is the wetted length (sample perimeter), and γ is the surface tension of the wetting liquid. Thus, the CA and the wettability of a solid can be determined from the known surface tension and the measured weight of the liquid meniscus. F/L is the so-called wetting tension and equals to the product of $\cos\Theta$ and γ, which is itself part of the fundamental Young equation for sessile drops in thermodynamic equilibrium. The difference between the advancing and the receding CAs is referred to as CA hysteresis. [45]

The hysteresis force loops are qualitatively described in terms of thermodynamic and kinetic hysteresis. For CA calculations, linear portions of the respective F/L lines are extrapolated to zero immersion depth by linear regression. Before each tensiometry wetting experiment, water surface tension γ was measured by means of the Wilhelmy method using a standard roughened platinum Wilhelmy plate. [46]

8. Influence of surface cleaning on contact angle analyzing

Surface cleaning method has a quantitative and qualitative influence on the results of contact angle (CA) measurements. An author studied the evolution of contact angle values versus the roughness for the three different cleaning methods [47]:

Type 0: water rinsing followed by nitrogen drying

Type I: successive soakings in ultrasonic baths of acetone, cyclohexane, and acetone, followed by water rinsing and nitrogen drying

Type II: "Type I" cleaning followed by an argon plasma cleaning

Based on the study, CAs around 150° was observed with type 0 cleaning, and no trivial correlation with the roughness was found. Type 0 cleaned surfaces are still covered by usual organic contaminants from ambient air, and no significant influence of the roughness on CA values is observed. The contact angle remained even around 140° in the roughness range between 2.5 and 12 μm, before it begin to reduce for *Ra* higher than 12 μm. Finally, type II cleaning strongly decreased contact angle values [48] compared to type I and type 0 ones. When the roughness increased up to a threshold value *Ra* = 10 μm, an increase in contact angle was detected. Above this threshold, the contact angle remained constant at 120°.

It thus clearly appears from these results that the more efficient the surface cleaning, the more strong and measurable the correlation between CA and roughness. To determine the Young equilibrium angle, CAs were measured on mirror polished titanium surfaces in the "two liquid phase" configuration, after the three different cleaning methods.

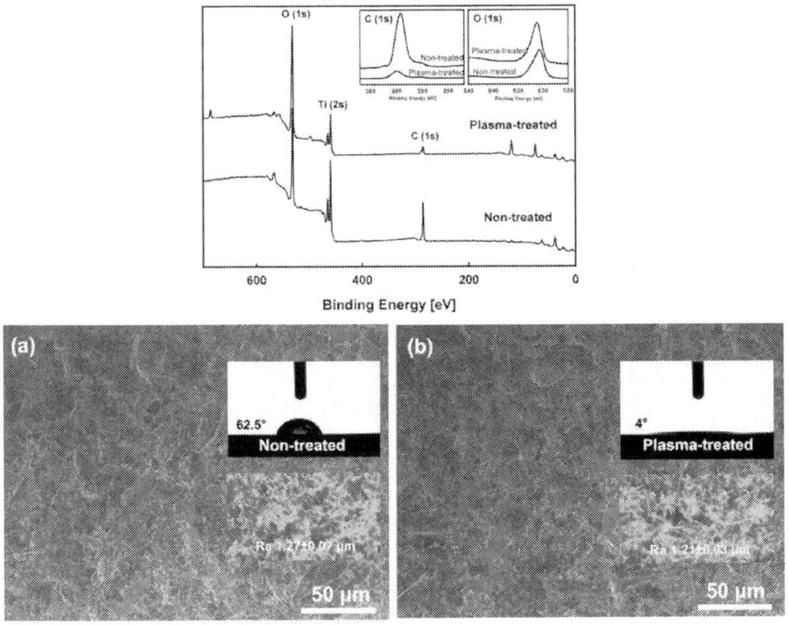

Figure 2. Ti surfaces of the untreated (a) and plasma-treated (b) for cleaning. Reproduced with permission from Kim et al. (49).

Plasma treatment also can be used as a cleaning method. Appropriate plasma processes render the surfaces more hydrophilic and modify the oxide layer. [49] Also, the application of plasma to metal implants can clean the surface of materials as shown in Figure 2. Among the wide range of plasma techniques, atmospheric plasma is one of the simplest and most efficient processes. We studied to evaluate the effects of atmospheric pressure plasma on the Ti surface. In this study, the plasma treatment did not affect the surface roughness. Therefore, atmospheric plasma is a powerful way of creating a functionalized hydrophilic surface of Ti implants as a simple and highly efficient method. An atmospheric plasma treatment has the potential as a surface modifying technique to clean the Ti implant surface.

9. Surface roughness and contact angle

The surface roughness is also an important parameter to be considered. To enhance our understanding of liquids in contact with rough surfaces, a systematic study was carried out in which water contact angle measurements were performed on a wide variety of rough surfaces with precisely controlled surface chemistry.

A uniform surface (Θ^Y) does have a unique value only if perfectly flat. On genuine surfaces, depending on how a drop is deposited, the contact angle can differ from advancing and receding contact angles. This hysteresis can be attributed to inhomogeneities in the division of adsorbents or the existence of contaminants, to surface roughness, or to time-dependent surface reorganizations. [50] The value of Θ is strongly influenced on the surface morphology on rough surfaces. On rough and hydrophobic surfaces, the liquid can either go after the surface topography or show strong pinning or can connect from sharpness to asperity while surrounding air below and presenting almost no hysteresis in the contact angle.

9.1. Wenzel

A roughness factor, describing the roughness influenced on Θ (Eq. 11), was introduced by Wenzel, as follows [51]:

$$\cos\Theta^W = r \cdot \cos\Theta^r \qquad (14)$$

where r is calculated by dividing the actual roughness-enhanced surface area by its projection. This behavior is often referred to as Wenzel-type wetting.

9.2. Cassie–Baxter

For the second case, Cassie and Baxter [52] modified Wenzel's equation by introducing the fractions f_1 and f_2, where f_1 corresponds to the area in contact with the liquid divided by the projected area and f_2 to the area in contact with the air trapped beneath the drop, also divided by the projected area:

$$\cos\Theta^{CB} = f_1 \cos\Theta^{Y} - f_2 \tag{15}$$

Structured surfaces that exhibit superhydrophobicity can also show an effect known as hemiwicking [53] or superwetting if they are surface-chemically functionalized to be hydrophilic. Hemiwicking is a complete wetting due to the presence of capillary forces in two dimensions. [54]

When the surface energy is high, the surface roughness indeed enhances wettability, causing hemiwicking in many cases caused by capillary forces. The pinning of the contact line results in a move to more hydrophobic θ values at lower surface energies. It was found that the surface topography outlines the pinning strength and with it the energy barrier working against the wetting behavior of the drop.

10. Contact angle and roughness of modified Ti implant surfaces

Although the increasing contact angle is in accordance with roughness on nontreated surfaces, modified surfaces have shown different consequences. In physical states, grit-blasting and etching treatments decrease the contact angle of the surfaces, except sandblasting and the acid-etching (SLA) treatment (Table 2). SLA-treated surface has nanosized features. This surface also contains two major roughness scales. The microscale roughness originates from the sandblasting step, leading to troughs. The superimposed nanoscale roughness was created by the acid-etching process. Thus, the apparent contact angle on the SLA surface is hydrophobic. This phenomenon can be explained by pinning the contact line. [48] On the other hand, anodizing treatment makes surface hydrophilic.

	Modifications of surface	Degrees of water contact angle (°)	Surface roughness (μm)	Ref.
Physical state	Nontreated (ground surfaces)	85.2 ± 3.6	0.65	[38] [7]
		76.3 ± 3.0	0.45	
		55.4 ± 4.1	0.26	
		43.0 ± 2.0	0.23	
	Grit-blasting	32.5 ± 3.5	1.64	[55]
	Etching	69.3 ± 3.0	0.37	[38]
		96.2 ± 9.2	0.51	
	Sandblast with large grit and acid etch (SLA)	138.3 ± 4.2	2.40	[6]
		120.1 ± 15.2	3.12	
	Thermal spray	57.4 ± 3.2	1.06	[56]
		0.0	-	
Chemical state	Electro chemical deposition	75.0 ± 1.0	3.50	[37]

	Modifications of surface	Degrees of water contact angle (°)	Surface roughness (μm)	Ref.
		86.0 ± 5.0	2.50	
	Ion implantation	81.5 ± 1.5	-	[57]
	Anodizing	47.25 ± 2.9	0.87	[38]
	Plasma-based fluorine ion implantation	90.0 ± 1.5	-	[58]
	Hydrothermally oxidation	38.5 ± 10.8	-	[12]
Biological state	rhBMP-2-immobilizing	13.7 ± 0.2	-	[59]
	RGDS-coated anodized Ti	29.9 ± 3.3	0.38	[42]
	Heparin and fibronectin adsorption	17.3 ± 3.5	0.31	[60]

Table 2. Surface contact angle and roughness value of modified titanium surfaces.

11. Conclusions

Surface composition and hydrophilicity are parameters that play a major role in implant–tissue interaction and osseointegration. In biological state, interfacial reactions *in vivo* change relevant physical and chemical surface parameters, such as the surface energy, affecting the long-term stability of implants. [61] In addition to surface topography, the properties of implant materials that affect cellular behavior include mechanical rigidity and wettability (SFE). The wettability of the surface plays an important role with respect to protein adsorption, cell attachment, and spreading. [62, 63] In some recent works, surfaces with a high surface free energy are reported to be more adhesive than those with a low surface free energy. [36] Thus, the understanding of surface factors, particularly surface wettability, is mandatory for better understanding of the bone implant biomaterial interface.

Author details

In-Hye Kim[1*], Tae-Yup Kwon[2,3] and Kyo-Han Kim[2,3]

*Address all correspondence to: hotsoul7@knu.ac.kr

1 Department of Dental Science, Graduate School of Kyungpook National University, Daegu, Korea

2 Department of Dental Biomaterials, School of Dentistry, Kyungpook National University, Daegu, Korea

3 Institute for Biomaterials Research and Development Dental Materials Testing and Evaluation Center, Kyungpook National University, Daegu, Korea

References

[1] den Braber ET, de Ruijter JE, Smits HT, Ginsel LA, von Recum AF, Jansen JA. Effect of parallel surface microgrooves and surface energy on cell growth. *Journal of Biomedical Materials Research* 1995;29:511–518.

[2] Rautray TR, Narayanan R, Kwon TY, Kim KH. Surface modification of titanium and titanium alloys by ion implantation. *Journal of Biomedical Materials Research Part B: Applied Biomaterials* 2010;93:581–591.

[3] Schwarz F, Wieland M, Schwartz Z. Potential of chemically modified hydrophilic surface characteristics to support tissue integration of titanium dental implants. *Journal of Biomedical Materials Research Part B: Applied Biomaterials* 2009;88:544–557.

[4] MacDonald DE, Rapuano BE, Deo N, Stranick M, Somasundaran P, Boskey AL. Thermal and chemical modification of titanium–aluminum–vanadium implant materials: effects on surface properties, glycoprotein adsorption, and MG63 cell attachment. *Biomaterials* 2004;25:3135–3146.

[5] Ferraz MP, Monteiro FJ, Serro AP, Saramago B, Gibson IR, Santos JD. Effect of chemical composition on hydrophobicity and zeta potential of plasma sprayed HA/CaO-P2O5 glass coatings. *Biomaterials* 2001;22:3105–3112.

[6] Rupp F, Scheideler L, Rehbein D, Axmann D, Gels-Gerstorfer J. Roughness induced dynamic changes of wettability of acid etched titanium implant modifications. *Biomaterials* 2004; 25: 1429–1438.

[7] Le Guehennec L, Soueidan A, Layrolle P, Amouriq Y. Surface treatments of titanium dental implants for rapid osseointegration. *Dental Materials* 2007;23:844–854.

[8] Zhao G, Schwartz Z, Wieland M. High surface energy enhances cell response to titanium substrate microstructure. *Journal of Biomedical Materials Research Part A* 2005;74A:49–58.

[9] Redey SA, Nardin M, Bernache-Assolant D. Behavior of human osteoblastic cells on stoichiometric hydroxyapatite and type A carbonate apatite: role of surface energy. *Journal of Biomedical Materials Research* 2000;50:353–364.

[10] Kennedy SB, Washburn NR, Simon CG, Jr., Amis EJ. Combinatorial screen of the effect of surface energy on fibronectin-mediated osteoblast adhesion, spreading and proliferation. *Biomaterials* 2006;27:3817–3824.

[11] Hallab NJ, Bundy KJ, O'Connor K, Moses RL, Jacobs JJ. Evaluation of metallic and polymeric biomaterial surface energy and surface roughness characteristics for directed cell adhesion. *Tissue Engineering* 2001;7:55–71.

[12] Zhang YM, Bataillon-Linez P, Huang P. Surface analyses of micro-arc oxidized and hydrothermally treated titanium and effect on osteoblast behavior. *Journal of Biomedical Materials Research Part A* 2004;68A:383–391.

[13] Barthlott W, Neinhuis C. Purity of the sacred lotus, or escape from contamination in biological surfaces. *Planta* 1997;202:1–8.

[14] Neinhuis C, Barthlott W. Characterization and distribution of water-repellent, self-cleaning plant surfaces. *Annals of Botany* 1997;79:667–677.

[15] Zhao G, Schwartz Z, Wieland M. High surface energy enhances cell response to titanium substrate microstructure. *Journal of Biomedical Materials Research Part A* 2005;74:49–58.

[16] Rupp F SL, Olshanska N, de Wild M, Wieland M, Geis-Gerstorfer J. Enhancing surface free energy and hydrophilicity through chemical modification of microstructured titanium implant surfaces. *Journal of Biomedical Materials Research Part A* 2006;76:323–334.

[17] Ponsonnet L, Reybier K, Jaffrezic N. Relationship between surface properties (roughness, wettability) of titanium and titanium alloys and cell behaviour. *Materials Science and Engineering C: Biomimetic and Supramolecular Systems* 2003;23:551–560.

[18] Adamson AW. The physical chemistry of surfaces. *Abstracts of Papers of the American Chemical Society* 2001;221:U320–U320.

[19] Janssen MI, van Leeuwen MB, van Kooten TG, de Vries J, Dijkhuizen L, Wosten HA. Promotion of fibroblast activity by coating with hydrophobins in the beta-sheet end state. *Biomaterials* 2004;25:2731–2739.

[20] Baier RE. Surface behaviour of biomaterials: the theta surface for biocompatibility. *Journal of Materials Science: Materials in Medicine* 2006;17:1057–1062.

[21] Combe EC, Owen BA, Hodges JS. A protocol for determining the surface free energy of dental materials. *Dental Materials* 2004;20:262–268.

[22] Strobel M, Lyons CS. An essay on contact angle measurements. *Plasma Processes and Polymers* 2011;8:8–13.

[23] Lim YJ OY. Initial contact angle measurements on variously treated dental/medical titanium materials. *Bio-Medical Materials and Engineering* 2001;11:325–341.

[24] Elias CN OY, Lima JH, Muller CA. Relationship between surface properties (roughness, wettability and morphology) of titanium and dental implant removal torque. *Journal of the Mechanical Behavior of Biomedical Materials* 2008;1:234–242.

[25] Bagno A, Di Bello C. Surface treatments and roughness properties of Ti-based biomaterials. *Journal of Materials Science-Materials in Medicine* 2004;15:935–949.

[26] Giavaresi G, Fini M, Cigada A. Mechanical and histomorphometric evaluations of titanium implants with different surface treatments inserted in sheep cortical bone. *Biomaterials* 2003;24:1583–1594.

[27] Fowkes FM. Attractive forces at interfaces. *Industrial and Engineering Chemistry* 1964;56:40–52.

[28] Owens DK. Estimation of the surface free energy of polymers. *Journal of Applied Polymer Science* 1969;13:1741–1747.

[29] van Oss CJ, Good RJ, Chaudhury MK. The role of van der Waals forces and hydrogen bonds in "hydrophobic interactions" between biopolymers and low energy surfaces. *Journal of Colloid and Interface Science* 1986;111:378–390.

[30] van Oss CJ, Chaudhury MK, Good RJ. Interfacial Lifshitz–van der Waals and polar interactions in macroscopic systems. *Chemical Engineering and Technology* 1988;86:927–940.

[31] Chibowski E, Ontiveros-Ortega A, Perea-Carpio R. On the interpretation of contact angle hysteresis. *Journal of Adhesion Science and Technology* 2002;16:1367–1404.

[32] Chibowski E. Surface free energy of a solid from contact angle hysteresis. *Advances in Colloid and Interface Science* 2003;103:149–172.

[33] Zenkiewicz M. Some effects of corona discharge treatment of biaxially-oriented polypropylene film. *Journal of Adhesion* 2001;77:25–41.

[34] Zenkiewicz M. Effects of electron-beam irradiation on wettability and surface free energy of a polypropylene film. *International Journal of Adhesion and Adhesives* 2005;25:61–66.

[35] Yang MW, Lin SY. A method for correcting the contact angle from the theta/2 method. *Colloids and Surfaces A: Physicochemical and Engineering Aspects* 2003;220:199–210.

[36] Stalder AF, Melchior T, Muller M, Sage D, Blu T, Unser M. Low-bond axisymmetric drop shape analysis for surface tension and contact angle measurements of sessile drops. *Colloids and Surfaces A: Physicochemical and Engineering Aspects* 2010;364:72–81.

[37] Extrand CW. A thermodynamic model for contact angle hysteresis. *Journal of Colloid and Interface Science* 1998;207:11–19.

[38] Lu X, Zhao ZF, Leng Y. Biomimetic calcium phosphate coatings on nitric-acid-treated titanium surfaces. *Materials Science and Engineering C: Biomimetic and Supramolecular Systems* 2007;27:700–708.

[39] Groessner-Schreiber B, Neubert A, Muller WD, Hopp M, Griepentrog M, Lange KP. Fibroblast growth on surface-modified dental implants: An in vitro study. *Journal of Biomedical Materials Research Part A* 2003;64A:591–599.

[40] Chien CC, Liu KI, Duh JG, Chang KW, Chung KH. Effect of nitride film coatings on cell compatibility. *Dental Materials* 2008;24:986–993.

[41] Elias CN, Oshida Y, Lima JHC, Muller CA. Relationship between surface properties (roughness, wettability and morphology) of titanium and dental implant removal torque. *Journal of the Mechanical Behavior of Biomedical Materials* 2008;1:234–242.

[42] Rosales-Leal JI, Rodriguez-Valverde MA, Mazzaglia G. Effect of roughness, wettability and morphology of engineered titanium surfaces on osteoblast-like cell adhesion. *Colloids and Surfaces A: Physicochemical and Engineering Aspects* 2010;365:222–229.

[43] Mabboux F, Ponsonnet L, Morrier JJ, Jaffrezic N, Barsotti O. Surface free energy and bacterial retention to saliva-coated dental implant materials—an in vitro study. *Colloids and Surfaces B: Biointerfaces* 2004; 39: 199–205.

[44] J.Y Chen, G.J Wan, Y.X Leng, P Yang HS, J Wang NH. Behavior of cultured human umbilical vein endothelial cells on titanium oxide films fabricated by plasma immersion ion implantation and deposition. *Surface and Coatings Technology* 2004;186:270–276.

[45] Balasundaram G, Shimpi TM, Sanow WR, Storey DM, Kitchell BS, Webster TJ. Molecular plasma deposited peptides on anodized nanotubular titanium: an osteoblast density study. *Journal of Biomedical Materials Research Part A* 2011;98A:192–200.

[46] Giljean S, Bigerelle M, Anselme K, Haidara H. New insights on contact angle/roughness dependence on high surface energy materials. *Applied Surface Science* 2011;257:9631–9638.

[47] Kim MC, Song DK, Shin HS. Surface modification for hydrophilic property of stainless steel treated by atmospheric-pressure plasma jet. *Surface and Coatings Technology* 2003;171: 312–316.

[48] Findenegg GH, Herminghaus S. Wetting: Statics and dynamics. *Current Opinion in Colloid and Interface Science* 1997; 2: 301–307.

[49] Kim IH, Son JS, Kwon TY, Kim KH. Effect of Atmospheric Plasma Treatment to Titanium Surface on Initial Osteoblast-Like Cell Spreading. *Journal of Nanoscience and Nanotechnology* 2015;15: 134–137.

[50] Findenegg GH, Herminghaus S. Wetting: Statics and dynamics. *Current Opinion in Colloid and Interface Science* 1997;2:301–307.

[51] Wenzel RN. Resistance of solid surfaces to wetting by water. *Industrial and Engineering Chemistry* 1936;28:988–994.

[52] Cassie ABD, Baxter S. Wettability of Porous Surfaces. *Transactions of the Faraday Society* 1944;40:0546–0550.

[53] Quere D. Rough ideas on wetting. *Physica A: Statistical Mechanics and its Applications* 2002;313:32–46.

[54] Extrand CW, Moon SI, Hall P, Schmidt D. Superwetting of structured surfaces. *Langmuir* 2007;23:8882–8890.

[55] Bathomarco RV, Solorzano G, Elias CN, Prioli R. Atomic force microscopy analysis of different surface treatments of Ti dental implant surfaces. *Applied Surface Science* 2004;233:29–34.

[56] Shi XL, Nakagawa M, Kawachi G, Xu LL, Ishikawa K. Surface modification of titanium by hydrothermal treatment in Mg-containing solution and early osteoblast responses. *Journal of Materials Science: Materials in Medicine* 2012;23:1281–1290.

[57] Luo Y, Ge S, Jin Z, Fisher J. Effect of surface modification on surface properties and tribological behaviours of titanium alloys. *Proceedings of the Institution of Mechanical Engineers, Part J: Journal of Engineering Tribology* 2009;223:311–316.

[58] Arita NK, Shinonaga Y, Nishino M. Plasma-based fluorine ion implantation into dental materials for inhibition of bacterial adhesion. *Dental Materials Journal* 2006;25:684–692.

[59] Ren XS, Wu YZ, Cheng Y, Ma HW, Wei SC. Fibronectin and bone morphogenetic protein-2-decorated poly (OEGMA-r-HEMA) brushes promote osseointegration of titanium surfaces. *Langmuir* 2011;27:12069–12073.

[60] Li GC, Yang P, Qin W, Maitz MF, Zhou S, Huang N. The effect of coimmobilizing heparin and fibronectin on titanium on hemocompatibility and endothelialization. *Biomaterials* 2011;32:4691–4703.

[61] Rupp F, Scheideler L, Geis-Gerstorfer J. Effect of heterogenic surfaces on contact angle hysteresis: Dynamic contact angle analysis in material sciences. *Chemical Engineering and Technology* 2002;25:877–882.

[62] Grinnell F. Cellular adhesiveness and extracellular substrata. *International Review of Cytology* 1978;53:65–144.

[63] Schakenraad JM, Busscher HJ, Wildevuur CR, Arends J. The influence of substratum surface free energy on growth and spreading of human fibroblasts in the presence and absence of serum proteins. Journal of Biomedical Materials Research 1986;20:773–784.

Chapter 10

Influence of Wettability and Reactivity on Refractory Degradation – Interactions of Molten Iron and Slags with Steelmaking Refractories at 1550°C

R. Khanna, M. Ikram-ul-Haq and V. Sahajwalla

Additional information is available at the end of the chapter

http://dx.doi.org/10.5772/61271

Abstract

Refractories, materials that can withstand high temperatures, play an important role in the iron and steel sector which alone accounts for ~70% of total refractories produced. In this chapter, detailed wettability and interfacial phenomena investigations on alumina-carbon and zirconia-carbon refractories at steelmaking temperatures. The wettability between refractory substrates and molten iron/slags was investigated at 1550°C using the sessile drop approach in a horizontal tube furnace equipped with a CCD camera. Detailed experimental results were obtained on alumina-carbon/molten iron system at high temperatures. Alumina is known to be non-wetting to molten iron while carbon can be easily wetted. Observed contact angles were found to depend strongly on the substrate composition and contact time. While the refractory substrates containing 50 and 60% carbon were found to be non-wetting to molten iron, the substrates containing higher amounts of C (≥ 70%) were found to become increasingly wetting. Molten iron droplets were seen to spread on these substrates.

The wettability of zirconia-carbon and alumina-carbon (C: 10-20 wt %) with two steelmaking slags (Slag 1: MnO: 50%, SiO_2: 25%, Al_2O_3:25%, and Slag 2: MnO: 40%, SiO_2:60%) was investigated at 1550°C. The wettability with Slag 1 was found to become poorer with increasing carbon content in the refractory; however a complete wetting was observed after 20 minutes of contact. Slag 2 showed a better wettability with higher carbon refractories. The extent of gasification of ZrO_2 – C system was found to be 9 times higher than the Al_2O_3 – C system. At ~5 minutes of contact, the slag appeared to get lifted away from the substrate leading to a slowing down of slag penetration. The maximum height of the gaseous gap decreased with increasing carbon in the substrate [880μm, 630μm & 440μm respectively for 20%, 15% & 10% carbon in the substrate]. The formation of the gaseous gap between slag 1 and the substrate was attributed to poor wetting and gas pressure. Slag 2 did not exhibit the formation of a gaseous gap due to lower contact angles between the slag and refractory. Alumina – carbon showed higher contact values but no gas gap formation was detected. The absence of gaseous gap was attributed to insufficient gas generation in this case.

Keywords: Carbon based refractories, alumina, zirconia, steelmaking

1. Introduction

Refractories—ceramics which can withstand very high temperatures—are used extensively by the steel industry in the internal linings of furnaces, in vessels for holding and transporting steel and slag, slide gate plates, shrouds and tundish rods [1,2]. Iron and steel producers are the largest consumers of refractories, accounting for the consumption of nearly 70% of all refractories. One of the greatest challenges lies in finding refractories that can withstand harsh environments for extended periods of time without degradation and associated product contamination. Carbon is added to a range of oxides such as alumina, magnesia, zirconia, silica, calcia etc. to enhance refractory resistance to slag corrosion, thermal shock and wear. However the degradation of carbon based refractories through interactions with molten iron and/or steelmaking slags is an important issue, as their failure/wear can lead to a loss of production time, equipment and even the steel product itself.

Globally, the industry generates several million tonnes of spent refractories annually. Refractories may last from just hours to several years, depending on service conditions and material performance [3]. When refractories reach the end of their service life, these are generally landfilled and replaced with new refractories. Carbon-bearing refractories are used extensively in steelmaking because of their excellent thermal shock, wear resistance, and strength at high temperatures. While the addition of even small amounts of carbon (< 5%) to oxide refractories has resulted in significant improvements in refractory performance, these refractories can undergo rapid carbon depletion on contact with molten steel [2, 4]. Refractory selection plays a critical role in optimizing productivity, enhancing energy efficiency and minimising labour costs.

High temperature wettability of carbon based refractories with molten iron and steelmaking slags is an important rate controlling factor influencing refractory degradation. The corrosion of alumina-graphite immersion nozzles at the slag/metal interface is caused by the oxide phase in the refractory dissolving preferentially in the slag, thereby producing a graphite rich layer [5]. Alumina itself has very poor wettability with molten iron; but the addition of even small amounts of carbon can affect its contact angle (a quantitative measure of wettability) with molten iron [6]. Atomistic computer simulations to investigate the wettability behaviour of alumina-carbon substrates with molten iron showed the presence of alumina inhibited the metal penetration in the refractory strongly depressing carbon dissolution [7].

The wettability of alumina by liquid iron has been investigated extensively; the contact angles have been reported to range from 100° to 141° in the temperature range 1530 -1600°C in inert atmosphere. However in an oxidising atmosphere, the contact angle was found to reduce, with liquid metal wetting the alumina substrate [8]. Kapilashrami et al [9] observed that in the static mode, the contact angles were found to decrease with increasing oxygen content of the metal phase. In the dynamic mode, the change of contact angle with time showed three stages at specific oxygen partial pressures; this phenomena was influenced by the diffusion of oxygen through the metal phase. A product layer consisting of $FeAl_2O_4$ was identified over the contact area between the metal drop and substrate.

Adsorption and chemical reaction are two types of chemical interactions during the wetting process: these occur when liquid is in mechanical and thermal equilibrium but not in chemical equilibrium with the solid and gaseous phases [10]. Adsorption involves the attraction of surface-active species to the two-phase interface; the time to reach equilibrium will depend on the diffusion rates of the active species to the surface. At high temperatures, the rate of diffusion would generally be quite high and the contact angle is expected to vary with time due to associated changes in surface tensions, energies and chemical composition.

Three types of chemical reactions can occur in a solid-liquid-vapour system: the formation of a new phase at the solid-liquid interface, the dissolution of some of the substrate constituents into the liquid or the dissolution of the liquid into the substrate [11]. The wetting behaviour can be characterised easily for the system in chemical equilibrium; the change in system behaviour is controlled by the system attaining mechanical and thermal equilibrium. When surface-active elements are present at the interface, these can alter the interfacial energy at that interface [12].

Understanding the wetting behaviour for non-reactive systems is relatively simple as compared to reactive systems. In non-reactive systems, characterized by the rapid kinetics of wetting and by weak temperature dependent variations of the contact angle (θ), the nature of the solid surface is not significantly changed by the contact with liquid phases. Solid and liquid contact phases under conditions of thermodynamic equilibrium with the chemical potential of each component; temperature and pressure in each of the contacting phases are the same. In a non-reactive wetting system, the process of adsorption could be observed at the interface, the interfacial energy in a non-reactive system depends on the nature and structure of the contacting phases [13].

The wettability and associated interfacial reactions in the metal/refractory system are influenced by the chemical compositions of species involved, the nature and concentration of impurities and additives, experimental conditions and the physical properties of the refractory [14]. In this chapter, we present detailed wettability and interfacial phenomena investigations on alumina-carbon and zirconia-carbon refractories at steelmaking temperatures during interactions with molten iron and a casting slag at 1550°C. The focus in this chapter is on the influence of reactions taking place in the interfacial region, production of gases through in-situ chemical reactions and subsequently the contact between the refractory and the molten steel or slag.

The chapter is organised as follows. Basic experimental details and contact angle measurement methodology is presented in section 2. Recent developments in the wetting behaviour of alumina-carbon refractories due to extended contact with molten iron are presented in section

3. Chemical interactions in the Al_2O_3-C/Fe system have recently been observed at 1550°C [5, 15-16] leading to low temperature carbothermic reduction of alumina in the presence of molten solvent. Implications of these reactions on the wetting behaviour are discussed. In section 4, the interactions of casting slags with carbon based refractories are presented as a function of refractory composition. The role of in-situ gas generation on interfacial contact and refractory corrosion is discussed. Section 5 summarises key findings and recent developments with implications for future developments in the field towards minimising refractory degradation and extending the operating life of steelmaking refractories.

2. Experimental

High purity (99.8%) fused alumina was mixed thoroughly with synthetic graphite with 5 wt % phenolic resin as a binder. The concentration of carbon in the substrate ranged between 50 and 90 wt%. Refractory substrates were prepared by compacting the mixture in a steel die using a hydraulic press and pressing to a pressure up to 10MPa. The compacted cylindrical substrates (25 mm diameter, 3-4 mm thick) were baked at 150 °C for 48 hours for enhancing their structural integrity. The weight of the metallic iron (0.03 %C) used was 1.0 gm. Sessile drop wettability investigations on the system were carried out at 1550°C in a laboratory scale, horizontal tube resistance furnace (Figure 1) [15]. Initially, the sample was held on a specimen holder, which could be pushed to the centre of the hot zone of the furnace with the help of a graphite rod.

The melting of iron marked the beginning of contact time. A high resolution charge-coupled device (CCD) camera fitted with an IRIS lens was used to capture the live in-situ phenomena in the furnace. The output from the CCD camera was channelled to a digital video recorder (DVD) and a television monitor to record the entire experimental process as a function of time. This feature allows specific images, displaying the contact between the metal and carbonaceous material to be captured as a function of time from the DVD, which were subsequently processed by a computer. A time-date generator was used in the system to display the duration of the experimental process. Specially designed software was used to measure contact angles on the basis of a curve fitting exercise. Details of this software have been given elsewhere [17].

Wettability investigations were also carried out on alumina-carbon (C: 10-20 wt %) refractories with a steelmaking slags (Slag 1: MnO: 50%, SiO_2: 25%, Al_2O_3:25%; Slag 2: MnO:40%, SiO_2:60%) at 1550°C. These studies were focussed on the role played by in-situ reduction reactions, gas generation and interfacial phenomena on the wettability between carbon-based refractories and molten slags.

The liquidus temperature of both meniscus slags was estimated using the thermodynamic package *Fact Sage* [18]. High purity reagent grade oxides MnO and SiO_2 powders supplied by Sigma Aldrich Chem., UK (Purity >99%) and, reactive Al_2O_3 powder supplied by Alcoa, Australia (Purity >99%)] were mixed uniformly in appropriate weight proportions and the melting of oxide mixture was carried out in a platinum crucible in an induction furnace

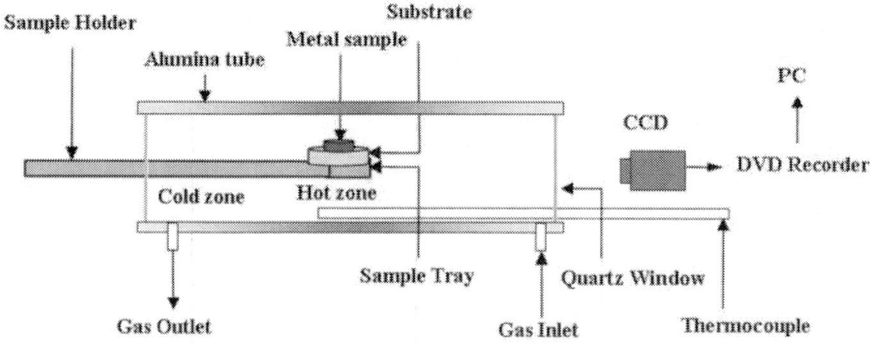

Figure 1. A schematic representation of the experimental arrangement

maintained at the melting temperature. The molten slag was water quenched, dried, grounded and remelted several times to ensure compositional homogeneity. The interfacial region between the slag and the refractory was analysed by optical and scanning electron microscopy (SEM) on Model S-3400X. Energy dispersive spectroscopy (EDS) was used to identify the elemental distribution in selected regions.

3. Interactions of Al_2O_3-C refractories with molten iron

3.1. Wettability of Al_2O_3 with molten iron

The wettability of alumina by liquid iron has been investigated extensively. The contact angles were found to be in the range 100° to 141° at 1530 -1600°C in the presence of inert gases such as helium, argon, hydrogen gases and under vacuum. Under oxidising atmosphere, the contact angle was found to reduce significantly and molten iron was found to wet alumina substrate [8]. Kapilashrami et al [19] also made a similar observation that in the static mode, the contact angles of liquid iron on alumina substrate were found to decrease with increasing oxygen content of the metal phase. Pitak and P'yanykh [20] investigated the wetting of refractories with molten steel; a change in wetting from 112° to 90° was observed starting from 7 min to 15 min at 1560 °C.

Alumina may therefore be expected to act in an inert manner, creating a physical barrier to chemical reactions and/or mass transfer and influence wettability when interacting with iron [19, 21-22]. Key experimental results on the wettability of alumina with molten iron have been summarised in Table 1.

The composition of molten iron used in these studies is given below:

ARMCO (M15) iron (ppm): 180[C]; 100[Si]; 1600[Mn]; 60[P]; 250[S]; 550[O]; 36[N]

St20K steel (%): 0.19C, 0.3Si, 0.65Mn, 0.03S, 0.016P, 0.02Ti, 0.025Al, 0.0075O

St35 steel (%): 0.35C, 0.29Si, 0.7Mn, 0.025S, 0.026P, 0.09Cr, 0.29Ni, 0.22Cu

Argon+%O: slightly oxidized Argon

St35L steel (%): 0.38C, 0.42Si, 0.56Mn, 0.03S, 0.04P

St60 steel (%): 0.61C; 0.67Mn; 0.27Si; 0.016P; 0.25S; 0.12Cu; 0.03Cr; 0.04Ni

Metal	Atmosphere	T (°C)	θ (°)
ARMCO (M15)	Vacuum	1550	141
Electrolytic iron	H_2	1550	121
Electrolytic iron	Helium	1550	129
Fe+3.9%C	Helium	1570	105
St50steel	Helium	1560	120
St60steel	Helium	1560	117
08Yu steel	Helium	1560	125
Fe+0.2%C	Argon	1550	123
St20K steel	Argon	1560	115
St35 steel	Argon	1500	112
St35 steel	Argon	1550	105
St35L steel	Argon	1560	130
Fe+0.45%C		1535	105
St60 steel	Helium	1530	109-100
St60 steel		1550	136

Table 1. The wettability of alumina with molten iron [23]

3.2. Effect of carbon concentration in the substrate

Experimental results in the form of video images of alumina-carbon (50-90%) substrates in contact with molten iron are shown in Figures 2 to 6. With 50% C in the substrate, molten iron is clearly non-wetting at T=0. There was not much change in the video images even after three hours of contact (Figure 2). The system continued to remain non-wetting; not much change was observed in contact angles.

With C increasing to 60% in the substrate (Figure 3), although the system was still non-wetting, the overall contact angles appear to have decreased. Once again, time did not have much influence on wettability. However, the system appeared to be reacting slightly as indicated by the gaseous whitish clouds rising from the substrate and tending to surround the melted iron droplet.

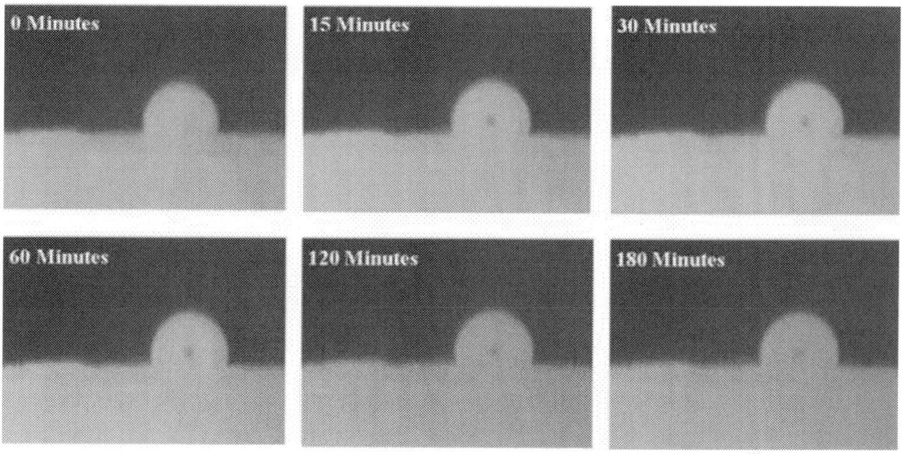

Figure 2. High temperature (1550°C) images of Alumina-50 wt% synthetic graphite substrate in contact with molten iron droplet after extended period of contact.

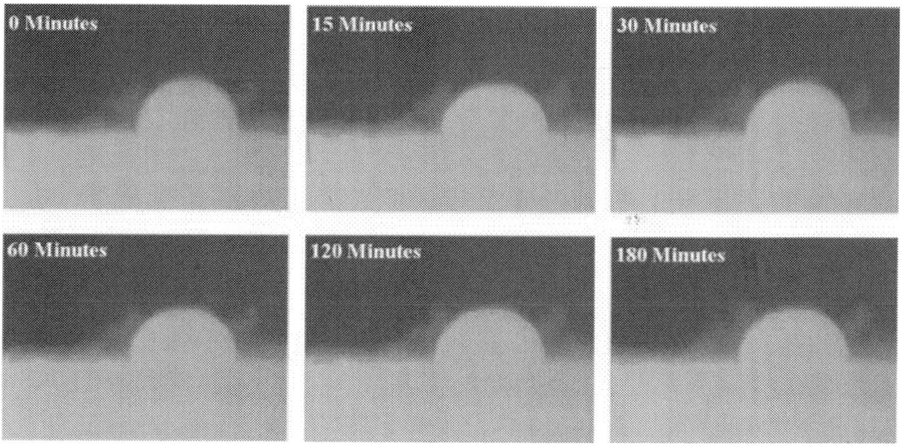

Figure 3. High temperature images of Alumina-60 wt% synthetic graphite substrate in contact with molten iron droplet after extended period of contact. A small amount of gaseous cloud can clearly be observed in the background.

With carbon concentration increasing to 70% in the substrate (Figure 4), although it was still non-wetting (θ> 90°) initially, the contact angle decreased rapidly within 15 minutes of contact. The droplet was seen to flatten out on the substrate. This increased level of contact was seen to persist even after three hours. The system also started to show high level of gaseous activity. Some gaseous clouds could be seen even at T=0; these spread right across the iron droplet after 2 and 3 hours. These gaseous products indicate chemical/reduction reactions occurring in the

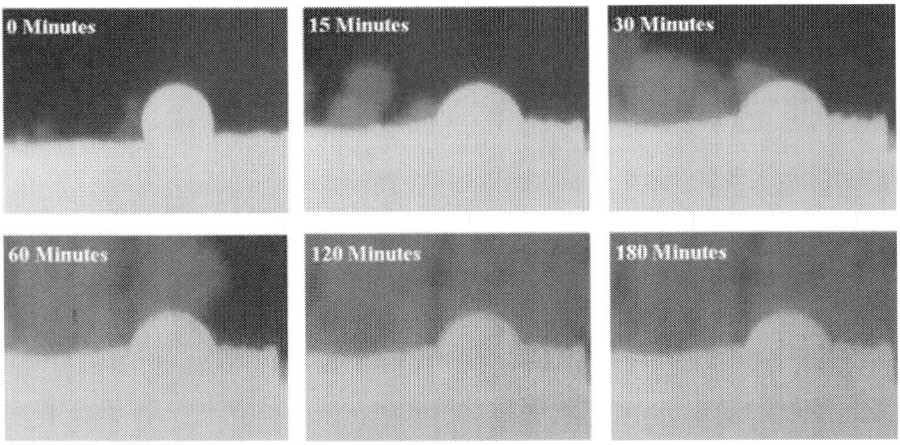

Figure 4. High temperature images of Alumina-70 wt% synthetic graphite substrate in contact with molten iron droplet after extended period of contact. A large amount of gaseous cloud can clearly be observed in the background.

system. Similar results were obtained for carbon contents of 80 and 90% in the substrate (Figures 5 and 6).

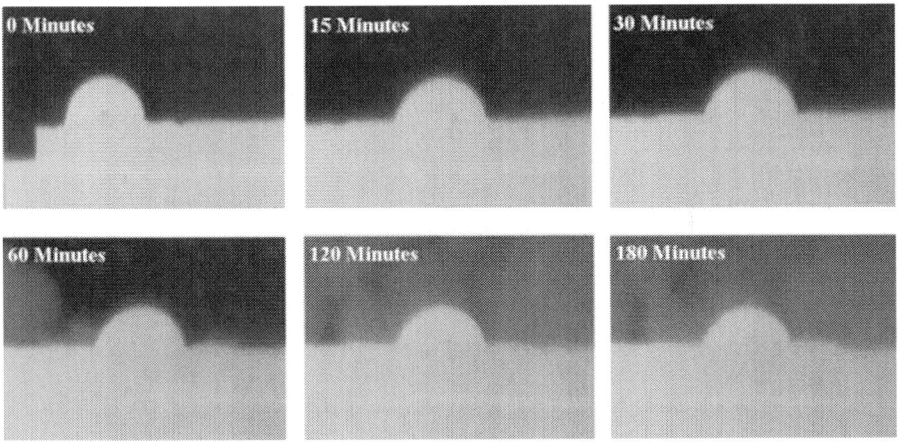

Figure 5. High temperature images of Alumina-80 wt% synthetic graphite substrate in contact with molten iron droplet after extended period of contact. A large amount of gaseous cloud can clearly be observed in the background.

Numerical values of contact angles as determined by especially designed software [17] have been plotted in Figure 7. For 50 wt% C in the substrate, the contact angles started at 96° initially and rose to 100° over extended contact. On the other hand, initial contact angles were higher

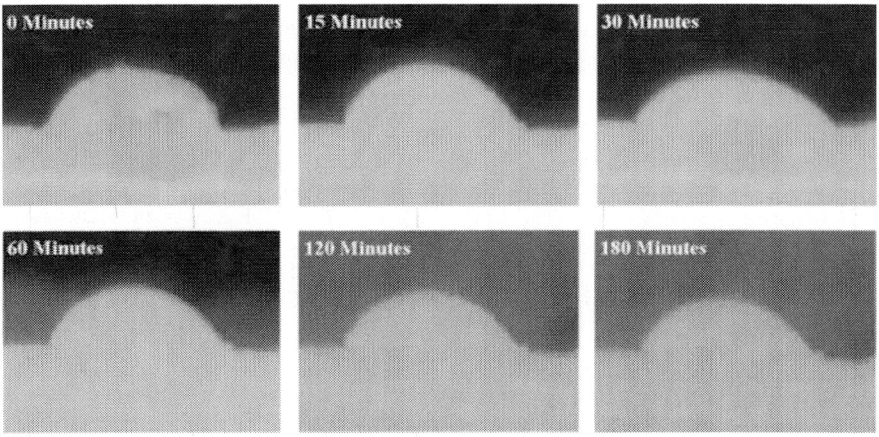

Figure 6. High temperature images of Alumina-90 wt% synthetic graphite substrate in contact with molten iron droplet after extended period of contact. A large amount of gaseous cloud and the flattening of metal droplet on the substrate surface can clearly be observed.

(109°) for the substrate containing 60 wt% C; these slowly reduced to 96° over extended period of contact. For substrates containing 70 wt% C, even though the contact angles showed poor wetting (92°) during initial contact, these showed a rapid decline and later settled down after 1 hour of contact. A similar behaviour was observed by substrates containing 80 and 90 wt% C, which showed a further improvement in wettability as represented by a reduction in contact angles.

These results indicate that the relative proportion of carbon in the refractory substrate had a very strong influence on their wettability with molten iron at 1550°C. Observed results can clearly be separated in to two groups: first group (C: 50%, 60%) which is non-wetting to molten iron and the second group ($C \geq 70\%$) which shows good wettability with molten iron. A clear transitional behaviour can be observed between the two groups. One of the key differences between the two groups was the high level of gaseous activity for the second group. The occurrence of chemical/reduction reactions between alumina-carbon substrates and molten iron can result in the generation of gaseous products (CO, CO_2, AlO, Al_2O etc.). These reactions appear to have a strong influence on wettability. While time had a strong influence for all substrates during initial stages of contact (up to 30 minutes), generally stable values of contact angles were observed over longer times up to 3 hours.

These studies have shown that the wettability of alumina-carbon refractories with molten iron is a strong function of substrate composition and carbon content. This study is expected to play a crucially important role in understanding degradation of carbon based refractories and the contamination of steel with reaction products. Extensive generation of gaseous products was observed after the system undergoing transition from poor wetting to good wetting; this study is therefore expected to help define limiting values of carbon that could help minimise refractory degradation.

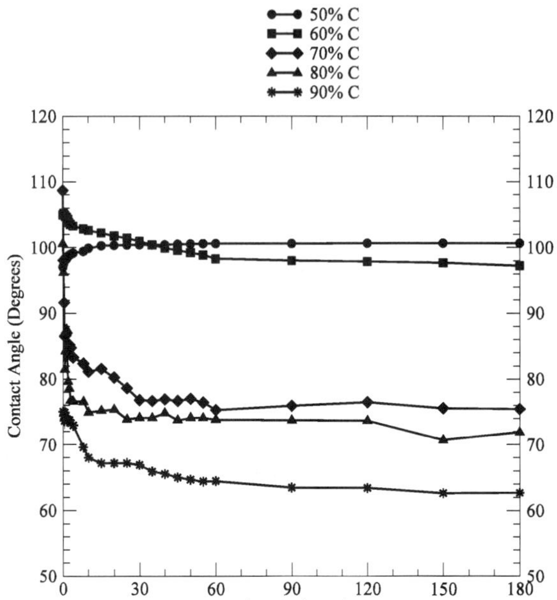

Figure 7. Contact angles as a function of substrate composition and time. The level of carbon in the refractory substrate has clearly been marked.

3.3. Effect of impurities

The effect of impurities present in carbons can have a significant influence on wettability. In this section, results on the use of synthetic as well as natural graphite are presented. The failure of the refractories depends on the interfacial behavior of the metal/refractory system and the ability of the liquid metal to 'wet' the refractory. The wettability investigations of liquid iron on alumina-synthetic graphite substrates were conducted at 1550°C sing the sessile drop approach. From the video images captured by the CCD camera, contact angles at various times were measured in different alumina-graphite substrates containing 10, 20 and 30% synthetic graphite labelled as AS1, AS2 and AS3 respectively. Figure 8 shows high temperatures sessile drop images for AS1, AS2 and AS3 refractory substrates containing 12.9 wt %C, 22.8 wt% C and 32.7 wt %C respectively at 1, 5 and 30 minutes time intervals.

From Figure 9, it can be seen that for all the substrates, there was a sharp drop in contact angles in the first five minutes followed by an increase in the next ten minutes with contact angles finally stabilizing after 15 min from the time of initial contact. With increasing carbon level from 10 to 30% affects wettability, but it does not have a strong influence and is not able to change the wetting behaviour from non-wetting to wetting. As with increased carbon content, the contact angle gets reduced, but still after 30 minutes of experimental time, where contact angle becomes stable, the final contact angle was more than 90°.

http://dx.doi.org/10.5772/61271

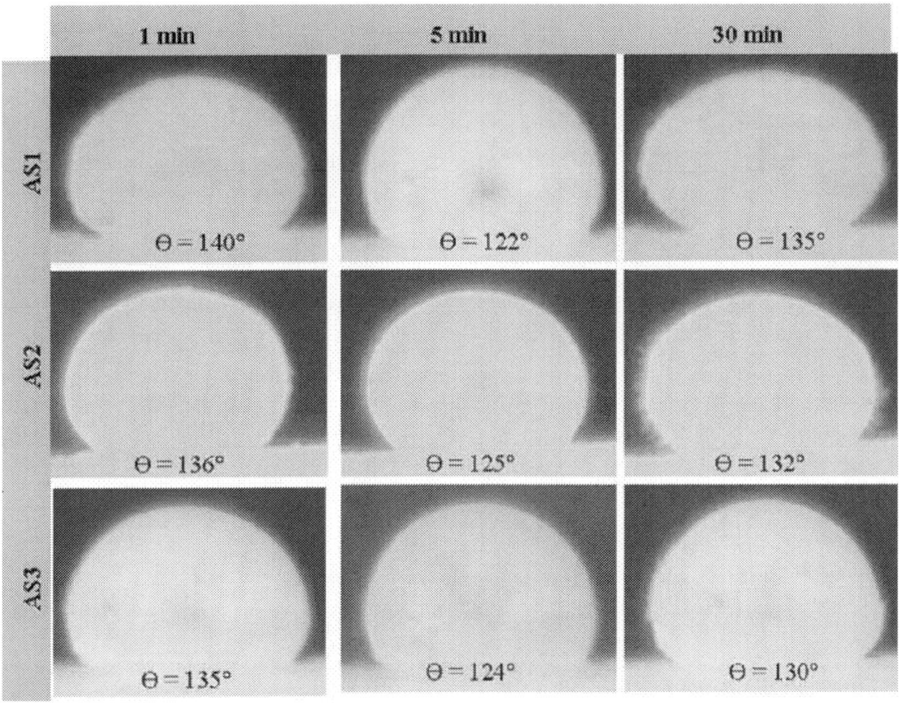

Figure 8. Molten iron droplet images at 1550 °C for different time intervals for alumina-synthetic graphite substrates

The wettability and contact angle measurements of liquid iron with natural graphite/alumina substrates were carried out at 1550°C under argon atmosphere to determine contact angles as a function of time. From the video images captured by CCD camera, the contact angles at various times were measured in different alumina-graphite substrates containing 10, 20 and 30% natural graphite labelled as AN1, AN2 and AN3 respectively.

Figure 10 represents high temperature images of molten iron droplets during reaction with alumina-natural graphite for AN1, AN2 and AN3 substrates containing 12.7wt%C, 22.4 wt% C and 32.1 wt% C respectively at 1, 5 and 30 minutes time interval. The measured data has been plotted in Figure 11 outlining the trend of the wetting behavior of all three alumina-natural graphite substrates during 30 minutes experimental time. It is seen that, there was first a drop in contact angles for all the substrates in the first five minutes which was followed by an increase over the next 10 minutes, after which the contact angles stabilized to around 130° for all the substrates within 15 min from the start of the experiments.

From the wettability data, it was concluded that the system was in chemical non-equilibrium till 15 minutes of reaction time and after that system achieved a state of chemical equilibrium from 15 to 30 minutes of reaction time. In alumina-natural graphite/iron refractory system, all

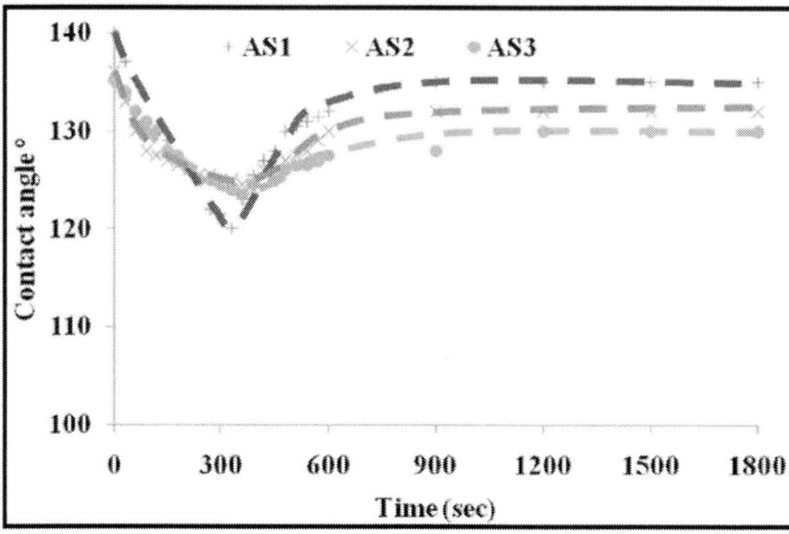

Figure 9. Variation of contact angles of molten iron with time for alumina-synthetic graphite substrates at 1550 °C.

three refractory substrates had stabilized at contact angles of 130°, while alumina-synthetic graphite substrates/iron refractory system, final angles were stabilized at 135°, 132° and 130° respectively for AS1, AS2 and AS3 substrates during 15 to 30 minutes of reaction time.

It can also be seen that increasing the carbon level from 10 to 30% had some influence on wettability, but it was not sufficient to change the wetting characteristics from non-wetting to wetting. With increased carbon content, the contact angles showed an increase; however throughout the course of the experiments, the angles were higher than 90°.

During extended contact of refractories with molten iron, some iron is known to penetrate the refractory substrates as carbon is being depleted. To study the influence of metal within the refractory substrate itself, the used refractory of AS1-Fe sample with 10, 20 and 30 % was mixed with virgin refractory composition of AS-1; the mixture termed as recycled refractory was labelled as AS1-10R, AS1-20R and AS1-30R. The wt% ratio of alumina and carbon content in these recycled refractories are given in Table 2.

Sample label	Used Refractory	Virgin Refractory	Wt. % Al_2O_3	Wt. % C
AS1-10R	10%	90%	87.3	12.7
AS1-20R	20%	80%	87.6	12.4
AS1-30R	30%	70%	87.9	12.1

Table 2. Composition of recycled refractory substrates

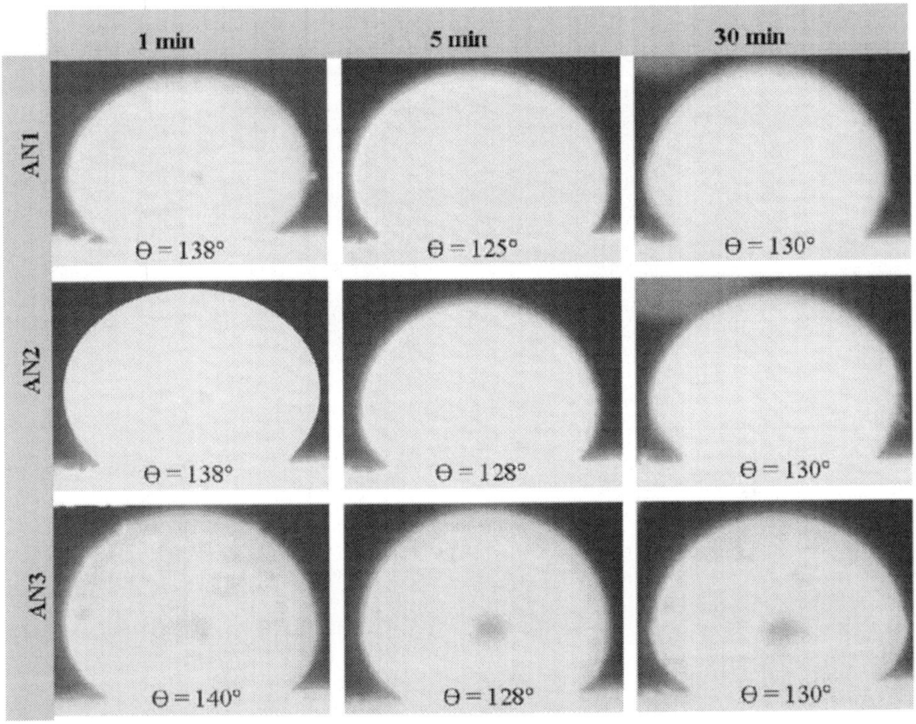

Figure 10. Molten iron droplet images at 1550 °C for different time intervals for alumina-natural graphite substrates

The wettability investigations of liquid iron droplet with recycled refractory substrates were conducted at 1550 °C under argon atmosphere and contact angles were measured as a function of time from the video images captured by a CCD camera. Figure 12 shows high temperatures sessile drop images for AS1-10R, AS1-20R and AS1-30R refractory substrates respectively containing 12.7 wt %C, 12.4 wt% C and 12.1 wt %C at 1, 5 and 30 minutes time intervals.

From Figure 12, the contact angles were determined as 118°, 119° and 130° during the first minute of the reaction for substrates AS1-10R, AS1-20R and AS1-30R respectively. After that contact angles dropped to 115°, 116° and 127° respectively for these substrates in the next 5 min. These then increased to 122°, 124° and 132° over the next ten minutes respectively and then these angles remained fairly unchanged during the reaction time (30 min).

It is seen from Figure 13 that for all the substrates, there was a sharp drop in contact angles in the first five minutes followed by an increase in the next ten minutes with the angles finally stabilizing within 15 min from the time of initial contact. This trend was quite similar to the one observed for virgin refractories as detailed above.

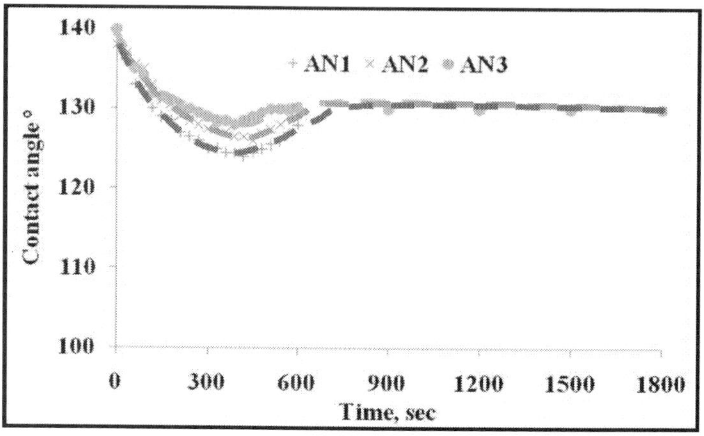

Figure 11. Variation of contact angles of molten iron with time for alumina-natural graphite substrates at 1550 °C.

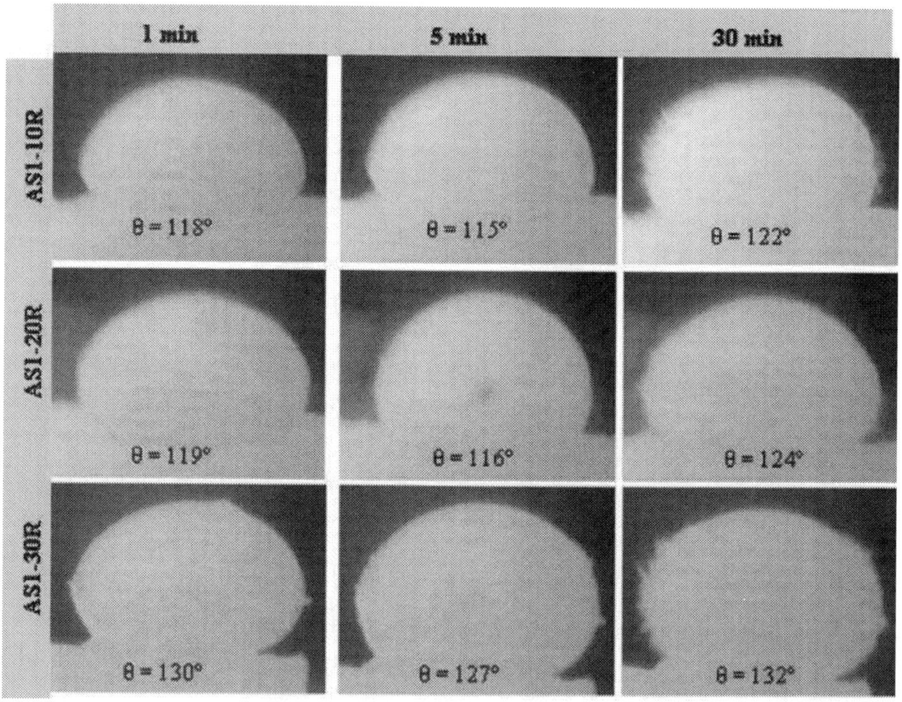

Figure 12. Molten iron droplet images at different times for recycled refractory substrates

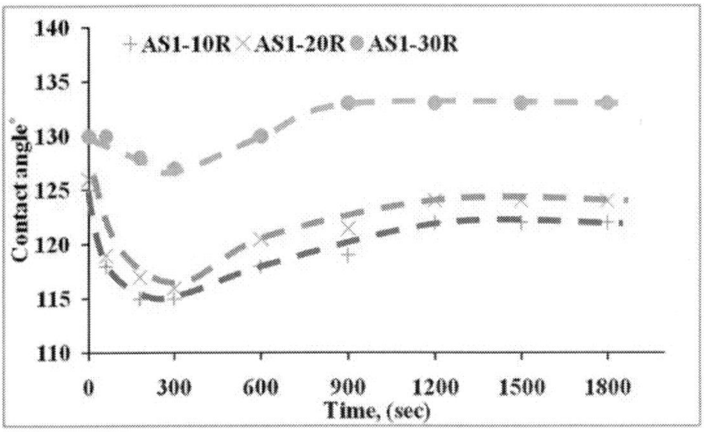

Figure 13. Dynamic contact angles of AS1-10R, AS1-20R and AS1-30R substrates at 1550 °C

4. Interactions of ZrO_2-C and Al_2O_3-C industrial refractories with molten slags

Detailed results presented in section 3 represent experimental results under controlled laboratory conditions. In this section, we present results from some industrial carbon based refractories with molten slags, typically encountered during casting. The composition of some of substrates investigated is given in Table 3. The chemical composition of slag 1 is given in Table 4.

Composition	Slag	Duration (min)	Shape	Surface
85%ZrO_2+15%C	1	60	Flat	Non-Oxidized
85%ZrO_2+15%C	1	60	Flat	Oxidized
85%Al_2O_3+15%C	1	60	Flat	Non-Oxidized

Table 3. Industrial flat ZrO_2-C & Al_2O_3 samples

		Slag 1
Slag composition, wt %	MnO	50
	SiO_2	25
	Al_2O_3	25
Estimated liquidus temperature, °C		1420
Melting temperature, °C		1600

Table 4. Chemical compositions of laboratory prepared slags

4.1. Industrial Flat ZrO_2-C Samples with Slag 1

In the sessile drop investigation, the contact angle between slag and refractory substrate can be used to investigate the slag wettability to the substrate samples. Video images of the reaction assembly during interactions at 1550 °C are given in Figure. 14. The time of interaction is indicated on the top right corner of these sessile drop images.

The contact angle measurement results of the non-oxidized ZrO_2-C sample (Figure. 14) shows that the contact angle between the substrate and slag 1 reached the highest value of 120° at 5 minute, then decreases dramatically to 60°and stable at around 60° for 15 minutes, after that the contact angle dropped quickly until the slag 1 entirely spreads onto the ZrO_2-C substrate. These results have been summarised in Figure. 15.

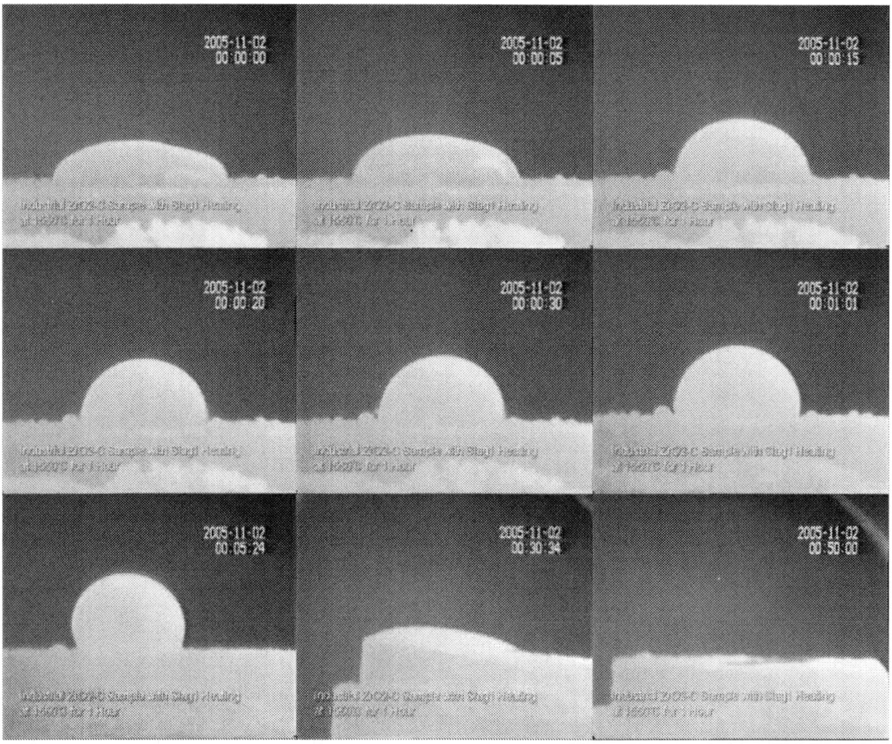

Figure 14. Industrial non-oxidized flat ZrO2-C samples with slag 1 (MnO 50%wt, SiO_2 25%wt, Al_2O_3 25%wt) heating at 1550°C for 1 hour

The condition of the refractory surface appeared to have a significant influence on the wettability behaviour of refractories. Figure 16 shows corresponding results when the refractory surface had been pre-oxidised. In contrast, the CCD images indicate good wetting

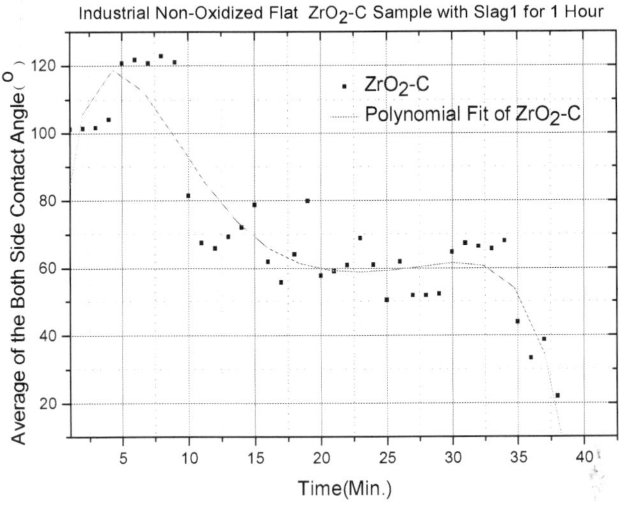

Figure 15. Industrial non oxidized flat ZrO_2-C sample with slag1 heating at 1550°C for 1 hour

between the oxidized ZrO_2-C substrate and slag 1. The slag 1 drop had spread on the top of the substrate, after the slag powder had fully melted in only 1 minute. Due to the limitation of the contact angle software, the quantitative contact angle trend is not available.

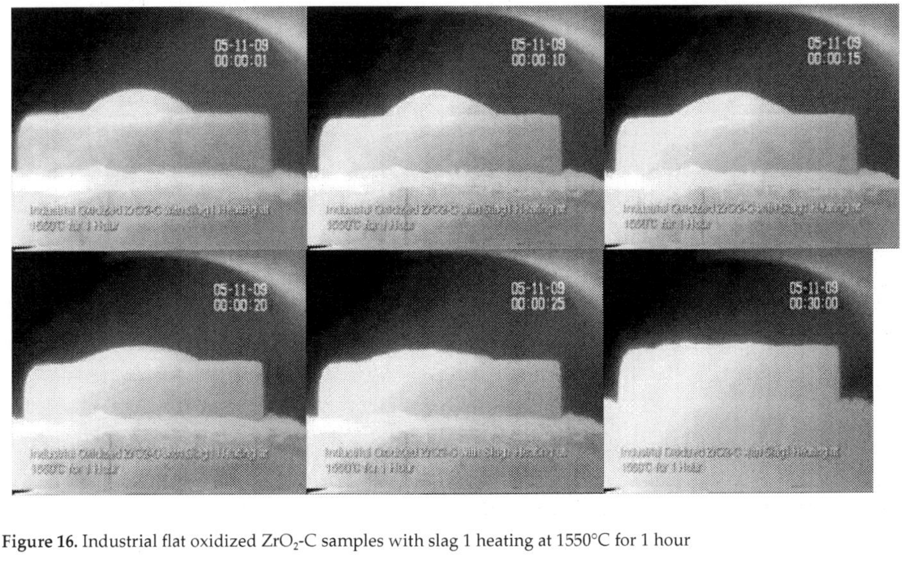

Figure 16. Industrial flat oxidized ZrO_2-C samples with slag 1 heating at 1550°C for 1 hour

4.2. Industrial Flat Al_2O_3-C Samples with Slag 1

Figure 17 and 18 show contact angles and video images for alumina-carbon refractories in contact with slag 1.

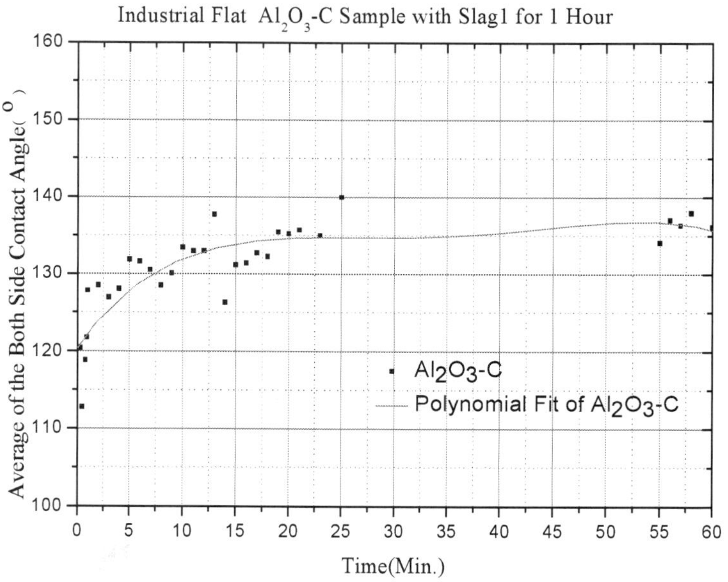

Figure 17. Industrial flat Al_2O_3-C sample with slag1 heating at 1550°C for 1 hour

These images show the contact angle between the Al_2O_3-C substrate and slag 1 was constantly more than 120° during 1 hour heating at 1550°C. Both industrial oxidized and non-oxidized flat ZrO_2-C samples had better wettability with slag 1 than the industrial flat Al_2O_3-C samples. Especially the oxidized ZrO_2-C sample was fully wetted by slag 1 in one minute.

Results for the gas data from the blank substrates is presented. Infrared analyser was used to analyse the outlet gas from the furnace. The IR results showed the outlet gases content of the CO and CO_2, and this can be converted to the removal amount of carbon. For the industrial non-oxide zirconia refractory, the cumulative volume of CO reached 1.8 litres after 60 minutes; similar volumes were recorded for the oxidised refractory (Figure 19). The corresponding volumes of CO_2 were negligibly small. Alumina-carbon refractories showed much lower levels of gas generation. Gases generated were generally higher in the presence of slag (Figure 20).

The generation of gases at the interface led to the production of gas bubbles in the slag, which in-turn had a significant influence on the contact between the slag and the refractory (Figure 21).

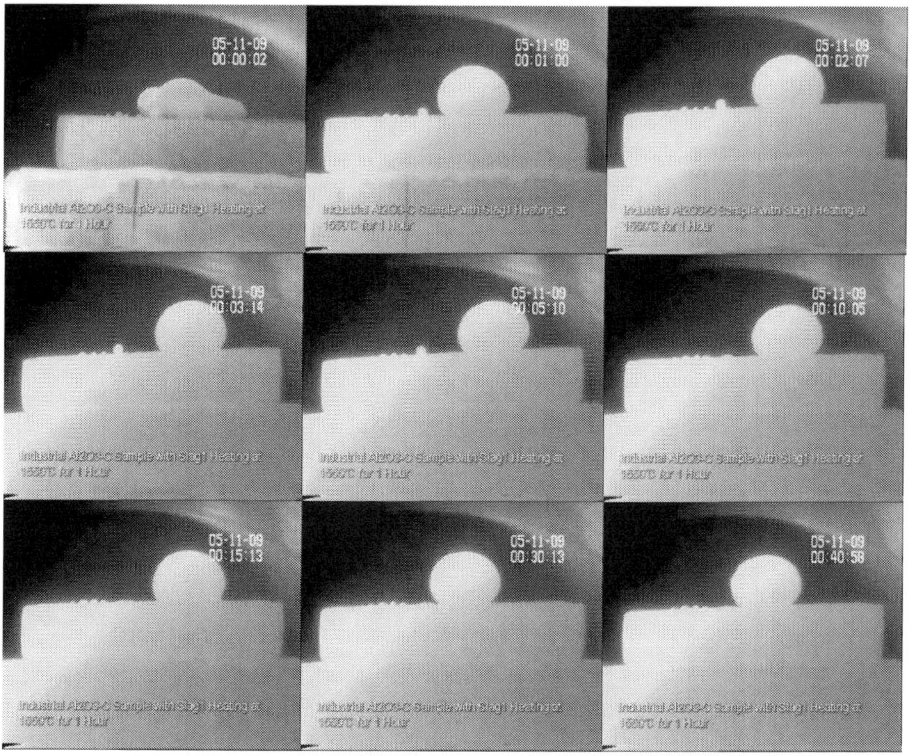

Figure 18. Industrial flat oxidized Al_2O_3-C samples with slag 1 heating at 1550°C for 1 hour

Detailed studies reported in this section have shown that reactions taking place in the interfacial region and associated gas generation can have a significant influence on the wettability of refractory substrates with strong implications for refractory degradation and corrosion.

5. Concluding remarks

These studies have shown that the wettability of alumina-carbon refractories with molten iron is a strong function of substrate composition and carbon content. Alumina and carbon are commonly used constituents of steelmaking refractories. Gases generated during the interaction make it easier for metal to penetrate in the substrate enhancing refractory degradation. As more and more gaseous products were generated after the system underwent transition from poor wetting to good wetting, this study will help define limiting values of carbon that could help minimise refractory degradation.

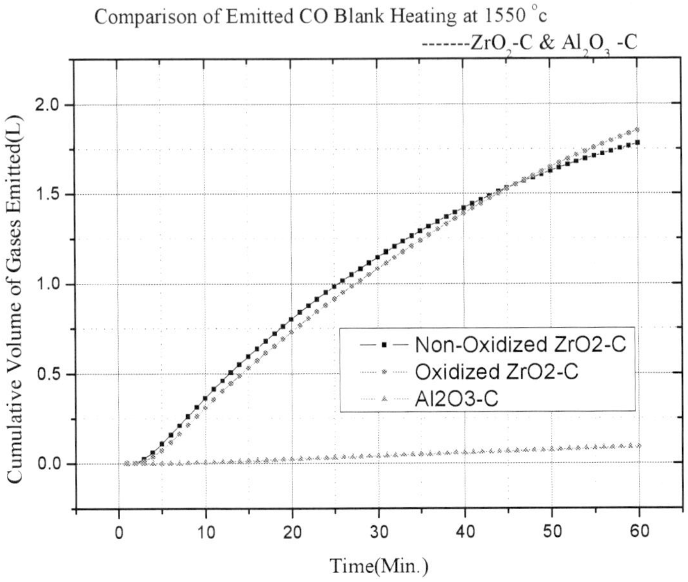

Figure 19. Comparison of emitted CO gas from blank ZrO_2-C & Al_2O_3-C refractories

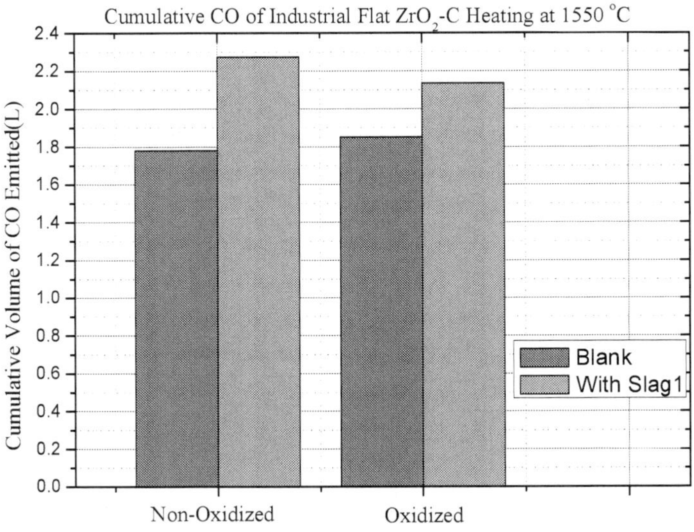

Figure 20. Cumulative CO of industrial flat ZrO_2-C sample blank & with slag

http://dx.doi.org/10.5772/61271

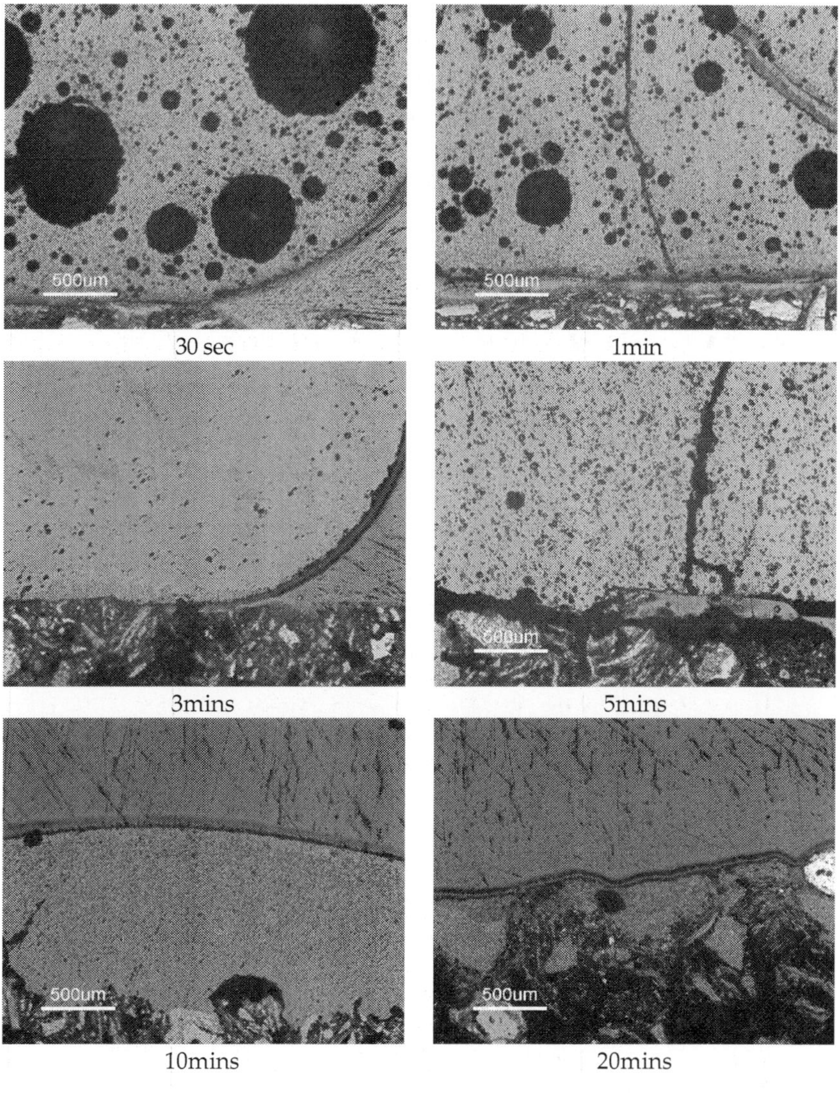

Figure 21. Interfacial phenomena between zirconia-carbon refractories and slag 1.

Alumina-synthetic graphite substrates showed a non-wetting behavior with liquid iron at 1550 °C. The initial contact angles were 140°, 136° and 135° respectively for substrates containing 12.9 wt% C, 22.8 wt% C and 32.7 wt% C, after which these experienced a 10-18º drop during first five minutes of the experiment, which was followed by a further increase in the angles to final values in the range of 130-135° for all the substrates. Alumina-natural graphite substrates also showed a non-wetting behavior with liquid iron at 1550 °C. The contact angles were initially at 138°, 138° and 140° respectively for substrates containing 12.7 wt% C, 22.4 wt% C and 32.1 wt% C, after which these experienced a drop of 10-15° during the first 5 minutes of reaction before stabilizing to the final contact angle value of approximately 130° for all the substrates for the rest of the experimental duration.

Recycled refractory substrates showed a non-wetting behavior with liquid iron at 1550 °C. The initial contact angles were 118°, 119° and 130° respectively for AS1-10R, AS1-20R and AS1-30R substrates, after which they experienced a 3º drop during first five minutes of the experiment, which was followed by a further increase in the angles to final values in the range of 122-132° for all the substrates. These values are much lower in comparison to virgin refractory system indicating the role played by the contamination in used refractory.

A number of experiments were reported on industrialized refractory substrates with casting slags. While zirconia based refractories generally showed good wettability, these were also associated with extensive in-situ gas generation and refractory degradation. Alumina based refractories however showed little gas generation and poor wettability. These studies have shown that a number of factors, including wettability, could play a significant role in the degradation of carbon based refractories and need to be taken into consideration.

Acknowledgements

This work was supported by the Australian Research Council (ARC) funds. We gratefully acknowledge their financial contribution. We thank Fuhai Liu and Tuanfeng Wang for some of the experimental work reported in this article.

Author details

R. Khanna*, M. Ikram-ul-Haq and V. Sahajwalla

*Address all correspondence to: ritakhanna@unsw.edu.au

Centre for Sustainable Materials Research and Technology, School of Materials Science and Engineering, The University of New South Wales, Sydney, Australia

References

[1] Sasai, K. and Mizukami, Y. Reaction rate between alumina graphite immersion nozzle and low carbon steel. ISIJ International. 1995; 35(1) 26-33.

[2] Schei, A., Tuset J.K., Tverit, H., "Production of High Silicon Alloys", Tapir Forlaug, Trondheim. 1998.

[3] Bennett, J., Kwong, K. An overview of recycling refractory materials. Industrial ceramics. 2004; 24(3) 165-171.

[4] Chen, CY, Lin, CI, Chen SH. Kinetics of synthesis of silicon carbide by carbothermal reduction of silicon dioxide. British Ceramic Transactions. 2000; 99(2) 57-62.

[5] K. Mukai, Interfacial phenomena in Materials processing edit ED Hondros, M McLean and KC Mills, Book 692 Royal Society, publisher: IOM, London, 1998, 201.

[6] Khanna, R., Sahajwalla, V. An atomistic model for the graphite alumina/liquid iron system: Monte-Carlo simulations on carbon dissolution. Acta Materialia. 2005; 53 (4) 1205-1214.

[7] Zhao, L., Sahajwalla, V. Interfacial Phenomena during Wetting of Graphite/Alumina Mixtures by Liquid Iron. ISIJ International. 2003; 43(1) 1-6.

[8] Poirier, DR., Yin, H., Suzuki, M., Emi, T. Interfacial properties of dilute Fe-OS melts on alumina substrates. ISIJ international. 1998; 38(3) 229-238.

[9] Kapilashrami, E., Seetharaman, S. Wetting characteristics of oxygen-containing iron melts on refractory oxides. Journal of Materials Science. 2005; 40(9) 2371-2375.

[10] McCarthy F. Interfacial Phenomena and Dissolution of Carbon from Chars into Liquid Iron During Pulverised Coal Injection in a Blast Furnace, PhD theis, University of New south wales Sydney; 2004.

[11] Aksay, IA., Hoge, CE, Pask, JA, Wetting under Chemical Equilibrium and Non-Equilibrium Conditions. Journal of Physical Chemistry. 1974; 78 (12) 1178-1183.

[12] Keene, BJ. Review of Data for the Surface Tension of Iron and its Binary Alloys. International Materials Review. 1988; 33 (1) 1-37.

[13] Naidich, JV. Wettability of Solids by Liquid Metals. Progress in Surface and Membrane Science. 1981; 14, 353-484.

[14] Bradley L., Li L., Stott, HF. Characteristics of the microstructures of alumina-based refractory materials treated with CO_2 and diode lasers. Applied Surface Science. 1999; 138, 233-239.

[15] Ikram-ul-Haq, M., Khanna R., Koshy, P., Sahajwalla, V. High-Temperature Interactions of Alumina-Carbon Refractories with Molten iron. ISIJ International. 2010; 50(6) 804-812.

[16] Ikram-ul-Haq, M., Khanna R., Kongkarat S., and Sahajwalla V. Chemical Interactions in Al_2O_3–C/Fe System at 1 823 K: Implications for Refractory Recycling. ISIJ International. 2012; 52(10) 1801–1808.

[17] Wu, C., Sahajwalla, V. Influence of melt carbon and sulfur on the wetting of solid graphite by Fe-C-S melts. Metallurgical and Materials Transactions B. 1998; 29(2) 471-477.

[18] Bale C.W., Pelton A.D., Thompson W.T, Eriksson G., Hack K., Chartrand P., Decterov S., Melancon J. and Peterson S. FactSage thermochemical software and databases. Factsage 6.0 GTT technology, Aachen Germany. 2009

[19] Kapilashrami, E., Investigation of Interactions between Liquid Iron Containing Oxygen and Aluminosilicate Refractories. Doctoral thesis, Royal institute of technology Stockholm; 2009

[20] Pitak, N., P'yanykh N. Wetting of refractories by molten steel and slag. Refractories and Industrial Ceramics. 1965; 6(5) 243-248.

[21] 2004 Udalov, YP, Lavrov B., Smirnov, VV., Sharov, DY, Sidorov, AS. Interaction of Molten Iron with Ceramics Based on Iron and Aluminum Oxides. Glass Physics and Chemistry. 2004; 30(1) 90-97.

[22] Nakashima, K. Mori, K. Interfacial Properties of Liquid Iron Alloys and Liquid Slags Relating to Iron- and Steel-Making Processes. ISIJ International. 1992; 32(1) 11-18.

[23] Zhao, L. Interfacial Phenomena and Chemical Reactions during Interactions of Iron and Graphite/Oxide Substrates for Refractory Applications. PhD Thesis, University of New South Wales Sydney; 2004.

Chapter 11

The Wetting of Leaf Surfaces and Its Ecological Significances

Huixia Wang, Hui Shi and Yanhui Wang

Additional information is available at the end of the chapter

http://dx.doi.org/10.5772/61205

Abstract

Leaf wettability, indicating the affinity for water on leaf surfaces, is a common phenomenon for plants in a wide variety of habitats. The contact angle (θ) of water on leaves measured at the gas, solid and liquid interface is an index of surface wettability. Leaves are termed as "super-hydrophilic" if $\theta < 40°$, "highly wettable" if $\theta < 90°$, and "wettable" if $\theta < 110°$. If $\theta > 110°$, the leaves are classified as being non-wettable, while $\theta > 130°$ for highly non-wettable and $\theta > 150°$ for super-hydrophobic. Both internal and external factors can influence leaf wettability. The chemical composition and structure of leaf surfaces are internal causes, but the external environment can also influence wettability by affecting the structure and composition of the surface. The main internal factors that affecting leaf wettability include the content and microstructure of the epidermal wax, the number, size and pattern of trichomes, stomatal density, the shape of epidermal cells, and leaf water status. The leaf contact angles increased with the increasing of leaf wax content. However, studies have shown that the contact angles were more dependent on the complexity of wax structure than on the absolute amount. For trichomes, there are three types of interaction between trichomes and water droplets, including (1) low trichomes density: no apparent influence of trichomes on the location of surface moisture, droplet formation and retention ; (2) medium trichomes density: trichomes appear to circle surface moisture into patches; (3) high trichomes density: trichomes appear to hold water droplets above the trichomes. In some cases, a higher stomatal density was accompanied with a higher contact angles. While, it was also observed that there was no significant correlation between contact angle and stomatal density for some species. For the effects of epidermal cells on leaf wettability, it was generally considered that the combination of a dense layer of surface wax and the convex epidermal cells was what created a hydrophobic leaf surface. However, the influence of leaf water content on contact angle of water droplets on different leaf surfaces was complex, e.g., contact angles increased with decreasing of leaf water content, contact angle remained to be constant with different leaf water content.

INTECH
open science | open minds

The ecological significances of leaf wettability include interception of precipitation, photosynthetic rate, pathogen infection and environmental quality. On leaf level, leaf surface wettability contributes to variability in interception between different plant species resulting different geometrical shapes of water on leaves (i.e., water film, patches, drops, and spherical droplets). For wettability leaves, water droplets form a layer of film on the surface that is relatively easy retained on leaves. For water repellent leaves, water on the surface produces droplets that are easily removed by wind and gravity. The spreading water film on leaves with high wettability can decrease photosynthesis due to the fact that diffusion of CO_2 is 10,000 times slower in water than in air. For pathogen infection, excess leaf wetness promotes pathogen infection in many species. However, bacterial spores did not attach firmly and did not germinate on super hydrophillic surfaces. Different leaf wettability can also influence the amount of air pollutants that can be captured, absorbed, and filtrated by leaf surfaces. However, these air pollutants may injure the leaf surface structure and result in a dramatic change in contact angles.

Keywords: Leaf surface, wettability, hydrophobic, microstructure, wax crystal, ecological significances

1. Introduction

Leaves are covered by a layer of cuticular wax, serving to decrease surface wetting and moisture loss. The epicuticular wax layer may be classified into two main types: a thin wax film that appears to be ubiquitous and a highly crystalline epicuticular wax consisting of wax crystals that is not present on all species [1, 2]. These outer layers control the wetting of leaves, which have been studied by many researchers in recent years because of the "lotus effect" (i.e., the self-cleaning properties that are a result of very high water repellence, as exhibited by the leaves of the lotus flower) [3–7].

Wetting is the ability of a liquid to maintain contact with a solid surface, resulting from intermolecular interactions when the two are brought together [8]. The degree of wetting is determined by a force balance between adhesive and cohesive forces, which is often characterized by contact angle of water measured at the gas, solid, and liquid interface [9–13]. Contact angle gives an inverse measurement of adhesion between a liquid and a solid. A lower contact angle indicates the liquid will spread over a larger area of the surface. A greater contact angle indicates the liquid will minimize contact with the surface and form a more spherical water droplet.

In a natural environment, leaf surfaces of a large variety of plants are frequently wetted by rainfall, dewfall, ground fog, and cloud mist, and both adaxial and abaxial surfaces are frequently affected. Depending on the tissue hygroscopicity, it may consist of individual drops, or of water films of thickness between a few nanometers and a few micrometers [11]. Several studies have shown that the contact angles between leaf surfaces and water droplets range from 0° to 180°, depending on plant species [3, 4, 9, 10, 12–25]. The differences in physical and chemical properties of leaf surfaces possibly lead to the different contact angles of leaves. These surface properties include the number and pattern of trichomes [12–14], the three-dimensional microstructures of epicuticular cells [3, 4, 15], the microstructures and compositions of wax crystals/films [3, 4, 16–18], and the number and distribution of stomata [12, 13, 19].

Wettability is a comprehensive response at the solid, gas, and liquid phase interface of leaves and it significantly affects physiological and ecological functions of plants. For example, leaf wettability influences pollutant deposition such as acid rain [20, 26, 27], ozone [20, 21], and particulate matter [4, 18, 28, 29], for foliar nutrient leaching [22], in the control of plant disease [23, 30–36], for plant photosynthesis and yield [24, 37, 38], and in the interception of precipitation [25, 39–42]. Recent advances in the area of the wetting of leaf surfaces and its ecological significances are reviewed at present. In Section 2, we discuss the contact angles on ideal and rough surfaces (the Young equation, Wenzel model, Cassie–Baxter model, and Cassie–Baxter to Wenzel transition), the classification of leaf surface wetting, and the methodologies used to measure leaf contact angles. In Section 3, we discuss how the wax content and structure; the number, shape, and pattern of trichomes; the stomatal density; the shape of epidermal cells; and the leaf water status, affect leaf wettability. In Section 4, we discuss the influence of leaf wettability on rainfall interception, photosynthesis rate, pathogen infection, and environmental quality.

2. Criteria and measurement of leaf wettability

2.1. Surface wetting and contact angle

2.1.1. The Young equation

A droplet on a solid surface wets the surface to a certain degree. To what extent a surface gets wet can be described by the contact angle. Contact angle is defined as the angle formed by a liquid at the three-phase boundary where the liquid, gas, and solid intersect (Figure 1). The contact angle directly provides information on three interfacial free energies involved: solid–gas/vapor (γ_{sg}), liquid–gas/vapor (γ_{lg}), and solid–liquid (γ_{sl}) [43]. The balance of forces of a water droplet on an ideal surface (i.e., the surface is flat, rigid, perfectly smooth, and chemically homogeneous, and has zero contact angle hysteresis) is explained by Young's equation [44]:

$$\gamma_{lg} \cos\theta = \gamma_{sg} - \gamma_{sl} \tag{1}$$

where θ is the intrinsic contact angle of a solid. Water droplets form contact angles of 90° when $\gamma_{sg} = \gamma_{sl}$ ($\cos\theta = 0$). Water droplets spread (i.e., $\theta < 90°$) on leaf surfaces when $0 < \gamma_{sg} - \gamma_{sl} < \gamma_{lg}$ ($0 < \cos\theta < 1$). Water droplets spread completely on leaf surfaces when $\gamma_{sg} - \gamma_{sl} = \gamma_{lg}(\cos\theta = 1)$. The contact angles between water droplets and leaf surfaces are greater than 90° when $\gamma_{sg} - \gamma_{sl} < 0$ ($\cos\theta < 0$).

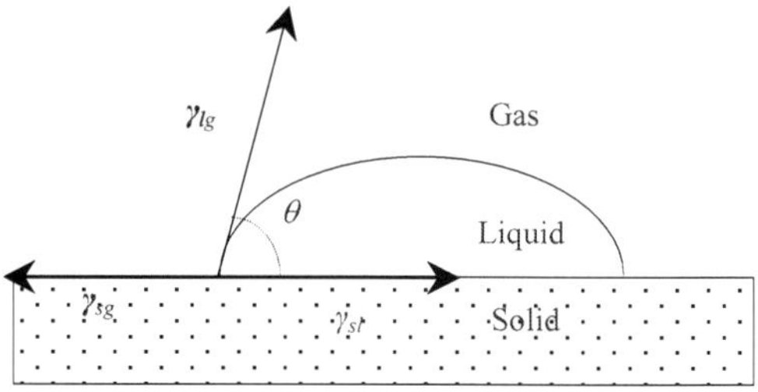

Figure 1. Wetting of a solid surface with water, with air as the surrounding medium.

2.1.2. Wenzel model

Young's equation applies strictly to an ideal surface. However, most leaf surfaces in nature are not perfect smooth, rigidity, or chemical homogeneity because of leaf surface properties [3, 4, 12–18, 45, 46]. Wenzel [47] modified Young's equation to account for contact angles formed on rough surfaces (θ_r) by adding a roughness factor (r) as follows:

$$\cos\theta_r = r\cos\theta \tag{2}$$

The roughness factor is defined as the actual contact area to apparent contact area, and acts as an amplification of the effect of the surface chemistry, i.e., smaller changes in θ become larger changes in θ_r. The value of $\theta = 90°$ is the changeover in sign of the cosine term. When $\theta > 90°$, θ_r is larger than θ, and the effect of increasing roughness is to further increase the θ_r toward to 180°. However, when $\theta < 90°$, θ_r is smaller than θ, and the effect of increasing roughness is to further reduce the θ_r toward to 0°. Thus, the Wenzel roughness emphasizes the intrinsic tendency of a surface toward either completely wetting or complete non-wetting [48].

2.1.3. Cassie–Baxter model

Wenzel's equation accounts for the surface roughness. It describes the homogeneous wetting regime in which water fills the roughness grooves on the surface, as seen in Figure 2a. However, it does not describe contact angle hysteresis that occurs on heterogeneous surfaces [49]. Contact angle hysteresis is defined as the difference between the advancing contact angle (i.e., the contact angle at the advancing edge of a liquid drop, θ_a) and the receding contact angle (i.e., the contact angle at the receding edge of a liquid drop, θ_r) (Figure 3). A more complex model is needed to measure how the apparent contact angle changes when various materials are involved. This heterogeneous surface, as shown in Figure 2b, is explained using the Cassie–Baxter equation [50]:

$$\cos\theta_c = f_1\cos\theta_1 + f_2\cos\theta_2 \tag{3}$$

where θ_1 is the contact angle for component 1 with areal fraction f_1, and θ_2 is the contact angle for component 2 with area fraction f_2 present in the composite material. This equation takes on special meaning when in a two-component system with one component being air with a contact angle of 180°. With cosine (180°) = −1 and $f_2 = 1 - f_1$, Eq. (3) can be reduced to

$$\cos\theta_c = f_1(\cos\theta_1 + 1) - 1 \tag{4}$$

Eq. (4) implies that with a small f_1 and a large θ_1, it is possible to create surfaces with a very large contact angle.

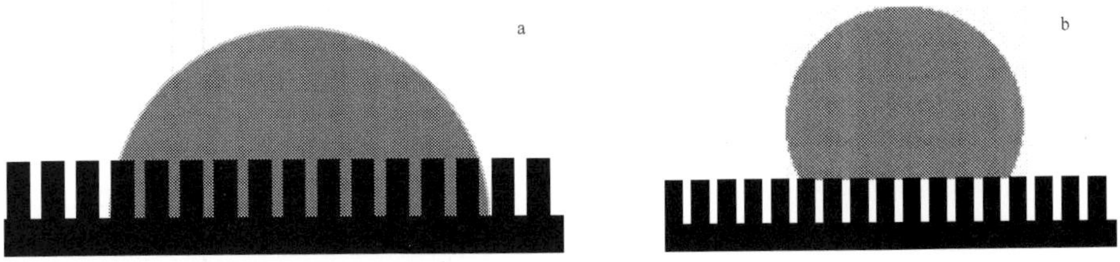

Figure 2. The sketch of Wenzel model (a) and Cassie–Baxter model (b).

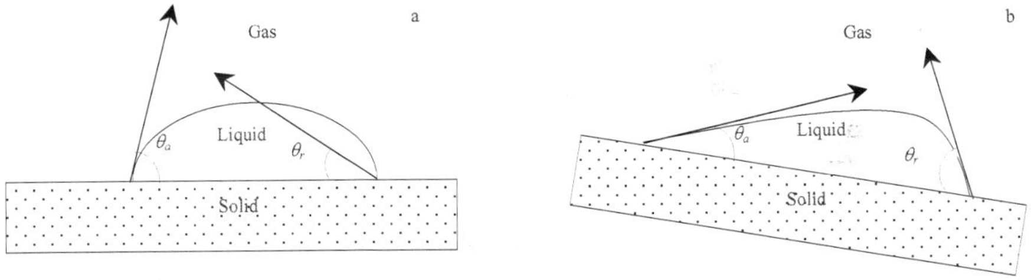

Figure 3. Schematic of advancing and receding contact angles (a) and advancing and receding contact angles determined by tilting experiment (b).

Figure 4. Sketch of a fluid drop on a hydrophilic surface where a film invades the solid texture.

In particular, Wenzel model applies in such case that the leaf surfaces with intermediate hydrophobic and intermediate hydrophilic characteristics, which remain dry beyond the water droplets. If the texture of surfaces is wettable, a film develops in the texture and the droplet sits upon a mixture of solid and liquid, consisting a composite surface (Figure 4) [51]. In this case, the contact angle between the liquid is 0°, and the equation can be written as

$$\cos\theta = f_s \cos\theta + 1 - f_s \tag{5}$$

where f_s is solid areal fraction of the composite surface.

2.1.4. Cassie–Baxter to Wenzel transition

Some researchers studied the stability of the composite interface of superhydrophobic surfaces and the transition from composite to homogeneous interface under pressure [52, 53] and vertical vibration [54]. The intermediate state between the Wenzel and the Cassie modes is shown in Figure 5. The penetration condition is given by

$$\cos\theta = \frac{\varphi_s - 1}{r - \varphi_s} \tag{6}$$

where φ_s is the fraction of the solid/liquid interface below the drop. The penetration front propagates to minimize the surface energy until it reaches the edges of the drop, thus arriving at the Wenzel state.

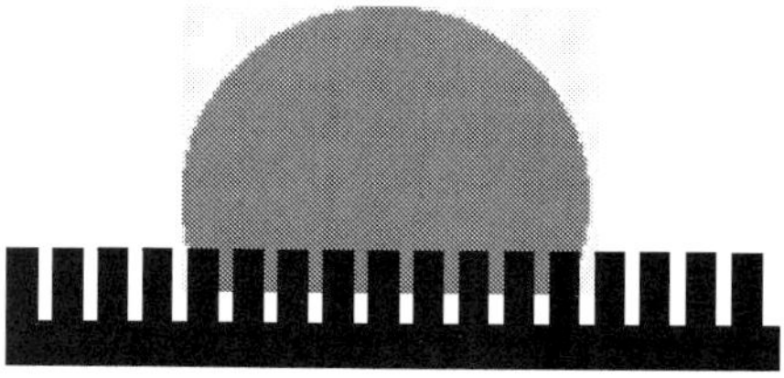

Figure 5. Intermediate state between the Wenzel and the Cassie modes.

2.2. Classification of leaf surface wetting

The larger the contact angle is, the more repellent a leaf surface shows (Figure 6). According to Aryal et al. [55] and Wang et al. [25], the judgment criteria for leaf wettability were as follows: if $\theta < 40°$, 40°– 90°, 90°–110°, 110°–130°, 130°–150°, and > 150°, leaves were termed as "super-hydrophilic," "highly wettable," "wettable," "highly non-wettable," and "superhydrophobic," respectively.

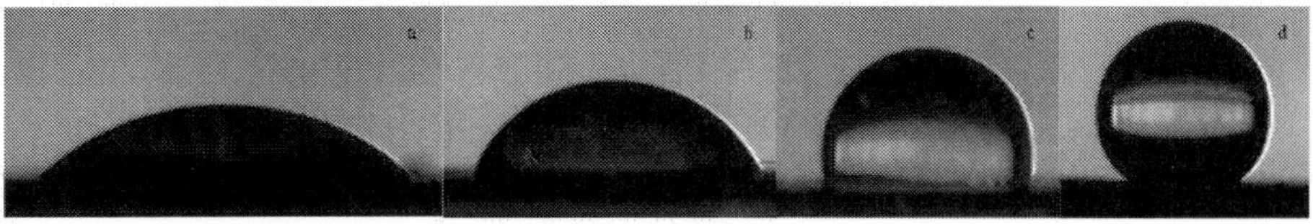

Figure 6. Images from a goniometer of a water droplet on (a) a superhydrophilic, (b) a highly wettable, (c) a non-wettable, and (d) a hydrophobic, leaf surfaces.

2.3. Measurement of leaf contact angle

The measuring methods of leaf contact angle include the static sessile drop method, the dynamic sessile drop method, the dynamic Wilhelmy method, and the Washburn equation capillary rise method [13–17, 28–32, 37–38, 56–78]. The sessile drop method is the most common method, which is measured by a contact angle goniometer using an optical subsystem to capture the profile of a pure liquid on a solid substrate. Older systems used a microscope optical system with a back light. Current-generation systems employ high-resolution cameras and software to capture and analyze the contact angle [77]. Water droplet sizes vary between studies from 0.2 to 15 μl (Table 1). Letellier et al. [79] considered that the contact angle is dependent on not only the nature and structure of the substrate but also the size of the drops. In their study, they found that if interface is a surface and with a thermodynamic dimension (m^{SL}) of 2/3, the contact angle does not depend on the drop mass. However, if interface is a fuzzy interface (i.e., a nonextensive phase), the contact angle depends on the surface wettable characteristics (hydrophilic vs hydrophobic), thermodynamic dimension, and water droplet volume. When $m^{SL} > 2/3$, the contact angle decreases as the volume increases if $\theta < 90°$, while the contact angle value increases with increasing volume if $\theta > 90°$. If $m^{SL} < 2/3$, the contact angle value increases with increasing volume when $\theta < 90°$, while the contact angle value decreases as the volume increases if $\theta > 90°$. However, both Knoll and Schreiber [31] and Schreiber [80] found that contact angles were independent of the droplet volumes between 1 and 10 μl. Therefore, the effects of water droplet volumes on leaf contact angles need further studies to provide relatively accurate information for comparing the results among researchers.

Methods	Water droplet size (μl)	Number of species	The measured contact angle (°)	Location	Reference
Digital camera and image-processing software	7.5	5	52–122	Belgium	17
Microscope and image-processing software	Not reported	1	64.7–138.0	Polish	56
Digital camera and geometric analysis	0.2	1	50–85	Japan	57

Methods	Water droplet size (μl)	Number of species	The measured contact angle (°)	Location	Reference
Binocular microscope with protractor graticule	1	1	72–115	UK	58
Stereomicroscopic photo and image-processing software	5	18	109–136	United States	30
Stereomicroscopic photo and image analysis	5	38	<15–>170	United States	59
Digital camera and image-processing software	10	36	40.33–144.25	Guatemala and United States	9
Digital camera and image-processing software	10	36	40.3–144.3	Guatemala and United States	60
Digital camera and image-processing software	10	5	44.68–77.92	Brazil	61
Digital camera and image-processing software	10	33	<20–>150	Guatemala and United States	62
Automated tension meter	2	1	85–120	Greece	16
Goniometer	2	1	45–75	Germany	31
Goniometer	2	1	40.7–134.3	Spain	14
Goniometer	2	6		New Zealand	63
Goniometer	7	3	60–140	Germany	4
Goniometer	0.2	1	65–99	New England	20
Digital camera and image-processing software	7.5	1	51.2–97.7	Belgium	37
Microprojector and image analysis	2 mm diameter	1	120–150	Australia	64
Drop shape analyzer or digital camera and image-processing software	5	227	50–145	Nepal	55
Not reported	1.5 mm diameter	52	39.9–136.1	New Zealand	65
Geometric analysis with inspection microscope	5	3	85–105	United States	21
Geometric analysis with inspection microscope	5	37	0–180	Argentina	12
Bench microscope with a protractor graticule	Not reported	1	101–108	Italy	66

Methods	Water droplet size (μl)	Number of species	The measured contact angle (°)	Location	Reference
Geometric analysis with inspection microscope	2	1	75–97.2	India	67
Not reported	5	1		United States	38
Geometric analysis with inspection microscope	5	50	23.8–180	United States	68
Goniometer	5	2	66.3–129.4	Japan	23
Digital camera and image analysis	0.2	1	60–120	Japan	69
Digital camera and image analysis	1–42 mm diameter	1	15–100	Sweden	70
Goniometer	2–3 mm diameter	200	117–164	Germany	3
Goniometer and image-processing software	500 μm diameter	50	<50–141	Germany	71
Goniometer	10	11	34.57–120.38	United States	10
Geometric analysis with inspection microscope	5	5	71–130	United States	72
Goniometer and image-processing software	2 or 6	3	67.1–135.9	China	18
Goniometer and image-processing software	5	5	43–146	China	73
Goniometer	5	9	53–153.5	China	74
Goniometer and image-processing software	6	21	42.3–134.7	China	13
Goniometer and image-processing software	6	21	41.5–136.0	China	28
Goniometer and image-processing software	6	18	47.6–142.7	China	29
Not reported	10	33	Not reported	Germany	15
Goniometer	5	2	62–153	United States	75
Goniometer	2	1	50.7–86.1	Germany	32
Goniometer	15	3	100–160	Germany	45
Microprojector and image analysis	3 mm diameter	7	29–152	UK	76

Table 1. Measurement methods and water droplet sizes used to calculate leaf contact angles.

3. Main factors influencing leaf wettability

3.1. Leaf wax

Leaves of higher plants are covered by a cuticle consisting of a cutin matrix with waxes embedded in and deposited on the surface of the matrix [80, 81]. The cutin fraction is a polyester-type biopolymer composed of hydroxyl- and hydroxyepoxy fatty acids, whereas the cuticular waxes are a complex mixture of long-chain aliphatic and cyclic compounds [45]. The major compound classes of plant cuticular waxes are *n*-alkanes (chain-length C_{21-35}) and smaller proportions of iso- and anteiso-homologues, primary alcohols (C_{22-40}), fatty acids (C_{16-34}), aldehydes (C_{21-35}), secondary alcohols (C_{21-35}), with a tendency for mid-chain hydroxylation, ketones (C_{21-35}), β-diketones (C_{22-36}), and *n*-alkyl esters (C_{32-64}) resulting from the combination of long-chain primary alcohols and fatty acids [1, 81, 82]. The cuticular waxes vary enormously between and among different species, different leaf developmental stages, and even between leaf sides [1, 13, 18, 25, 81, 83]. The main components of epicuticular wax always have hydrophobic properties, having contact angles of 94°–109° [84]. However, the contact angles of the investigated leaves cover a wide range (0° –180°), suggesting that the physicochemistry of the cuticular wax (e.g., wax content, composition, and microstructure of cuticular wax) influence leaf surface wettability [4, 17, 24, 62, 63, 75]. Meanwhile, the external environments (e.g., precipitation, temperature, water stress, ozone, and acid rain) can influence the wettability through altering the leaf structures [20, 21, 45, 85–87].

Studies have shown that wax content had distinct effects on leaf wettability [13, 19, 45, 88]. A study conducted by Koch et al. [45] showed that the contact angle increased with the increasing of leaf wax content for the three investigated species (*Brassica oleracea*, *Eucalyptus gunnii*, and *Tropaeolum majus*). Wang et al. [13] investigated the contact angles of water droplets on leaves of 21 plant species and their relation with leaf wax contents. They found that leaf contact anglesincreased with increasing of wax contents. However, the correlation was not significant. Burton and Bhushan [88] measured the contact angles of *Nelumbo nucifera* and *Colocasia esculenta* for both with wax and without wax (removed using acetone). And they found that the contact angle dramatically reduced to less than 90° when the wax was removed from the surface. Their results suggested that the leaf material itself was a hydrophilic material, and the combination of the wax and the roughness of the leaf is what creates the hydrophobic surface. Wang et al. [13] also found that contact angles decreased after wax removal for most species, especially for hydrophobic leaves (e.g., *Ginkgo biloba*, *Rosa chinensis*, and *Berberis thunbergii*). But, an increase in contact angles for a few hydrophilic species (e.g., *Populus canadensis*, *Prunus persica*, and *Koelreuteria paniculata*) after wax removal was observed, and the observed contact angles were always less than 110°. Scanning electronic microscopy (SEM) showed that removal of epicuticular wax by organic solvents could affect the physical appearance and perhaps structure of the foliar surface [89]. The studies of Wang et al. [13], Koch et al. [45], Kumar et al. [19], and Burton and Bhushan [88] suggested that leaf contact angle was more dependent on the complexity of wax structure than on the absolute amount.

Since leaf contact angle was more dependent on the wax structure, recent papers have highlighted the importance of wax structure on leaf wettability [4, 18, 24, 25, 39, 45, 46, 65, 73–

75]. Neinhuis and Barthlott [4] found that the leaves of *G. biloba* were densely covered by wax crystals, resulting in high contact angles of 130°–140° throughout the whole growing season. However, the leaves of *Quercus robur* were partly covered by "amorphous" wax within a few weeks, serving as water repellent for only a few weeks after the wax crystals were produced. Then the contact angle decreased until autumn. The different seasonal changes in leaf wettability of leaves of *G. biloba* and *Q. robur* during the whole growing season maybe due to the chemical composition and regeneration rate of the waxes. The wax forms (crystals vs films) and the regeneration rate of leaves of *Sophora japonica*, *Platanus acerifolia*, and *Cedrus deodara* may result in the significant differences in leaf contact angles and different seasonal changes in wettability of three species [18]. A study conducted in Shaanxi, China, by Wang et al. [25] showed that the presence of wax crystals led to higher contact angles as compared with leaves with wax films. Besides, they found that the leaves densely covered by wax crystals had higher contact angles than those with only sporadic distribution with wax crystals or "amorphous" waxes. Haines et al. [39] observed the relation between leaf surface microstructure and leaf wettability. They reported that the leaves of *Robinia pseudoacacia* and *Liriodendron tulipifera*, having contact angles of 131.6° ± 2.3° and 124.8° ± 1.6°, respectively, were covered by a densely layer of wax crystals. However, the leaves of *Erechtites hieracifolia* were glabrous and waxless, leading to lower contact angles. Hanba et al. [24] have demonstrated that the leaves of *Pisum hieracifolia* were water repellent because of the wax crystals.

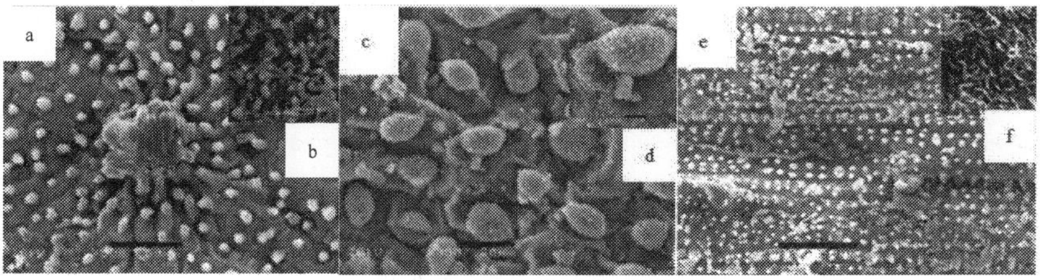

Figure 7. SEM images with different magnifications of lotus (a, b, from references [46, 90]; a: scale bar 20 μm; b: scale bar 1 μm), taro (c, d, from reference [91]; c: scale bar 20 μm, d; scale bar 5 μm), and rice (e, f, from reference [91]; e: scale bar 50 μm, f: scale bar 1 μm).

Barthlott et al. [2] observed surface micromorphology of at least 13,000 species, representing all major groups of seed plants. In total, 23 wax types are classified, for example, granule, platelet, plate, rodlet, thread, and tubule. For water repellency, a classical example is the lotus leaf, which has a very superhydrophobic surface (contact angle over 150°). The SEM images (Figure 7a, b) showed that the surface of the lotus leaf comprises randomly distributed, almost hemispherically topped papillae with sizes 5–10 μm (height to basal radius aspect ratio ~1) decorated with branchlike protrusions with sizes of about 150 nm [46, 90, 91]. Many elliptic protrusions with an average diameter of about 10 μm were uniformly distributed in the nestlike caves, forming a microstructure on taro leaf (*Colocasia*). Many nanoscale pins were also harmoniously disseminated on the whole surface, resulting in a hierarchical structure on its surface together with the formed microstructure (Figure 7c, d). This binary structure resulted in a superhydrophobic leaf surface, with a contact angle of about 159° ± 2° [91]. A binary structure (Figure 7e, f) on surface of rice (*Oryza sativa*) was observed, and leaf surface had a contact angle of 157° ± 2°. On top of the surface, the papillae with average diameter of about 5–8 μm were arranged similarly. Innumerable pins about 20–50 nm were proportionally distributed on the sublayer of surface [91].

3.2. Trichomes

Trichomes, also known as hairs, are fine outgrowths or appendages on plants, which are of diverse structure (e.g., puberulent, hispid, strigose, villous, pilose, strigillose, tomentose, pubescent, downy, and articulate) and function. Plant hairs may be unicellular or multicellular, branched or unbranched. Hairs on plants are extremely variable in their presence across species and even within a species, such as their location on plant organs, size, density. Trichomes can reflect the sunlight, absorb water and nutrients, and reduce transpiration. Besides, trichomes affect the leaf wettability, which has been analyzed in some investigations [12–14, 19, 38, 68].

Some papers have demonstrated that different trichome density and structure may result in different leaf wettability [3, 59, 67, 68, 92]. In a study conducted in the center Rocky Mountain (USA), a positive correlation was observed between contact angle and trichome density for 50 subalpine/montane species [68]. Brewer et al. [59] and Pandey and Nagar [67] considered that the leaves with trichomes were more water repellent, especially where trichome density was greater than 25/mm^2, at which they may develop trichome canopy. The adaxial surface of leaves of *Callistephus chinensis*, having a contact angle of 139°, was densely covered with conoid trichomes, whereas the adaxial surface of leaves of *Cucurbita pepo*, having a contact angle of 70°, was sparsely covered by villous [92]. The study of Wang et al. [92] also found that the variation in structure of trichomes had great influence on leaf wettability. The adaxial surface of *N. nucifera* with waxy trichomes was extremely water repellent (high contact angle of 150°). However, the abaxial surface of *C. chinensis* with nonwaxy trichomes only had a contact angle of 97°. The wettability of hairy leaves strongly depended on the presence or absence of wax crystals on the trichomes. Leaves with nonwaxy trichomes were only water repellent for a short time after a water droplet had been applied. In contrast, leaves with waxy trichomes are extremely water repellent, although the trichomes were up to 2 mm high and only loosely distributed over the leaf surface [3].

Three types of interaction between trichomes and water droplets were evident [38, 68]. On one group of leaf surfaces (Figure 8a), trichomes appeared to have no influence on the location of surface moisture, droplet formation, or retention. In this group, the trichome density was relatively low, and usually a film of water formed on the leaf surface. The leaves of *Abutilon theophrasti*, *Helianthus tuberosus*, and *Prunus triloba* had pilose, and the trichomes may penetrate water droplet deposited on leaves, resulting in contact angles of 43° ± 2°, 46° ± 2°, and 84.8° ± 12.3°, respectively [13, 73]. The enhanced wetting observed in some plant species with "open" trichome pattern caused by capillary action, which segregated water into patches based on water drawn along the trichomes [25, 84]. In a second group, leaf surfaces showed a "segregating strategy" (Figure 8b). In this group, the trichome density was relatively low, but not as low as that on leaf surfaces with no trichome–water interaction, and the hydrophilic trichomes encircled patches of water. Trichomes on the third group of leaf surfaces exhibited the "lifting strategy" (Figure 8c), which water droplets deposited above the trichomes. In this group, the high trichome density might develop trichome canopy. Leaves of *S. japonica* and *P. acerifolia* were densely covered by tiny trichomes that created hierarchical roughness (microbumps superimposed with a nanostructure). Water droplets on these surfaces readily sit on the apex of nanostructures because air bubbles fill in the valleys of the structure under the droplet (Figure 9) [13, 93]. Therefore, these leaves exhibit considerable superhydrophobicity, just like the water droplet on lotus leaf (Figure 9) [93, 94].

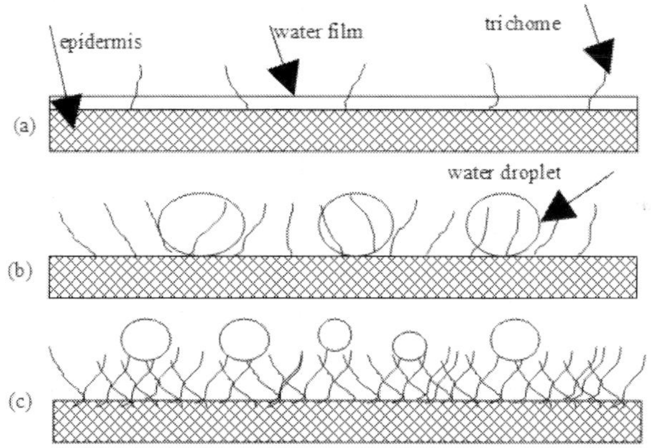

Figure 8. Types of trichome interactions with water. (a) No apparent influence of trichomes on the location of surface moisture, droplet formation or retention. (b) Segregating strategy—trichomes appear to circle surface moisture into patches. (c) Trichomes appear to hold water droplets above the trichomes. (From references [38, 68].)

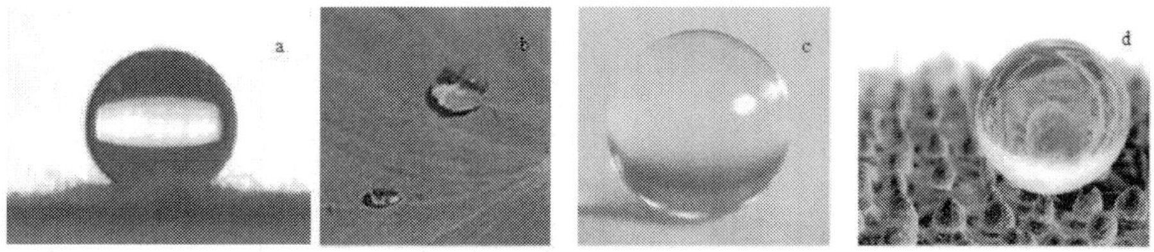

Figure 9. Water droplet on *P. acerifolia* (a, From reference [13]) and lotus leaf (b–d, From references [93, 94].)

3.3. Stomatal density

Stomatal density and aperture (length of stomata) vary under a number of environmental factors such as atmospheric CO_2 concentration, light intensity, air temperature, photoperiod (daytime duration), and pollutants [85, 95–98]. Previous studies have shown that the contact angle on abaxial surface was higher than that on adaxial surface [9, 12, 13, 65, 68], which was consistent with the stomatal distribution on leaf surfaces. The results of Brewer and Nuñez [12] showed that the surface with the greatest concentration of stomata was the least wettable. In a study conducted in Xi'an by Wang et al. [13], they found that a higher stomatal density was accompanied with a higher contact angle. However, the relationship between leaf contact angle and stomatal density was not shown a simple linear relationship. The studies of Juniper and Jeffree [99], Brewer and Smith [38], and Kumar et al. [19] indicated that no significant correlation was observed between contact angle and stomatal density.

3.4. Epidermal cells

A gradient in the cell shape directly related to the contact angle of leaf surfaces was observed in some studies conducted by Wagner et al. [15] (Figure 10), Haines et al. [39], Neinhuis and Barthlott [4], Wang et al. [18], and Wang et al. [25]. Wang et al. [25] found that the leaves with convex epidermal cells with wax crystals, for example, *Cynanchum chinense*, *Agropyron mongolicum*, and *Anemone vitifolia*, had higher contact angles and lower water droplet adhesion. However, the leaves with smooth epidermal cells, such as *Populus simonii* and *Cynanchum komarovii*, had lower contact angles and higher water droplet adhesion. Haines et al. [39] found that the combination of a dense layer of granular wax and the convex epidermal cells was what created a hydrophobic surface of *L. tulipifera*. Wang et al. [18] and Neinhuis and Barthlott [4] both found that the leaf surfaces with convex epidermal cells had higher contact angles than those with flat epidermal cells, and the differences in leaf microstructure could also lead to the differences in leaf wettability during the whole growing season. These studies suggested that the combination of the wax and the convex epidermal cells can create a higher leaf contact angle. However, Neinhuis and Barthlott [3] observed the micromorphological characteristics of 200 water-repellent plant species by using SEM. They found that the scale of the epidermal relief ranged from 5 μm in multipapillate epidermal cells up to 100 μm in large epidermal cells. These variations in scale had almost no effect on leaf wettability of the investigated plant species.

3.5. Leaf water status

Leaf water status of plants is also a key factor in the wetting of leaf surfaces. Quantitative studies on the wetting by water of the exterior surfaces of leaves of *Sinapis arvensis* and *Triticum vulgare* showed that contact angles varied markedly on the detached leaves during the wilting process. This change was reversible on the recovery of turgor, and the magnitude of the contact angle of water droplets on a leaf surface showed a diurnal fluctuation characteristic with a range of as high as 30°. The change of contact angle could be explained by a function of leaf water status in rapidly wilting leaves. In addition, this change of contact angle was related to the corrugation of the leaf surfaces [76]. Weiss [100] studied the relationship between the contact angle of distilled water droplets and the leaf water potential as a function of time of a day on three different types of leaf surface: alfalfa (*Medicago sativa*), a smooth, waxy surface; dry edible beans (*Phaseolus vulgaris*), a corrugated, relatively waxless surface; and soybeans (*Glycine max*), a dense pubescent, corrugated, relatively waxless surface. The results showed

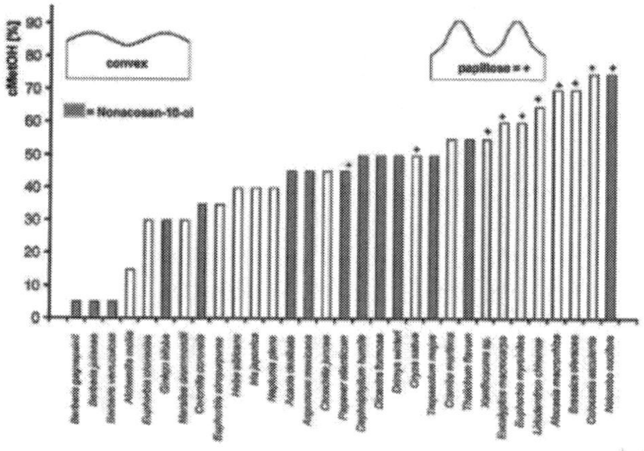

Figure 10. Resistance of leaf surfaces against wetting with water ± methanol mixtures. Leaf surfaces without papillose epidermal cells (on the left) are more easily wetted than those with prominent papillae (right). Gray columns mark wax tubules composed of nonacosan-10-ol, indicating that the high methanol resistance is independent of the individual fine structure of the wax layer but mainly depends on the sculpturing of the outer epidermal cell wall. (From reference [15].)

that although the contact angle of the droplets on dry edible beans and soybeans displayed a diurnal variation, the contact angle remained to be constant at approximately 140° in 1 day for alfalfa leaves with smooth and waxy surface. There were no significant differences among the means of contact angle for alfalfa, dry edible beans, and soybeans under the well-watered condition. These results conflict with those of Fogg [76], who found a diurnal trend in the contact angle of water droplets on leaf surfaces which he attributed to changes in the cuticular structure of the leaves. Therefore, the effect of leaf water content on contact angle of water droplets on different leaf surfaces was complex and needs further studies in the future.

4. Ecological significance of leaf wettability

4.1. Interception of precipitation

Rainfall interception of forest areas is an important hydrological process that alters the quantity, timing, and distribution of water input and output on a catchment. On leaf level, leaf surface characteristics contribute to variability in interception between different plant species, resulting in different geometrical shapes of water on leaves (i.e., water film, patches, drops, and spherical droplets) [65]. Wang et al. [25] investigated leaf water drop adhesion of 60 plant species from Shaanxi, northwest China. The adhesion of water droplets to leaves covered a wide range of area, from 4.09 to 88.87 g/m^2 on adaxial surfaces and 0.72 to 93.35 g/m^2 on abaxial surfaces. The combined values for adaxial and abaxial surfaces in a single species ranged from 5.67 to 159.59 g/m^2. Wilson et al. [40] found that the leaf maximum water storage capacity of potato was 150 g/m^2. They also reported that leaf water accumulation in the upper position of the canopy always exceeded that in the lower canopy, and clumping caused less water accumulation in the upper canopy and greater accumulation in the lower half compared to the random case. Tanakamaru et al. [41] compared leaf water retention by young and old leaves of *Cryptomeria japonica* and used values of 56 and 128 g/m^2, respectively. They attributed this difference in leaf water retention to the epicuticular wax of this species, which was very susceptible to erosion by rainfall. Haines et al. [39] explored the relation between water storage capacities and leaf wettability. They reported that *L. tulipifera*, having a low wettability, captured less water compared to those captured by *E. hieracifolia* and *Platanus occidentalis*, having a high wettability. Hanba et al. [24] reported values of 202, 210 g/m^2 and 120, 116 g/m^2 for the adaxial and abaxial surface of bean and pea, respectively. A survey of 50 subalpine/montane species indicated that moisture accumulation differed among species and habitats [68]. Wohlfahrt et al. [42] investigated the water storage capacities of nine plant species in Stubai Valley and found that the maximum water storage capacities covered a wide range from 13.2 to 314.0 g/m^2. They found the correlations between leaf maximum water storage capacity and leaf perimeter, hemisurface area, shape factor, and specific leaf area were all not significant and suggested that other parameters than those investigated were responsible for determining the maximum water storage capacity for plant species. Therefore, Wang et al. [25] explored leaf physical (roughness) and physicochemical (surface free energy, its dispersive and polar components, and work of adhesion for water) properties in relation to the adhesion of water droplets on leaves. The adhesion of water droplet on leaves decreased as leaf roughness increased but positively correlated with surface free energy, its dispersive component, and work of adhesion for water. However, a significant power correlation was observed between adhesion of water droplet and the polar component of surface free energy. These results indicated that leaf roughness, surface free energy, its components, and work of adhesion for water played important roles in leaf water droplet adhesion.

4.2. Photosynthetic rate

The effect of leaf surface wettability on plant photosynthesis stems from the fact that diffusion of CO_2 is 10,000 times slower in water than in air [9, 12, 24, 38, 72, 101]. Surface wetness induces different changes in leaf photosynthesis among species because leaf surface wettability varied greatly (i.e., from being covered almost completely by water to being water repellent) [24, 68]. For alpine and subalpine plants, natural dew depressed assimilation by 77% in species having wettable leaves, whereas assimilation was stimulated by 14% in species having nonwettable leaves [101]. For bean and pea, a 22% stimulated assimilation rate was obtained for nonwettable pea leaves in the 72-h mist-treated artificial surface, but the wettable bean leaves were on the contrary, which decreased 28%. They postulated that the photosynthetic responses to wetness are due to the change in stomatal regulation [24]. The results of Brewer and Smith [72] indicated that surface wetting, either from natural events or spraying irrigation might lead to significant reduction in CO_2 exchange and growth potential in agricultural species. Leaf surface wetness caused the greatest decline in photosynthesis for the surfaces with the lowest contact angle, which was due to the fact that water physically blocked stomatal pores.

4.3. Pathogen infection

Different plant species vary widely in pathogen infection, perhaps due to leaf micromorphology, surface chemistry, and degree of leaf wetness [19, 81, 102–104]. Water droplets on the leaf surface can be an important source of water for pathogen infection. Excess leaf wetness promotes pathogen infection in many native and agricultural species [19]. Kuo and Hoch [33] found that pycnidiospores of *Phyllosticta ampelicida* could only germinate on substrata on which they were firmly attached. Such surfaces had contact angles of >80°. When pycnidiospores were deposited on more wettable surfaces, they did not attach firmly and did not germinate. Such surfaces had contact angles of < 40°. A significant increase in leaf wettability of wheat (*Triticum aestivum* L. cv. 'Minaret') was observed when the leaves were infected by *Pseudomonas putida*. The densities of *P. putida* decreased when leaf wettability increased [82]. The degradation of epicuticular wax crystals was observed with leaf age [7, 10, 18] and under environmental conditions [56, 87], which offered, in general, a more suitable microhabitat for most phyllosphere organisms [31–32]. In this way, these features increased the coverage by epiphytic microorganisms [31–32]. Huber and Gillespie [34] studied the relation between leaf surface wettability and pathogen infection. They found that the factors (e.g., micrometeorology, leaf wettability, and plant structure) that influenced the duration of dew all influence the pathogen infection. Pinon et al. [35] found that infection by *Melampsora larici-populina* on poplar leaves was dependent on the duration of leaf moisture. They also suggested that leaf wettability should be considered as an additional trait when breeding poplar for durable resistance to *M. larici-populina,* as this characteristic is likely to be a useful defense against all pathotypes of the pathogen. In addition, Cook [36] considered that the difference in leaf wettability provided a quick and discriminating technique for the preliminary screening of cultivars and lines within cultivars of peanut for resistance to *Puccinia arachidis*.

4.4. Environmental quality

It is universally accepted that trees and other vegetation are effective at trapping and absorbing many pollutants, such as particulate matters, CO, NO_2, and SO_2 [17–18, 56, 105–108], and they can act as biological absorbers or filters of pollutants [18, 105–107]. Leaf as the multifunctional interface between plants and environment is also continuously exposed to high levels of varieties of air pollutants. Air pollutants may cause plants both acute and chronic damages on anatomical and morphological characteristics [17–18, 56, 109, 110], leading variations in leaf wettability. Therefore, leaf wettability is potentially a good indicator to point out differences in urban habitat quality [18, 56]. Adams and Hutchinson [26] investigated the ability of four species (cabbage: *Brassica oleracea*; sugar beet: *Beta vulgaris*; radish: *Raphanus sativus*; sunflower: *Helianthus annuus*) to neutralize acid rain with contrasting leaf surfaces (e.g., wettability, droplet retention, hairiness, and thickness of epidermal waxes). Droplets were neutralized on the leaves of radish, and sunflower which had lower contact angles, causing a pH increase of between 0.3 and 1.5 pH unites. On leaves of cabbage and sugar beet, which had larger contact angle, however, droplets were generally acidified. The wettability of leaves of *P. vulgaris, Vicia faba, Pisum sativum,* and *Brassica napus* from emergence to full expansion exposed to simulated acid rain at pH values between 5.6 and 2.6 were investigated by Percy and Baker [27, 111]. Leaf contact angles in all species decreased on leaves exposed to simulated acid rain at pH $\leq$ 4.6 relative to those exposed at pH 5.6. The variations in epidermal wax and epidermal membrane could have important consequences for leaf wettability [111]. Ozone exposure decreased contact angles of *Populus nigra*, *Populus euramericana*, and *Populus menziesii* foliage, and the effects of ozone on cuticular interactions with liquid water varied with the ozone exposure regime and species [21]. Neinhuis and Barthlott [4] and Wang et al. [18] observed the seasonal changes in leaf wettability, the relation between the particulate matter accumulation and leaf wettability. They reported that *G. biloba* and *S. japonica*, having nonwettable leave surfaces, did not show significant seasonal changes in contact angle and particulate matter ability during the growing season. However, *Q. robur*, *Fagus sylvatica*, *C. deodara*, and *P. acerifolia*, having wettable leaf surfaces, showed significant seasonal variations in contact angle and particulate matter capturing ability during the growing season. Besides, they found that the particulate matter capturing ability of leaves of *Q. robur*, *F. sylvatica*, *C. deodara*, and *P. acerifolia* increased with the decrease in contact angle.

5. Conclusions

Leaf surfaces represent the key interfaces between plants and their environment, which influence the biodiversity and biomass, nutrient and water balance, biogeochemical cycle, and productivity of ecosystems. Leaf surface wettability, indicating the affinity for water on the leaf surface, is a common phenomenon for plants in a wide variety of habitats, which directly affect leaf photosynthesis, canopy interception, pathogen infection, and environmental quality. Many studies concentrated on the differences in leaf surface wettability and its relation with leaf microstructure. Leaf surface wettability has been considered to be of great theoretical and academic importance, and studies focused on the ecological significances of leaf wettability should be encouraged in the future.

Acknowledgements

This work was supported by the National Natural Science Foundation of China (grant no. 41230852).

Author details

Huixia Wang[1*], Hui Shi[1] and Yanhui Wang[2]

*Address all correspondence to: dasenlin-80@163.com

1 School of Environmental and Municipal Engineering, Xi'an University of Architecture and Technology, Xi'an, China

2 Institute of Forest Ecology, Environment and Protection, Chinese Academy of Forestry, Beijing, China

References

[1] Taylor P. The wetting of leaf surfaces. Current Opinion in Colloid and Interface Science, 2011, 16(4): 326–334.

[2] Barthlott W, Neinhuis C, Cutler D, Ditsch F, Meusel I, Theisen I, Wilhelmi H. Classification and terminology of plant epicuticular waxes. Botanical Journal of the Linnean Society, 1998, 126(3): 237–260.

[3] Neinhuis C, Barthlott W. Characterization and distribution of water-repellent, self-cleaning plant surfaces. Annals of Botany, 1997, 79: 667–677.

[4] Neinhuis C, Barthlott W. Seasonal changes of leaf surface contamination in beech, oak, and ginkgo in relation to leaf micromorphology and wettability. New Phytologist, 1998, 138: 91–98.

[5] Fürstner R, Barthlott W, Neinhuis C, Walzel P. Wetting and self-cleaning properties of artificial superhydrophobic surfaces. Langmuir, 2005, 21: 956–961.

[6] Hansen W R, Autumn K. Evidence for self-cleaning in gecko setae. Proceedings of the National Academy of Sciences of the United States of America, 2005, 102(2): 385–389.

[7] Lotus effect. Wikipedia, the Free Encyclopedia. http://en.wikipedia.org/wiki/Lotus effect. (accessed 5 May 2015).

[8] Wetting. Wikipedia, the Free Encyclopedia. http://en.wikipedia.org/wiki/Wetting (accessed 12 December 2014).

[9] Holder C D. Leaf water repellency of species in Guatemala and Colorado (USA) and its significance to forest hydrology studies. Journal of Hydrology, 2007, 336: 147–154.

[10] Holder C D. The relationship between leaf hydrophobicity, water droplet retention, and leaf angle of common species in a semi-arid region of the western United States. Agricultural and Forest Meteorology, 2012, 152: 11–16.

[11] Sutton M A, Spindler G, van Putten E. A climatology of leaf surface wetness. Theoretical and Applied Climatology, 2002, 71: 107–117.

[12] Brewer C A, Nuñez C I. Patterns of leaf wettability along an extreme moisture gradient in western Patagonia, Argentina. International Journal of Plant Sciences, 2007, 227, 168(5): 555–562.

[13] Wang H X, Shi H, Li Y Y. Leaf surface wettability of major plant species for urban greening in Xi'an and related affecting factors. Chinese Journal of Ecology, 2010, 29(4): 630–636. (in Chinese with English abstract)

[14] Fernández V, Sancho-Knapik D, Guzmán P, Peguero-Pina J J, Gil L, Karabourniotis G, Khayet M, Fasseas C, Heredia-Guerrero J A, Heredia A, Gil-Pelegrín E. Wettabili-

ty, polarity and water absorption of *Quercusilex* leaves: effect of leaf side and age. Plant Physiology, 2014, 166(1): 168-180.

[15] Wagner P, Fürstner R, Barthlott W, Neinhuis C. Quantitative assessment to the structural basis of water repellency in natural and technical surfaces. Journal of Experimental Botany, 2003, 54(385): 1295–1303.

[16] Kolyva F, Stratakis E, Rhizopoulou S, Chimona C, Fotakis C. Leaf surface characteristics and wetting in *Ceratoniasiliqua* L. Flora, 2012, 207: 551–556.

[17] Kardel F, Wuyts K, Babanezhad M, Wuytack T, Adriaenssens S, Samson R. Tree leaf wettability as passive bio-indicator of urban habitat quality. Environmental and Experimental Botany, 2012, 75: 277–285.

[18] Wang H X, Shi H, Li Y Y, Yu Y, Zhang J. Seasonal variations in leaf capturing of particulate matter, surface wettability and micromorphology in urban tree species. Frontiers of Environmental Science and Engineering, 2013, 7(4): 579–588.

[19] Kumar N, Pandey S, Bhattacharya A, Ahuja P S. Do leaf surface characteristics affect *Agrobacterium* infection in tea [*Camellia sinensis* (L.) O Kuntze]? Journal of Biosciences, 2004, 29(3): 309–317.

[20] Percy K E, Jensen K F, McQuattie C J. Effects of ozone and acidic fog on red spruce needle epicuticular wax production, chemical composition, cuticular membrane ultrastructure and needle wettability. New Phytologist, 1992, 122: 71–80.

[21] Schreuder M D J, van Hove L W A, Brewer C A. Ozone exposure affects leaf wettability and tree water balance. New Phytologist, 2001, 152: 443–454.

[22] Adriaenssens S, Staelens J, Wuyts K, de Schrijver A, Van Wittenberghe S, Wuytack T, Kardel F, Verheyen K, Samson R, Boeckx P. Foliar nitrogen uptake from wet deposition and the relation with leaf wettability and water storage capacity. Water, Air, and Soil Pollution, 2011, 219(1–4): 43–57.

[23] Burch A Y, Zeisler V, Yokota K, Schreiber L, Lindow S E. The hygroscopic biosurfactant syringafactin produced by *Pseudomonas syringae* enhances fitness on leaf surfaces during fluctuating humidity. Environmental Microbiology, 2014, 16(7): 2086–2098.

[24] Hanba Y T, Moriya A, Kimura K. Effect of leaf surface wetness and wettability on photosynthesis in bean and pea. Plant, Cell and Environment, 2004, 27, 413–421.

[25] Wang H X, Shi H, Li Y Y, Wang Y H. The effects of leaf roughness, surface free energy and work of adhesion on leaf water drop adhesion. PLoS One, 2014, 9(9): e107062. doi:10.1371/journal.pone.0107062.

[26] Adams C M, Hutchinson T C. Comparative abilities of leaf surfaces to neutralize acidic raindrops II. The influence of leaf wettability, leaf age and rain duration on changes in droplet pH and chemistry on leaf surfaces. New Phytologist, 1987, 106(3): 437–456.

[27] Percy KE, Baker EA. Effects of simulated acid rain on leaf wettability, rain retention and uptake of some inorganic ions. New Phytologist, 1988, 108(1): 75–82.

[28] Wang H X, Shi H, Li Y Y. Relationships between leaf surface characteristics and dust capturing capability of urban greening plant species. Chinese Journal of Applied Ecology, 2010, 21(12): 3077–3082. (in Chinese with English abstract)

[29] Liu L, Guan D S, Chen Y Q D. Morphological structure of leaves and dust-capturing capability of common street trees in Guangzhou Municipality. Acta Ecologica Sinica, 2013, 33(8): 2604–2614. (in Chinese with English abstract)

[30] Bradley D J, Gilbert G S, Parker I M. Susceptibility of clover species to fungal infection: the interaction of leaf surface traits and environment. American Journal of Botany, 2003, 90(6): 857–864.

[31] Knoll D, Schreiber L. Influence of epiphytic micro-organisms on leaf wettability: wetting of the upper leaf surface of *Juglans regia* and of model surfaces in relation to colonization by micro-organisms. New Phytologist, 1998, 140: 271–282.

[32] Knoll D, Schreiber L. Plant–microbe interactions: wetting of ivy (*Hedera helix* L.) leaf surfaces in relation to colonization by epiphytic microorganisms. Microbial Ecology, 2000, 41: 33–42.

[33] Kuo K C, Hoch H C. Germination of *Phyllosticta ampelicida* pycnidiospores: prerequisite of adhesion to the substratum and the relationship of substratum wettability. Fungal Genetics and Biology, 1996, 20(1): 18–29.

[34] Huber L, Gillespie T J. Modeling leaf wetness in relation to plant disease epidemiology. Annual Review of Phytopathology, 1992, 30: 553–577.

[35] Pinon J, Frey P, Husson C. Wettability of poplar leaves influences dew formation and infection by *Melampsora larici-populina.* Plant Disease, 2006, 90: 177–184.

[36] Cook M. Peanut leaf wettability and susceptibility to infection by *Puccinia arachidis.* Phytopathology, 1980, 70(8): 826–830.

[37] Van Wittenberghe S, Adriaenssens S, Staelens J, Verheyen K, Samson R. Variability of stomatal conductance, leaf anatomy, and seasonal leaf wettability of young and adult European beech leaves along a vertical canopy gradient. Trees, 2012, 26(5): 1427-1438.

[38] Brewer C A, Smith W K. Leaf surface wetness and gas exchange in the pond lily *Nuphar polysepalum* (Nymphaeaceae). American Journal of Botany, 1995, 82(10): 1271–1277.

[39] Haines B L, Jernstedt J A, Neufeld H S. Direct foliar effects of simulated acid rain II. Leaf surface characteristics. New Phytologist, 1985, 99: 407–416.

[40] Wilson T B, Bland W L, Norman J M. Measurement and simulation of dew accumulation and drying in potato canopy. Agriculture and Forest Meteorology, 1999, 93: 111–119.

[41] Tanakamaru S, Takehana T, Kimura K. Effect of rainfall exposure on leaf wettability in near-isogenic barley lines with different leaf wax content. Journal of Agricultural Meteorology, 1998, 54(2): 155–160.

[42] Wohlfahrt G, Bianchi K, Cernusca A. Leaf and stem maximum water storage capacity of herbaceous plants in a mountain meadow. Journal of Hydrology, 2006, 319: 383–390.

[43] Rosado B H P, Holder C D. The significance of leaf water repellency in ecohydrological research: a review. Ecohydrology, 2013, 6(1): 150–161.

[44] Young T. An essay on the cohesion of fluids. Philosophical Transactions of the Royal Society of London, 1805, 95: 65–87.

[45] Koch K, Hartmann K D, Schreiber L, Barthlott W, Neinhuis C. Influences of air humidity during the cultivation of plants on wax chemical composition, morphology and leaf surface wettability. Environmental and Experimental Botany, 2006, 56: 1–9.

[46] Koch K, Bhushan B, Barthlott W. Multifunctional surface structures of plants: an inspiration for biomimetics. Progress in Materials Science, 2009, 54: 137–178.

[47] Wenzel R N. Resistance of solid surfaces to wetting by water. Industrial and Engineering Chemistry, 1936, 28: 988–994.

[48] Shirtcliffe N J, McHale G, Atherton S, Newton M I. An introduction to superhydrophobicity. Advances in Colloid and Interface Science, 2010, 161: 124–138.

[49] Marmur A. Wetting on hydrophobic rough surfaces: to be heterogeneous or not to be? Langmuir, 2003, 19(20): 8343–8348.

[50] Cassie A B D, Baxter S. Wettability of porous surfaces. Transaction of the Faraday Society, 1944, 40: 546–551.

[51] Bico J, Thiele U, Quéré D. Wetting of textured surfaces. Colloids and Surfaces A: Physicochemical and Engineering Aspects, 2002, 206(1): 41–46.

[52] Yoshimitsu Z, Nakajima A, Watanabe T, Hashimoto K. Effects of surface structure on the hydrophobicity and sliding behavior of water droplets. Langmuir, 2002, 18(15): 5818–5822.

[53] Liu B, Lange F F. Pressure induced transition between superhydrophobic states: configuration diagrams and effect of surface feature size. Journal of Colloid and Interface Science, 2006, 298(2): 899–909.

[54] Bormashenko E, Pogreb R, Whyman G, Erlich M. Cassie–Wenzel wetting transition in vibrating drops deposited on rough surfaces: is the dynamic Cassie–Wenzel wetting transition a 2D or 1D affair? Langmuir, 2007, 23(12): 6501–6503.

[55] Aryal B, Neuner G. Leaf wettability decreases along an extreme altitudinal gradient. Oecologia, 2010, 162: 1–9.

[56] Pogorzelski S J, Rochowski P, Szurkowski J. *Pinus sylvestris* L. needle surface wettability parameters as indicators of atmospheric environment pollution impacts: novel contact angle hysteresis methodology. Applied Surface Science, 2014, 292: 857–866.

[57] Takamatsu T, Sase H, Takada J. Some physiological properties of *Cryptomeria japonica* leaves from Kanto, Japan: potential factors causing tree decline. Canadian Journal of Forest Research, 2001, 31: 663–672.

[58] Cape J N. Contact angles of water droplets on needles of Scots pine (*Pinus sylvestris*) growing in polluted atmosphere. New Phytologist, 1983, 93: 293–299.

[59] Brewer C A, Smith W K, Vogelmann T C. Functional interaction between leaf trichomes, leaf wettability and the optical properties of water droplets. Plant, Cell, and Environment, 1991, 14: 955–962.

[60] Holder C D. Leaf water repellency as an adaptation to tropical montane cloud forest environments. Biotropica, 2007, 39(6): 767–770.

[61] Rosado B H P, Oliveira R S, Aidar M P M. Is leaf water repellency related to vapor pressure deficit and crown exposure in tropical forests? Acta Oecologica, 2010, 36: 645–649.

[62] Holder C D. The relationship between leaf water repellency and leaf traits in three distinct biogeographical regions. Plant Ecology, 2011, 212: 1913–1926.

[63] Nairn J J, Forster W A, van Leeuwen R M. Quantification of physical (roughness) and chemical (dielectric constant) leaf surface properties relevant to wettability and adhesion. Pest Management Science, 2011, 67(12): 1562–1570.

[64] Netting A G, von Wettstein-Knowles P. The physico-chemical basis of leaf wettability in wheat. Planta, 1973, 114(4): 289–309.

[65] Hall D M, Burke W. Wettability of leaves of a section of New Zealand plants. New Zealand Journal of Botany, 1974, 12: 283–298.

[66] Paoletti E, Raddi P, La Scala S. Relationships between transpiration, stomatal damage and leaf wettability in declining beech trees. Chemosphere, 1998, 36(4–5): 907–912.

[67] Pandey S, Nagar P K. Leaf surface wetness and morphological characteristics of *Valeriana jatamansi* grown under open and shade habitats. Biological Plantarum, 2002, 45(2): 291–294.

[68] Brewer C A, Smith W K. Patterns of leaf surface wetness for montane and subalpine plants. Plant, Cell and Environment, 1997, 20: 1–11.

[69] Sase H, Takahashi A, Sato M, Kobayashi H, Nakata M, Totsuka T. Seasonal variation in the atmospheric deposition of inorganic constituents and canopy interactions in a Japanese cedar forest. Environmental Pollution, 2008, 152: 1–10.

[70] Sundberg M, Månsson A, Tågerud S. Contact angle measurements by confocal microscopy for non-destructive microscale surface characterization. Journal of Colloid and Interface Science, 2007, 313: 454–460.

[71] Hauck M, Jürgens S R, Brinkmann M, Herminghaus S. Surface hydrophobicity causes SO_2 tolerance in lichens. Annals of Botany, 2008, 101: 531–539.

[72] Brewer C A, Smith W K. Influence of simulated dewfall on photosynthesis and yield in soybean isoline (*Glycine max* [L.] merr. cv Williams) with different trichome densities. International Journal of Plant Sciences, 1994, 155(4): 460–466.

[73] Yang X D, Shang G R, Li Y T, Xuang M. Surface morphological characteristics comparisons of serious plant leaves with hydrophobic or hydrophilic function. Journal of Northwest Normal University (Natural Science Edition), 2006, 38(3): 91–95. (in Chinese with English abstract).

[74] Han Z W, Qiu Z M, Wang S J, Ren L Q. Relationship between non-smooth appearance and wettability of plant leaf surface. Journal of Jilin University (Engineering and Technology Edition), 2008, 38(1): 110–115. (in Chinese with English abstract)

[75] Bhushan B, Jung Y C. Wetting, adhesion and friction of superhydrophobic and hydrophilic leaves and fabricated micro/nanopatterned surfaces. Journal of Physics: Condensed Matter, 2008, 20. doi:10.1088/0953-8984/20/22/225010.

[76] Fogg G E. Quantitative studies on the wetting of leaves by water. Proceedings of the Royal Society of London. Series B, Biological Sciences, 1947, 134(877): 503–522.

[77] Contact angle. Wikipedia, the Free Encyclopedia. http://en.wikipedia.org/wiki/Contact angle. (accessed 12 December 2014).

[78] Shen Q, Ding H G, Zhong L. Characterization of the surface properties of persimmon leaves by FT-Raman spectroscopy and wicking technique. Colloids and Surfaces B: Biointerfaces, 2004, 37: 133–136.

[79] Letellier P, Mayaffre A, Turmine M. Drop size effect on contact angle explained by nonextensive thermodynamics. Young's equation revisited. Journal of Colloid and Interface Science, 2007, 314: 604–614.

[80] Schreiber L. Wetting of the upper needle surface of *Abies grandis*: influence of pH, wax chemistry and epiphyllic microflora on contact angles. Plant, Cell and Environment, 1996, 19: 455–463.

[81] Müller C, Riederer M. Plant surface properties in chemical ecology. Journal of Chemical Ecology, 2005, 31(11): 2621–2651.

[82] Kunst L, Samuel A L. Biosynthesis and secretion of plant cuticular wax. Progress in Lipid Research, 2003, 42: 51–80.

[83] Faini F, Labbé C, Coll J. Seasonal changes in chemical composition of epicuticular waxes from the leaves of *Baccharis linearis*. Biochemical Systematics and Ecology, 1999, 27: 673–679.

[84] Holloway P J. Chemistry of leaf waxes in relation to wetting. Journal of the Science of Food and Agriculture, 1969, 20(2): 124–128.

[85] Pal A, Kulshreshtha K, Ahmad K J, Behl H M. Do leaf surface characters play a role in plant resistance to auto-exhaust pollution? Flora, 2002, 197: 47–55.

[86] Cameron K D, Teece M A, Bevilacqua E, Smart L B. Diversity of cuticular wax among Salix species and *Populus* species hybrids. Phytochemistry, 2002, 60(7): 715–725.

[87] Xu S J, Jiang P A, Wang Z W, Wang Y. Crystal structures and chemical composition of leaf surface wax depositions on the desert moss *Syntrichia caninervis*. Biochemical Systematics and Ecology, 2010, 37(6): 723–730.

[88] Burton Z, Bhushan B. Surface characterization and adhesion and friction properties of hydrophobic leaf surfaces. Ultramicroscopy, 2006, 106: 709–719.

[89] Boyce R L, McCune D C, Berlyn G P. A comparison of foliar wettability of red spruce and balsam fir growing at high elevation. New Phytologist, 1991, 117(4): 543–555.

[90] Zobra V, Stratakis E, Barberoglou M, Spanakis E, Tzanetakis P, Anastasiadis S H, Fotakis C. Biomimetic artificial surfaces quantitatively reproduce the water repellency of a lotus leaf. Advanced Materials, 2008, 20: 4049–4054.

[91] Guo Z G, Liu W M. Biomimic from the superhydrophobic plant leaves in nature: binary structure and unitary structure. Plant Science, 2007, 172(6): 1103–1112.

[92] Wang S J, Ren L Q, Han Z W, Qiu Z M, Zhou C H.. Non-smooth morphology of typical plant leaf surface and its anti-adhesion and hydrophobicity. Transactions of the Chinese Society of Agricultural Engineering, 2005, 21(9): 16–19. (in Chinese with English abstract)

[93] Bhushan B, Jung Y C. Natural and biomimetic artificial surfaces for superhydrophobicity, self-cleaning, low adhesion, and drag reduction. Progress in Materials Science, 2011, 56: 1–108.

[94] Computer graphics of the lotus-effect. http://wthielicke.gmxhome.de/bionik/index-uk.htm (accessed 10 January 2015).

[95] Doheny-Adams T, Hunt L, Franks P J, Beerling D J, Gray J E. Genetic manipulation of stomatal density influences stomatal size, plant growth and tolerance to restricted

water supply across a growth carbon dioxide gradient. Philosophical Transactions of the Royal Society B: Biological Sciences, 2012, 367(1588): 547–555.

[96] Rashidi F, Jalili A, Kafaki S B, Sagheb-Talebi K, Hodgson J. Anatomical responses of leaves of Black Locust (*Robinia pseudoacacia* L.) to urban pollutant gases and climatic factors. Trees, 2012, 26(2): 363–375.

[97] Haworth M, Elliott-Kingston C, Gallagher A, Fitzgerald A, McElwain J C. Sulphur dioxide fumigation effects on stomatal density and index of non-resistant plants: implications for the stomatal palaeo-[CO_2] proxy method. Review of Palaeobotany and Palynology, 2012, 185: 44–54.

[98] Carins Murphy M R, Jordan G J, Brodribb T J. Acclimation to humidity modifies the link between leaf size and the density of veins and stomata. Plant, Cell and Environment, 2014, 37(1): 124–131.

[99] Juniper B E, Jeffree C E. Plant Surfaces. London: Edward Arnold, 1983.

[100] Weiss A. Contact angle of water droplets in relation to leaf water potential. Agricultural and Forest Meteorology, 1988, 43(3): 251–259.

[101] Smith W K, McClean T M. Adaptive relationship between leaf water repellency, stomatal distribution, and gas exchange. American Journal of Botany, 1989, 76(3): 465–469.

[102] Rottstock T, Joshi J, Kummer V, Fischer M. Higher plant diversity promotes higher diversity of fungal pathogens, while it decreases pathogen infection per plant. Ecology, 2014, 95: 1907–1917.

[103] Boyette C D, Hoagland R E. Bioherbicidal potential of a strain of *Xanthomonas* spp. for control of common cocklebur (*Xanthium strumarium*). Biocontrol Science and Technology, 2013, 23(2): 183–196.

[104] Szechyńska-Hebda M, Hebda M, Mierzwiński D, Kuczyńska P, Mirek M, Wędzony M, van Lammeren A, Karpiński S. Effect of cold-induced changes in physical and chemical leaf properties on the resistance of winter triticale (× *Triticosecale*) to the fungal pathogen *Microdochium nivale*. Plant Pathology, 2013, 62(4): 867–878.

[105] Beckett K P, Freer-Smith P H, Taylor G. Urban woodlands: their role in reducing the effects of particulate pollution. Environmental Pollution, 1998, 99: 347–360.

[106] Nowak D J, Hirabayashi S, Bodine A, Hoehn R. Modeled $PM_{2.5}$ removal by trees in ten US cities and associated health effects. Environmental Pollution, 2013, 178: 395–402.

[107] Nowak D J, Crane D E, Stevens J C. Air pollution removal by urban trees and shrubs in the United States. Urban Forestry and Urban Greening, 2006, 4(3–4): 115–123.

[108] Shi H, Wang H X, Li Y Y. Wettability on plant leaf surfaces and its ecological significance. 2011. ActaEcologicaSinica, 2011, 31(15): 4287–4298. (in Chinese with English abstract)

[109] Kayode J, Otoide J E. Environmental pollution and leaf cuticular variations in *Newbouldia laevis* Seem. ex Bureau. Asian Journal of Plant Sciences, 2007, 6(6): 1024–1026.

[110] Gostin I N. Air pollution effects on the leaf structure of some *Fabaceae* species. Notulae Botanicae Horti Agrobotanici Cluj-Napoca, 2009, 37(2): 57–63.

[111] Percy K E, Baker E A. Effects of simulated acid rain on production, morphology and composition of epicuticular wax and on cuticular membrane development. New Phytologist, 1987, 107(3): 577–589.

Chapter 12

Wettability and Other Surface Properties of Modified Polymers

Nikola Slepickova Kasalkova, Petr Slepicka,
Zdenka Kolska and Vaclav Svorcik

Additional information is available at the end of the chapter

http://dx.doi.org/10.5772/60824

Abstract

Surface wettability is one of the crucial characteristics for determining of a material's use in specific application. Determination of wettability is based on the measurement of the material surface contact angle. Contact angle is the main parameter that characterizes the drop shape on the solid surface and is also one of the directly measurable properties of the phase interface. In this chapter, the wettability and its related properties of pristine and modified polymer foils will be described. The wettability depends on surface roughness and chemical composition. Changes of these parameters can adjust the values of contact angle and, therefore, wettability. In the case of pristine polymer materials, their wettability is unsuitable for a wide range of applications (such as tissue engineering, printing, and coating). Polymer surfaces can easily be modified by, e.g., plasma discharge, whereas the bulk properties remain unchanged. This modification leads to oxidation of the treated layer and creation of new chemical groups that mainly contain oxygen. Immediately after plasma treatment, the values of the contact angles of the modified polymer significantly decrease. In the case of a specific polymer, the strongly hydrophilic surface is created and leads to total spreading of the water drop. Wettability is strongly dependent on time from modification.

Wettability plays a key role, e.g., in the development of biomaterials in tissue engineering and regenerative medicine. Biocompatibility tests of the cell adhesion, proliferation and viability are performed in an aqueous medium, and it is necessary to control the surface wettability. Various cell types have different requirements on surface properties, but while maintaining suitable parameters, the optimal value of water contact angle for cell adhesion is in the interval of 50-70°. In the case of polymers, which have usually higher values of contact angles, the decrease in water contact angle and adjustment of the surface may be accomplished by introducing selected chemical groups (by plasma treatment), chemical compounds (by grafting, i.e., chemical bath deposition), or nanoparticles. In the case of grafting of polyethylene glycol (PEG) on the plasma activated high-density polyethylene was under "properly" selected conditions of modification (molecular weight of PEG, concentration of PEG in solution) prepared layers that positively influenced the cell adhesion. In addition, low concentration of gold nanoparticles increased the number of adhered cells without decreasing cell viability.

Whether the surface of the polymeric substrate is attractive for the adhesion and proliferation of cells is determined not only by wettability but also by a range of other surface properties of the substrate. Some of these properties, e.g., morphology, chemical composition, and electrokinetic potential (zeta potential), will be discussed in this chapter. These properties will also be discussed in relation to the cytocompatibility of surface.

Keywords: wettability, polymer, surface modification, plasma, morphology

1. Introduction

Polymeric materials can be used in many applications due to their mechanical (e.g., elastic) characteristics, chemical stability, and light weight, as well as their several design possibilities. Polymers are very well applied in tissue engineering, in particular, due to a wide range of physical, mechanical, and chemical properties, cytocompatibility, microstructure, and biodegradability [1, 2]. These properties can precisely be controlled by the composition and processing of polymers. Many of these polymers were originally developed as plastics, elastomers, or fibers for medical and industrial applications, but the latter have been adopted for biomedical use [3–5]. Polymeric materials differ in their molecular weight, polydispersity, crystallinity, thermal transitions, and decomposition rate, which can significantly affect the properties of the polymer matrix [6]. The chemical composition, structure, and morphology of polymers influence the surface-free energy and thus have a great influence on wettability [7, 8]. Wettability is a fundamental property of solid-state surfaces that plays important roles in the industry [9]. The wettability of a material surface is generally characterized in terms of the contact angle (CA) between solid and liquid interfaces [10]. Wettability can be described as a property governed by intermolecular interactions that characterizes the degree of wetting of a solid surface by a liquid droplet. One common way of characterizing wettability is through the measurement of the static CA θeq that results from the equilibrium between the interfacial tensions between solid, liquid, and vapor phases [11]. A surface is typically defined as water wetting for (approximately) $\theta int<60°$ to 75° (measured through the aqueous phase), NAPL (non-aqueous phase liquid) wetting for $\theta int>105°$ to 120°, and intermediate wetting for

75°<θint<105° [12]. The exact boundaries between each class are quite inexact, particularly regarding the limits that separate hydrophilicity/hydrophobicity and hydrophobicity/ superhydrophobicity. Wettability is a very important parameter of materials that specifies and limits the next material's application [13, 14]. Assessment of the wettability, namely, measurement of the CA, can also be used to determine the surface stabilization of the modified polymer [15].

Summarized results of our scientific group in the field of research
Ing. Nikola Slepičková Kasálková, Ph.D., born 1981; assistant professor; plasma modification of polymers; surface structure characterization; study of polymer surface cytocompatibility (cell adhesion and proliferation) and goniometry; according to Web of Science - publications 36, H index 8, citations 188. **References:** [15, 57, 78, 79, 80, 92, 106, 107, 109, 110, 112]
Doc. Ing. Petr Slepička, Ph.D., born 1980; associate professor of Materials Science and Engineering; plasma and laser beam modification of polymers; modification of surface structure; coatings patterning; microscopy (AFM, confocal and FIB SEM) characterization; according to Web of Science - publications 94, H index 17, citations 874). **References:** [15, 16, 18, 19, 38, 40, 42, 43, 57, 78, 79, 92, 106, 107, 109, 110, 112, 114]
Doc. Ing. Zdeňka Kolská, Ph.D., born 1969; associate professor of Physical Chemistry; modification of surface structure of materials; physicochemical characterization of material surfaces (electrokinetic and surface area analyses, goniometry); according to Web of Science - publications 76, H index 14, citations 543. **References:** [15, 16, 18, 38, 40, 41, 78, 92, 111, 112]
Prof. Ing. Václav Švorčík, DrSc., born 1957; professor of Materials Science and Engineering; ion, plasma and laser beam modification of polymers; modification of surface structure; surface, chemical, physical properties and cytocompatibility of polymers, metal nanostructures, preparation and characterization; according to Web of Science - publications 324, H index 28, citations 3051. **References:** [15, 16, 18, 19, 38, 40, 41, 42, 43, 57, 78, 79, 80, 81, 92, 95, 102, 106, 107, 109, 110, 111, 112, 114, 116]

2. Laser-treated surfaces

Surface modification can be applied with respect to the applicability of polymer surface and interface in the construction of surface-enhanced substrates, which, for instance, provide biocompatibility or anti-microbial properties, may be used in sensor applications or for support of metal layers deposition [16-20]. Surface properties, such as wettability, morphology, and surface chemistry, play a crucial role in polymer/material application. A compilation of surface and interface analysis techniques will be described in this chapter, since besides the surface laser treatment itself, it is of great importance to understand the surface changes induced by a particular laser type (semiconductor, excimer, etc.) or by set up of input irradiation parameters. The determination of the CA and surface-free energy is one the most widely used surface characterization techniques. It provides very valuable information on the wetting properties of the surface and is determined by the upper molecular layer at the surface. The interpretation of measured values is complex and also involves discussion of measured values for different liquids, surface roughness influence, and other factors that involve surface chemistry changes, which can significantly be modified by the laser beam treatment [16–24].

The surface and interfacial properties of polymers play a key role in many technological processes, such as polymer processing, metallization, or their cytocompatibility. All these processes are controlled by wetting and adhesion phenomena. Based on the conclusions of wettability measurement, we can estimate the surface chemistry with regard to polar or nonpolar functional groups as well as acidic or basic sites available at the modified surface region. In addition, the interaction of the solid surface with consequently applied (e.g., grafted and sputtered) surface-active substances can be influenced, or the biocompatibility is enhanced [16, 17, 20]. The importance of wettability in adhesion lies in the quality of adhesive bonding and coatings, which strongly depend on the spreading of these materials on the adherent or solid substrate. The laser surface treatment also significantly influences the surface energy (SE), and CA measurements are believed to be one of the simplest and most straightforward approaches to obtain surface-free energies of pristine or modified solids. The solid surfaces may very often be chemically heterogeneous, roughness can significantly be altered, or the reorientation of functional surface groups may occur during aging processes.

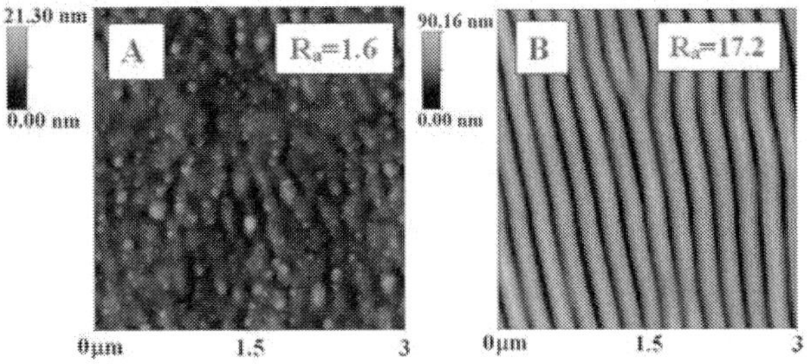

Figure 1. Detail of surface morphology (2D, 3x3 μm²) of PEN treated with 10 mJ cm⁻² and the number of laser pulses (A) 1000 and (B) 6000. R_a represents the average surface roughness in nanometers [18].

Several types of polymers may be changed by the laser treatment of the polymer surface. A broad spectrum of wavelengths, laser types, and pulse durations can be used. One of the most important types of polymer surface treatment is excimer irradiation. The excimer laser treatment of polymers can deliver periodic dimensional features in arrays across the surface of a substrate with lateral dimension close to 100 nm. Excimer laser can induce periodic nanopatterns on the surface of different polymers, which can absorb appropriate wavelengths [16, 17]. Besides polymers such as polyethylene naphthalate (PEN) [18], polyethyleneterephthalate (PET) [19], polystyrene (PS) [20], and polyethersulfone (PES) [21], surface changes (e.g., morphology and chemistry) can also be induced (Fig. 1) by excimer laser treatment on biopolymers or biocompatible polymer blends.

Laser surface modification of polyether-ether-ketone (PEEK) as a polymer with great mechanical and chemical properties was performed with the aim of wettability changes. A systematic approach of the laser irradiation wavelength influences the modification of surface layers of

http://dx.doi.org/10.5772/60824

PEEK under laser wavelengths (λ = 1064, 532, and 355 nm), with the focus on the determination of the optimal experiment parameters to increase the roughness and wettability, was executed [22]. It was found that a statistically significant factor during the processing using the 1064 nm laser radiation is the combination of pulse frequency, scanning speed, spot diameter, and pulse overlapping. In order to minimize the CA and improve wettability, the utilization of high-pulse-frequency, low-scanning-speed, low-overlapping, and small spot diameter was determined. On the other hand, the most influencing factors, using the 532 nm laser radiation, were the scanning speed and spot diameter [22]. Ultraviolet (UV) irradiation (172 nm) was used for the exposure of chosen polymer foils, i.e., polyethylene (PE), polyimide (PI), and polytetrafluoroethylene (PTFE), in air atmosphere [23]. The wettability (characterized by CAs and water drops) of the PE and PI films significantly increased after the exposure of excimer laser in air atmosphere. The subsequent X-ray photoelectron spectroscopy (XPS) study of the irradiated layers determined that the decrease of CAs was ascribed to the increase in the oxygen concentration at the modified surface. The surface characteristics of the PTFE film, however, was not considerably changed by the UV exposure. The aged UV exposed PE and PI films exhibited a significant decrease in wettability [23]. The influence of surface morphology on wettability was reported in [24], wherein a novel approach to creating hierarchically patterned PEEK surfaces composed of nanostructures and microstructures was introduced. The method was reported to be fast, easy, and applicable to large areas without using complicated lithography and mask-based techniques. Surfaces with a wide spectrum of wetting parameters ranging from a super-hydrophilic to a hydrophobic nature were reported [24]. The biological response on the performed surface materials modification is strongly dependent on the changes in surface properties (e.g., surface morphology, roughness, and composition). The modification of the surface of carbon-coated high-molecular-weight polyethylene (HMWPE) by different laser wavelengths (λ = 1064, 532, and 355 nm) was conducted [25]. The influence of experimental conditions (pulse frequency, scanning speed, and spot overlapping) on the changes in the surface properties of material was studied. It was found that the laser treatment of these carbon-coated samples always improved the wettability of the samples. The effect of the laser power on wettability was similar under the three laser treatments. The results suggested that the increment in the scanning speed reduces the CA during the surface treatment at 1064 nm. On the contrary, this parameter increments the CA during the surface treatment at 355 nm [25]. KrF laser was successfully applied for both laser-induced patterning and whole-area irradiative processing to modulate the wettability characteristics and osteoblast cell response following 24-h and 4-day incubation [26]. For the patterned samples, the CA was increased by up to 25°, which was attributed to a mixed-state wetting regime. For all samples, CA was a decreasing function of the SE [26]. The effects of using CO_2 and F_2 lasers to modify the wettability of a nylon,6-6 by producing 50 μm spaced (with depths between 1 and 10 μm) trench-like patterns using various laser parameters such as laser power for the CO_2 laser and the number of pulses for the excimer laser was studied in [27]. The CO_2 laser couples into the material through resonant coupling, which gives rise to bond vibrations, allowing the temperature to rise and melt the material. Upon cooling, the molten material resolidifies, and a protrusion away from the surface becomes evident on the surface. The CO_2 laser was proved to be capable of producing CAs that are slightly larger compared with the as-received sample.

In contrast, for the F_2 excimer laser-patterned samples, the apparent polar component and surface O_2 content decreased by up to 11.69 mJ.m^{-2} and 1.6 at %, respectively. It has been proposed that the increase in CA that results from the laser modifications is due to the patterned topographies [27].

The type of atmosphere during the modification process was reported to have a great influence on the wettability of a polymer [28]. The surface of poly(e-caprolactone) (PCL) fiber scaffolds was significantly altered by femtosecond laser irradiation. Variations in input parameters (e.g., laser fluence and scanning area) led to a change in the surface CA. Scaffolds that were treated in ambient air or oxygen exhibited increased atomic oxygen concentration and also ablation of material. The concentrations of oxygen were generally larger, and sessile drop CAs smaller for surfaces for which the ablation was more pronounced. This effect was independent of the fact that the ablation was caused by higher laser power, decreased scanning space, or smaller defocusing distance [28]. The influence of roughness changes induced by femtosecond laser on the polypropylene (PP) surface was investigated in [29], wherein the influence of multi-scale roughness on CA measurements was described. An experimental investigation of the static CA measured on PP textured surfaces by a femtosecond laser with multi-scale roughness shown the dependence of the 3D roughness parameters. The proposed physical approach showed that the drop is not in such a case as in the state described by Wenzel or Cassie–Baxter, even with the roughness parameters calculated at the relevant scales [29]. The use of CO_2 laser-induced surface patterns to modify the wettability of poly(methyl methacrylate)(PMMA) with respect to the surface roughness of patterns was reported [30].

Different types of polymers were treated with excimer lasers, with the aim of modifying the surface wettability of treated samples. Polystyrene (PS) substrates were modified with excimer laser (ArF excimer pulsed laser, λ=193 nm) argon, and oxygen plasmas (radio frequency (RF) glow discharge). The substrates were processed at different experimental conditions (number of pulses or treatment times) [31]. Contact angle determination showed higher wettability for the modified PS with both argon and oxygen plasmas and lower wettability for the laser-treated samples. It was also observed that, in the case of laser irradiation, wettability and surface morphology affect cell attachment and spreading [31]. Laser modifications of PS using suitable exposure doses and an appropriate choice of the processing gas (helium or oxygen) enabled a highly localized control of wetting. The hydrophilic and hydrophobic behavior was caused by chemical and topographical surface changes [32]. The wettability of PI surfaces that were microstructured using KrF laser radiation at fluences above the material ablation threshold was studied by static CA measurements. The morphology of laser-treated surfaces consisted of conical features whose dimensions and areal density depended on the fluence [33]. The effect of the cone geometry on the surface CA depended on the wetting regime. When wetting occurred in the homogeneous regime, the CA of the treated surfaces increased with the radiation fluence because the cone dimensions increased. In contrast, for the heterogeneous regime, an increase in CA was observed due to an increase in the average distance between cones [33].

Surface wettability changes may also influence the growth and adhesion of consequently deposited metal layers. The effect of an excimer laser treatment at low laser fluences on a PET surface was investigated, as well as its efficiency in improving the metal/polymer adhesion when the treated polymer is coated with aluminum [34]. For all treatments, an important increase in hydrophilicity was detected at low pulse numbers. When increasing the pulse number (>200), the evolution of the hydrophilicity depended on the laser process parameters. At 248 nm, the wettability either remained constant or even decreased at a high number of pulses due to the surface decarboxylation. At 193 nm, increasing the pulse number improved the hydrophilicity, and very low water contact angles (WCAs) were obtained [34].

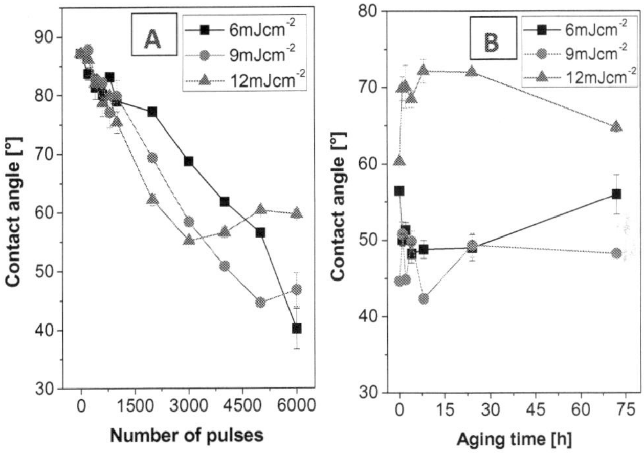

Figure 2. A) Dependence of the CA of laser-treated PS on several laser pulses treated with fluences of 6-12 mJ cm^{-2}. (B) Dependence of the CA on time from the laser treatment (aging time) for samples exposed with 6 000 pulses and laser fluence of 6, 9 and 12 mJ cm^{-2} [38].

Excimer laser and femtosecond laser were also used for the surface modification of PCL membranes for tissue engineering applications [35]. In the case of femtosecond laser, the results gathered from the wettability test revealed that the PCL membranes with holes perforated at lower pulse numbers had higher SEs. At lower pulse numbers, the degree of surface roughness was higher because there were more surface asperities due to a greater "splattering" effect [35]. If compared, the results from the KrF treatment suggested that increasing the pulse number (corresponding to a longer laser irradiation time) at both energies increased the surface hydrophilicity of the PCL. The improvement in surface hydrophilicity after laser irradiation was proposed to be due to the sufficient energy of the excimer laser provided to overcome the bond strength of the PCL chemical structure at these irradiated regions [35]. Advanced functional surface patterns with high lateral resolution can be obtained by an appropriate combination of laser microstructuring or nanostructuring and laser surface modification. A flexible and rapid surface functionalization of amorphous carbon films showed a great potential for various applications such as biological surfaces and tribological systems [36]. The formation of carboxyl groups at the surface was detected, which corresponds to an improved wettability of water. The SE has shown a linear dependence on the laser energy dose and laser pulse number, respectively. For laser fluences near the ablation threshold, a selective ablation of hydrogen-enriched domains seems to be responsible for the formation of micro-sized cones [36]. Recently, a modification of a method for CA measurement, an axisymmetric drop shape method, has also been used for wettability analysis [37]. Contact angle hysteresis phenomena on polymer surfaces have been studied through CA measurements using sessile liquid droplets and captive air bubbles in conjunction with a drop shape. This method was reported to be promising also for characterization of wettability of laser treated surfaces with controlled roughness and chemical composition [37].

Excimer laser is able to create regular or semi-regular patterns on the surface of polymer, which comply with specific parameters (e.g., ability to absorb the wavelength). A study of KrF laser treatment on a PS foil surface was performed. The influence of several laser pulses and laser fluence on surface properties (morphology, wettability, and chemistry) of PS was investigated [38]. The optimal input parameters for ripple nano-patterning with high regularity were determined. It was observed that increasing the number of laser pulses at all laser fluences applied led to a decrease in CA (Fig. 2A), which confirms the increase in wettability after laser exposure and is also with an excellent agreement with the work in [20]. Hydrophilicity is increased due to the formation of surface radicals after polymer bonds breakage caused by the impact and absorption of laser beam and consequent creation of new oxygen (polar) groups. Consequently, samples treated with 6-12 mJ cm^{-2} and 6 000 pulses remained at ambient atmosphere, so they can undergo aging procedures for an interval of 1–72 h (Fig. 2B).

When a laser fluence of 6 mJ cm^{-2} was applied, a mild decrease in CA was observed, down to 50°. The treatment with a fluence of 9 mJ cm^{-2} at the first stages of aging primarily resulted in the fluctuation of CA. After 24 h of consequent aging, the mild constant decrease was determined. A different situation was observed when a fluence of 12 mJ cm^{-2} was applied. At this laser fluence, a sharp increase of up to 70° was first observed, and after achieving its maximum, a slow decrease was detected. After 72 h, the polymer surfaces can be signed as "aged," i.e., with no further significant changes. The modification induced in PS by the ArF excimer laser radiation has also been investigated [39]. The disperse component of the SE approximately raised linearly with the R_a value, whereas the polar component remained approximately constant. Surprisingly, laser-induced oxidation of the PS surface layer had no appreciable impact on the SE polar component. Laser-induced surface roughness was reported to be the main factor that influences the SE.

The wettability of laser-induced nano-ripples on PEN foils was studied in [40]. Polyethylene naphthalate (PEN) is chemically "similar" to PET, which is a widely industrially used polymer. Due to the presence of condensed benzene rings PEN has "higher" properties (mechanical, chemical, and thermal) compared with PET. The consequent measurements of PEN foils treated with several laser pulses (1000–6000) and laser fluence of 6-12 mJ cm^{-2} were conducted from the point of wettability changes. A sharp decrease in surface CA was observed for all three studied laser fluences with an increasing number of laser pulses. The selected samples were chosen for the consequent aging studies. The CA of the laser-modified surface was evaluated for 1, 2, 4, 8, 24, and 72 h from the treatment. The lowest laser fluence first exhibited a small decrease during the process. After 8 h, no significant fluctuation could be observed. Rather, a different situation occurred for samples treated with higher laser fluences (9 and 12 mJ cm^{-2}). A sharp increase during the initial stage of the aging process was observed. After

http://dx.doi.org/10.5772/60824

approximately 8 h, a mild decrease was determined, and the surface CA was almost constant after 72 h. For longer aging periods, no more fluctuations have been observed [40].

Polymer films (PE, PP, PTFE, PS, and PET) were modified by a F_2 laser (157 nm) [41]. The modification led to a degradation of the polymer surface layer and ablation. Wettability, characterized by the measurement of the CA, was determined as a function of the number of laser pulses. The laser irradiation led to a decrease in contact in all cases, except for PP substrates. The values of the CAs are affected not only on surface morphology but also on chemical composition. After an initial fast increase in wettability, a saturation of the CA was achieved. For more than 2000 laser pulses the CA remained unchanged [41]. An initial fast increase in wettability was detected on PE and PET, and a much slower increase for PTFE and PS.

Biopolymer surfaces can also be significantly altered by the use of laser treatment. Excimer laser irradiation can induce significant morphology (Fig. 3) and wettability changes [42, 43], or even a regular nanopattern can be constructed [44]. Different types of laser treatment also lead to the significant degradation of biopolymers. Degradation of poly(L-lactide) under a CO_2 laser treatment above the ablation threshold was reported in [45], but ablation of the poly(L-lactide) by CO_2 laser did not affect the wettability of the surface of the material, which confirmed the lack of formation of polar functional groups. A different situation can be observed if excimer laser is used.

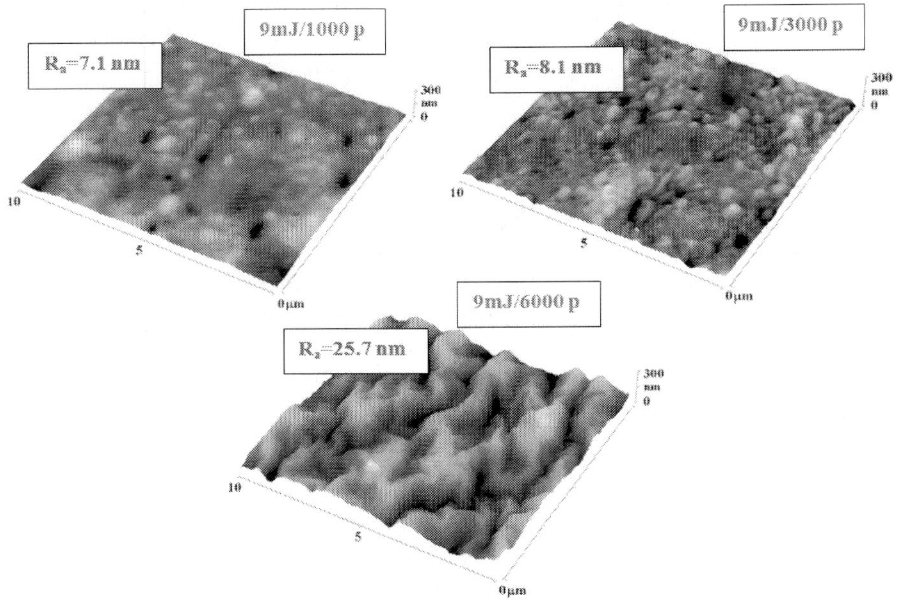

Figure 3. Surface morphology of plasma pretreated PLLA, consequently exposed with a laser beam of 9 mJ cm^{-2} and different numbers of pulses (1000, 3000, and 6000). R_a represents the arithmetic mean roughness in nanometers [43].

Surface ablation of poly(L-lactide acid) (PLLA) induced by excimer laser was studied in [42]. It was found that irradiation by KrF laser results in an increase in wettability of PLLA. This increase significantly depended on processing parameters (e.g., the number of pulses and laser fluence). The substrates treated with 500 pulses exhibited the smallest difference in wettability compared with the pristine PLLA. With an increasing count of laser pulses, the wettability increased (see Fig. 4). However, for laser fluences above 30 mJ cm^{-2}, the increase in wettability was neutralized, and a "constant" behavior was detected. This phenomenon is probably connected with both a significant change in surface morphology, roughness, and changes in structure and with the ablation of the biopolymer (Fig. 5).

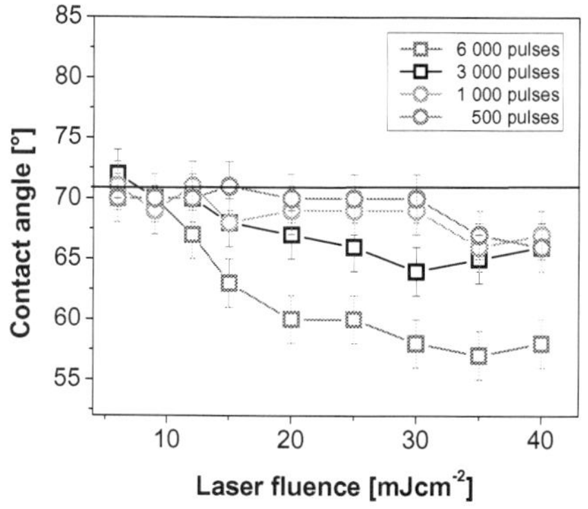

Figure 4. Dependence of the PLLA CA on the KrF laser fluence by different numbers of pulses (500–6 000) [42].

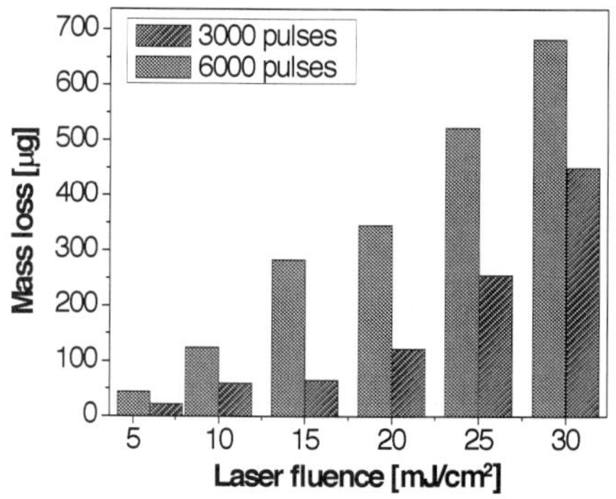

Figure 5. Dependence of the mass loss of PLLA on the applied KrF laser fluence. The treatment was accomplished at 3 000 and 6 000 pulses [42].

Plasma pretreatment strongly influences the CA of consecutively treated biopolymer [37]. The samples that were pretreated with plasma and aged for more than 72 h (which means that the input values of the PLLA CA before laser exposure were those of 90° (surface is stabilized)) were consecutively treated with KrF laser. The samples treated with 500 pulses exhibited the highest values of CA compared with the samples exposed to a higher number of pulses. With increasing laser fluence, the surface morphology was strongly influenced, and surface chemistry was significantly altered. The macromolecular chains' rearrangement ability and the chemical structure of the modified surface were influenced by the laser fluence. Slight fluctuations of CA appeared, probably due to the ablation of the polymer surface. The most pronounced decrease in CA was observed for samples treated for 6 000 pulses. However, a slight increase with higher laser fluence (40 mJ cm^{-2}) was observed, probably due to more pronounced ablation [43].

3. Plasma-treated polymer

Many polymers have good bulk properties, which are used in a wide range of applications. In many application areas, surface properties besides bulk characterization are equally important for their successful use. However, their surface properties, especially adhesion, wettability, and cytocompatibility, are unsuitable for many applications. The effective method leading to the modification of the surface and changes of the chosen parameters is modification by plasma discharge [46, 47].

The polymer generally has a low surface reactivity. Using physical modification requires the use of high-energy particles, by which an inert polymer surface is bombarded. Most of the modifications are performed using low-temperature or "cold" plasma [48]. When we speak of plasma, we mean the highly ionized gas consisting of neutral and charged particles (ions, radicals, excited molecules, and electrons), wherein the total charge remains neutral [49]. Plasma can provide modification of the upper layer of polymer surface without using solvent or generating chemical waste and with less degradation and roughening of the material compared with many wet chemical treatments [50]. This procedure enables the modification of heat-sensitive materials, e.g., polymers [48, 51]. This type of plasma can be generated in many ways, e.g., by corona discharge treatment, UV radiation, X-ray beam, and gamma irradiation.

The plasma modification of polymers is very often used in surface cleaning or etching [52], changes in surface wettability and adhesion [53], reducing friction [54], or regulation of the cell adhesion in cytocompatibility studies [55]. The creation of the polar groups on the modified surface reaction by the reaction of activated polymer surface with gas atmosphere leads to the augmentation of adhesion, increasing hydrophilicity, or modifying the surface morphology (Fig. 6). It is known that modification by plasma treatment leads to changes in the surface layers (approximately 10 nm or less), whereas the bulk properties are not affected. The choice of the gas used to modification may affect the chemical composition of the "newly created" functional groups introduced to the polymer surface [56,57].

Figure 6. Surface morphology of pristine PLLA and PLLA treated in Ar plasma discharge (discharge power: 3 W; time of modification: 300 s). R_a represents the arithmetic mean roughness in nanometers.

The plasma treatment is a complex process, and a variety of chemical and physical reactions can occur. Plasma processing is generally used for film deposition or etching and may also be used for resist development and removal [58]. The type of functionalization can be varied by a selection of plasma gas (Ar, N_2, O_2, CO_2, and NH_3) and processing parameters (power, time, pressure, and gas flow rate). Plasma treatment with oxygen or nitrogen plasma provides polar functional groups on the surface, which alters the SE of polymers [59–61] and enhances the attachment of consequently applied chemical substances or particles. It is known that oxygen plasma can react with a wide range of polymers to produce a variety of oxygen functional groups. Oxygen plasma provides different functional groups on the polymer surface, e.g., the C–O, C=O, and O=C–O groups [61–63]. The wettability of surface as well as its biocompatibility can significantly be altered by plasmas containing nitrogen [64]. Cell affinity was enhanced using ammonia plasma treatment on PS [65]. Nitrogen- or ammonia-based plasmas were applied to create amino-binding sites [66] or to increase the peel strength between polymers [67]. The plasma is able to increase adhesive strengths on the polymer surface [68].

3.1. Oxygen in plasma gas and the surface wettability of modified polymer

Although PLLA and polyhydroxybutyrate with 8 % polyhydroxyvalerate (PHBV) can be found in biomedical applications, both are hydrophobic, making their interaction with cells difficult. Thus, the surface modification of these polymers is important for all applications involving their contact with physiological components [69, 70]. The influence of the oxygen and nitrogen plasmas on the hydrophilicity of the PLLA and PHBV was studied. The oxygen plasma treatments on both PLLA and PHBV samples, in general, caused a reduction in the CA and an increase in the SE. This increase in the SE reflects an increase in surface hydrophilicity. Changes in the surface morphology were observed, as well as a decrease in CA, and an increase in SE and polar components was detected. These modifications were associated with the plasma parameters (pressure, power, and time) and made both polymers more hydrophilic [49]. In a study aimed at the evaluation of plasma treatment effects on the surface properties of PLLA, film air plasma was used at a pressure of 10–20 Pa, and a discharge current of 50 mA for 60 s was applied. The CA of wettability measurements of the films showed that plasma modification led to an increase in hydrophilicity and SE [71]. The use of oxygen plasma treatment on fibril tip surfaces in polymer elastomeric polyurethane (PU) and polydimethylsiloxane (PDMS) fibrillar adhesives improved their wettability and adhesion capacity. Surface modifications of fibril tips involved UV/ozone and oxygen plasma treatments to make the fibril tips more hydrophilic. Treatment of fibril tip surfaces renders them more hydrophilic, and the adhesion capacity of treated fibrillar arrays is sensitive to the relative humidity of the environment [72]. Surface effects induced by oxygen plasma exposition (time of treatment: 60 s; constant pressure: 15 Pa; and at different process powers, ranging from 20 W to 200 W) on PET wettability were investigated by WCA measurements. Plasma treatment produced surface modifications with the incorporation of hydrophilic functional groups such as carbonyl (C=O) and carboxylic (–COOH) groups. At a low power level (20 W), the plasma was mainly rich of radical species that react with methylene (C–O) and ester groups (O–C=O) present in the polymeric chains, producing a slight etching of the material and new hydrophilic species as carbonyl groups are introduced. At a high power level (200 W), a more efficient incorporation of hydrophilic species into the PET surface was observed. Contact angle measurements confirmed the enhancement of the surface wettability as a function of plasma power [73]. The degree of hydrophilicity of the treated HMWPE surface was determined by measuring the CA of a water droplet on the surface. The plasma was generated using compressed air as the ionization gas. It was found that plasma power and scan speed affect the grafting process. The lower power and lower scan speed generally resulted in a greater change in CA from the HMWPE [74]. Experiments on plasma treatment of PP and PET foils were carried out with the use of atmospheric pressure plasma sources operating with ambient air as plasma forming gases. The atmospheric pressure plasma source was ambient airflow as a plasma-forming gas. Airflow was perpendicularly directed to the electric current. The velocity of airflow was about 20 m s^{-1}. The average power of this dc plasma source was 35 W. The results of measurements made within about 1 min after plasma processing indicated that the air plasma processing of PP and PET resulted in a decrease in CA till a processing time value of 1 s, followed by a phase of a nearly constant value of 36° for PP and 30° for PET. The effect was noticeably stronger for the PP surface. This was explained by different chemical compositions of these polymers [75]. The effect of low-pressure O_2 plasma treatments on the surface-free energy and morphology of PP films was studied in [76]. The treatment was performed at a pressure of 0.35 bar. Each PP sample was treated at 10, 30, and 50 W for 10, 20, 30, 60, 120, 180, and 300 s, respectively. Based on the surface-free energy calculation and Fourier transform infrared spectroscopy-attenuated total reflection (FTIR–ATR) spectroscopy, it was evident that the changes in the SE PP film significantly increased already after a very short treatment time (20 s). The surface modifications with O_2 plasma proceeded only with hydrophilic modification. The SE reached the maximum value for a short treatment time and then leveled off [76]. For improving wettability, RF plasma treatment in the air atmosphere was reported to be a very effective tool. The high-density polyethylene (HDPE) was modified by air and oxygen plasmas (operating pressure from 0.1 to 1 Torr, power ranging between 5 W and 10 W). The CA measurement, ATR FTIR, and XPS analysis showed that plasma treatment leads to a dramatic increase of the hydrophilic character of the HDPE surfaces. A hydrophilic surface was created mainly due to the generation of the carboxylic, carbonyl, and peroxide groups at the surface interface [77].

In all the above-mentioned cases, the oxygen-containing plasma treatment leads to a decrease in WCA and increase in hydrophilicity of the polymer surface.

3.2. Inert plasma treatment and wettability of modified polymers

The influence of Ar plasma on the surface polarity and wettability on the PLLA and poly-4-methyl-1-pentene was studied by measuring the CA. The samples were modified in diode plasma discharge for 0–240 s, using DC Ar plasma (gas purity: 99.997%; power: 3 W and 8 W). Exposed samples exhibited a sharp decrease in CA immediately after modification. It was found that the values of CA depend on both exposure time and discharge power. As a reset of the plasma treatment, the polarity of the polymer surface increased, the CA decreased, and the SE increased. With an increasing aging time, the surface polarity was spontaneously reduced. The increase in CA of the polymer surface and decrease in wettability may be caused by the reorientation of polar groups into the polymer bulk [78]. The dependence of the WCA and surface-free energy on exposure time at different discharge powers was determined on plasma-modified PHBV. The samples were modified in diode plasma discharge on Balzers SCD 050 device for 0–240 s, using dc Ar plasma (gas purity: 99.997%; power: 3 W, 5 W, and 8 W). Based on the CA measurement using a goniometry technique, it was concluded that the treatment of PHBV in Ar plasma leads to a decrease in the measured WCA (increase in wettability). Exposure to the lowest plasma discharge power of 3 W led to fast decrease in CA after 10 s of treatment. For a 40-s exposure, WCA achieved a minimum value of about 30°. For exposure times above 40 s, WCA slowly increased with increasing exposure times. It was proposed that this turnout in WCA dependence may be due to the surface ablation and change in surface roughness. A similar trend was observed at a higher discharge power of 5 W, where the lowest WCA (36.6°) was observed for an exposure time of 10 s. At 8 W, the discharge power was monotonous, but a slow WCA decrease with an increasing exposure time was shown. Differences in WCA and SE during the aging of the plasma-modified PHBV were observed; however, after some time, a saturated, constant value was achieved. Both effects strongly depended on the plasma discharge power [79]. The work focused on the modification of PHBV and PLLA foil with Ar plasma and presented the influence of different conditions of plasma treatment on the wettability of the two biopolymers. Both biopolymers were treated with Ar plasma discharge (discharge power: 3 W; time of modification: 25–400 s). For both polymers, it was determined that immediately (5 min) after irradiation, all samples exhibit a sharp decrease in CA to a minimum value. With an increasing aging time, the CA increases. The stabilization of the polymer surface was observed after 10 days (for PLLA) and 13 days (for PHBV) from the treatment (Fig. 7). Finally, the CA was strongly dependent on the exposure time. Substrates modified for a shorter times (5–75 s) had a significantly lower CA than substrates modified for longer times (300–400 s) [57].

The changes in wettability of 2-hydroxy-ethyl-methacrylate (HEMA) and 2-ethyl-oxy-ethylmethacrylate (EOEMA) induced by Ar plasma treatment were measured using goniometry. The experiment was performed on a Balzers SCD 050 device, and the processing parameters were as follows: the time of modification was 0–400 s, and the discharge power was 1.7 W [80]. Based on the analyses, the creation of the "new" groups containing oxygen on HEMA and EOEMA chains leads to changes in wettability. The CAs were immediately evaluated (in 6 min) and 386 h after the plasma irradiation. It was found that the plasma modification leads to the rapid decrease in CA in the first few minutes after modification. The

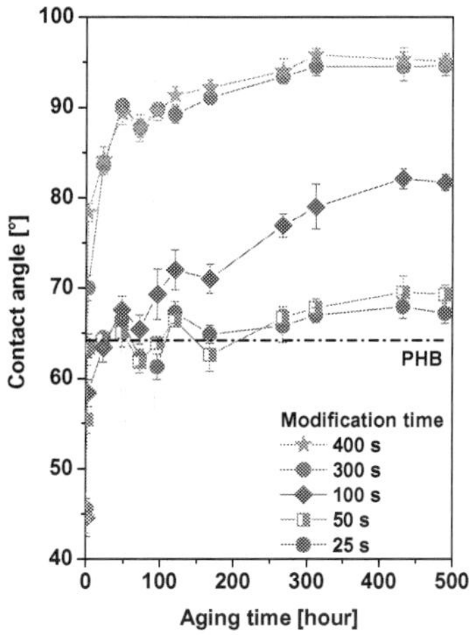

Figure 7. Dependence of the CA of PHB plasma treated by 3 W on the aging time (after modification). Plasma exposure times were 25, 50, 100, 300, and 400 s. The CA value for the pristine PHB is shown as a dotted line [57].

decrease was higher for longer modification times. Comparable dependence was observed for both polymers. Then, 386 h after plasma modification, the CA increased compared with the CA measured immediately after the plasma treatment. The experiment in [81] proved that the concentration of oxygen groups created by Ar plasma irradiation does not change with time from modification. This result [81] and the finding from goniometry analyses [80] can be explained by the reorganization of surface layers and rotation of degraded macromolecules and newly created structures to the "free volume of polymer" [80, 82]. The studies of plasma treatments processed by Ar on the drug release from the poly(ethylene-co-vinyl acetate) (EVA) polymer films demonstrated the effects of plasma modification on the wettability of a typical biocompatible polymer. Plasma treatment was carried out using RF plasma-enhanced chemical vapor deposition: power was 200 W, and the total pressure was kept at 13.3 Pa. The plasma processing time was changed from 0 s (untreated) to 45 s in order to evaluate the plasma processing time effects on the drug eluting characteristics. After stabilizing the sample in air for 1 day, the CA measurements of the samples were performed. The plasma treatment by Ar resulted in a significant reduction in WCA, even by a short plasma treatment time of only 5 s. By longer than 5 s of the plasma treatment, the WCA became relatively low and stayed almost constant thereafter, regardless of the plasma processing time. This behavior led to a reduction of the drug release amount from EVA by only a short plasma processing time [83]. The aging behavior of the plasma-irradiated PP and PET polymer foils was studied by goniometry. Measurement was performed by the determination of the WCA. Plasma modification causes the creation of oxidized structures on the polymer surface. The exposure of the irradiated samples to the air atmosphere occurs at the reorientation of the oxygen groups from the surface into the bulk of the material. The influence of aging time on the CA showed the change of the values of CAs of the plasma-modified polymer foils (PP and PET) as a function of storage time.

Based on the results, it is evident that the aging is characterized by a fast decrease in wettability during the first few hours of storage. At longer storage times, the wettability decreases more slowly, and the CA finally reaches a plateau value [84].

4. Grafting of plasma-treated polymers

The plasma treatment of polymers leads to a modification of polymer surface due to the production of reactive sites or changes in cross-linking or molecular weight. The bombardment of polymer surfaces by energetic particles breaks the covalent bonds, and this process results in the formation of surface radicals. The typical gas used for activating the polymer surface by forming radicals is Ar or He. These active sites (radicals) can later react with other chemical substances. The polymer surface can also be functionalized by "plasma-induced grafting," which is the combination of plasma activation and conventional chemistry [46]. Grafting is a method in which new compounds are bounded from the gas or liquid phase on the activated polymer surface [85]. Grafting may be realized to the polymer surface with the use of one or more substances. In the first case, modification is performed in one step, and in the latter case, it may occur for networking several substances simultaneously or in consecutive steps. Not only radicals but also the functional groups created due to plasma treatment, such as carboxyl, hydroxyl, or amine groups, can be used for subsequent modification of the polymer surface.

For many applications, it is necessary to bind suitable substances on the polymer surface. In tissue engineering, development of biosensors and other biodiagnostic devices is of great importance, and immobilizing of proteins onto a surface is a key step in many biotechnological processes and applications. Various immobilization techniques such as physical adsorption, covalent immobilization, and bioaffinity immobilization have widely been practiced [86]. For polymers, a technique for covalently binding of proteins, i.e., pretreating the polymer surface in plasma discharge combined with the bath deposition, may be used. On plasma-activated surfaces, proteins can be immobilized through a reaction with the amine and carboxyl groups [87].

In the development of, e.g., biomarkers and biosensors or in molecular design, nanoparticles are very often applied [88-90]. These may be prepared from organic materials (e.g., polymers) and inorganic materials (e.g., metal nanoparticles). In addition, one can design nanoparticles with controlled hydrophilic/hydrophobic characteristics, with positive or negative charge, with response to external pH or thermal stimuli, and, finally, with high resistivity or conductivity. It is therefore clear that combining nanoparticles with different nature, shape, chemistry, and physical properties together with polymer substrates permits the design of an infinity of systems for a plethora of applications [91]. Metal nanoparticles can be prepared through physical (mechanical separation of metal aggregates) or chemical (nucleation and growth nucleus) methods. Colloidal solutions of metal nanoparticles and nanoparticle composites with ceramic or various organic materials are currently prepared in particular. The most widely used metals include gold, silver, and platinum, as well as iron and their compounds. Hence, a key aim of materials science is to understand roles that substances or particles bonded on material surface play at the interface of materials. It is clear that the grafting a new material on the surface leads to changes in chemical structures and composition, surface morphology (Fig. 8), roughness, and other parameters, and they affect surface wettability.

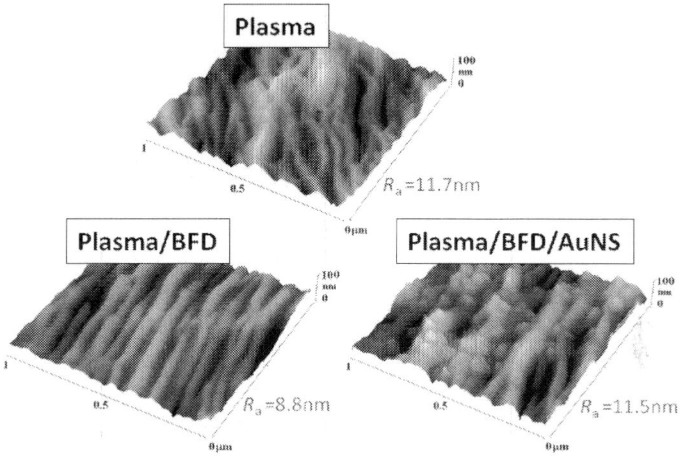

Figure 8. Surface morphology of the PE treated with plasma for 100 s (Plasma), PE treated with plasma and subsequently grafted with biphenyl-1,4-dithiol (Plasma/BFD), and PE grafted with biphenyl-1,4-dithiol and Au nanoparticles (Plasma/BFD/AuNS). R_a represents the arithmetic mean roughness in nanometers [92].

4.1. Wettability of grafted polymer surface and cytocompatibility

One of the most studied nanomaterials for biomedical applications are carbon materials (nanotubes as a carrier for biosensors, fullerenes), polymer nanofibers with high porosity (dressings, tissue engineering, and targeted drug delivery), hydrogels (dressings for open wounds and materials release medicines), or dendrimers (cancer treatment) and inorganic materials (silica and metal nanoparticles) [93, 94]. Fullerenes, spherical molecules composed solely of carbon atoms (e.g., C60), exhibit a broad spectrum of biological activity. Their unique shape and ability to "capture" some substances or compounds within their structure make fullerenes significant, especially in their potential use for targeted drug delivery or gene therapy [95, 96]. After irradiation with ultraviolet or visible light, fullerenes may convert molecular oxygen to become highly reactive and cause photodynamic disintegration of biological systems. This effect can be used in the therapy against cancerous growths, viruses, and bacteria that are resistive to broad-spectrum drugs. In their original unmodified state, fullerenes are highly hydrophobic and insoluble in water. On the other hand, they are relatively reactive, allowing modification of their structure. The modification leads to the increase in their hydrophilicity and wettability. They will also become soluble in aqueous solutions, which improves their interaction with biological systems [97, 98]. Deposition and grafting of gold–silver particles on polymer grafts has been the subject of a large number of studies for their potential application in electronic and optical detection systems, therapeutics, catalysis, and the environment [99-101]. Metallic nanoparticles such as iron oxide, gold, and silver have been developed and adopted for use in targeted drug delivery and diagnostic imaging. Super paramagnetic nanoparticles composed of nanoparticles of oxides, such as magnetite (Fe_3O_4) and hematite (Fe_2O_3), allow sensitive detection with high-resolution MRI. Elemental silver and silver salts are well known due to their relative non-toxicity in human cells and their antibacterial effects. Silver nanoparticles have been developed for antibacterial applications, because they more easily penetrate the cell wall of microbes and, thus, are more effective [102].

In the case of modification of the material that is in contact with cells, the nanoparticles or the other compounds grafted on the polymer surface change the surface wettability (Fig. 9) and thus significantly affect the adhesion and proliferation of living cells.

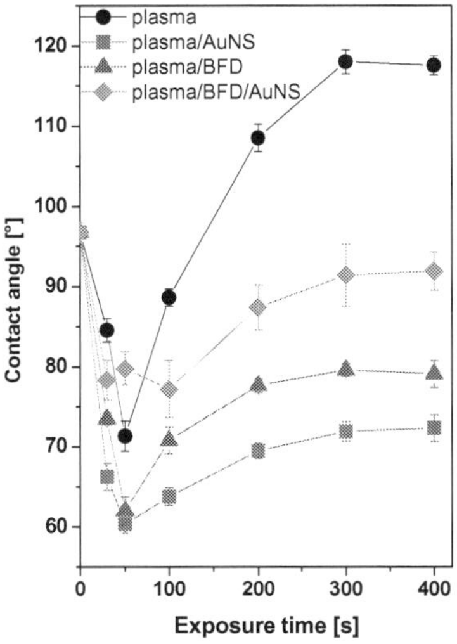

Figure 9. Dependence of the CA on the plasma exposure time for plasma-treated PE (plasma), plasma-treated and Au nanoparticle grafted PE (plasma/AuNS), plasma-treated and biphenyl-1,4-dithiol grafted PE (plasma/BFD), and plasma-treated and biphenyl-1,4-dithiol and Au nanoparticle grafted PE (plasma/BFD/AuNS) [92].

It was found that the unique properties of nanomaterials may not only affect the cell adhesion and interconnectivity across the cell structure on the surface of the sample but also represent an important factor for the structural and organizational stability of the cells [103]. The properties of the substrate significantly affect the subsequent establishment and organization of various cell types. Cells are "attached" to the substrate surface through the adhesive points that constitute the communication channels between the substrate's surface and the cellular cytoskeleton. Formation of this interface is influenced not only by the surface chemistry of the substrate but also by the surface charge, wettability, and modulus of elasticity [104, 105]. One widely accepted advanced "trend" in tissue engineering is the creation of surfaces that promote cell colonization. For this application, it is necessarily suitable to adjust not only chemical composition but also surface wettability. The decrease in surface wettability influenced by the grafting of adhesive molecules was studied in [106]. In this work, the changes in surface properties and biocompatibility of PE substrates after the plasma activation and subsequent grafting with chosen molecules glycine (Gly), tripeptide arginine–glycine–aspartate (RGD), fibronectine (FN), vitronectin (VN), and collagen (CG) were determined. It was found that plasma activation and grafting with molecules led to significant changes in the surface properties (polarity, chemical composition, morphology, and roughness) of the substrates.

Grafting of PE with molecules led to a strong decrease in CA. The biocompatibility of the substrates and cell–material interactions were studied by *in vitro* experiments. It was found that the surface activation of PE samples by Ar plasma discharge had a favorable effect on the adhesion, spreading, homogeneity of distribution, and, moderately, proliferation activity of mouse embryonic fibroblasts (NIH 3T3 cells). This effect was even more apparent on the samples coated with biomolecules. In the experiment, it was demonstrated that the treatment of PE by Ar plasma discharge and subsequent grafting with molecules dramatically increased the attractiveness of the presented polymer foils for cytocompatibility of NIH 3T3 cells [106]. The decrease in CA on polyolefins (e.g., PE) may also occur due to the interaction of their surfaces with Gly [80]. The grafting of this molecule on the PE surface leads to the increase in oxygen concentration on the modified surface and decrease in WCA. Combination with other surface properties (morphology and roughness) leads grafting with Gly to an increase in the attractiveness of PE for vascular smooth muscle cell (VSMC) colonization [107].

Grafting of poly(ethylene)glycol (PEG) on polymer surface grafts has been the subject of a large number of recent studies because of their properties applicable for creating polymer brushes. Polymer brushes are used as model systems in polymer science, in colloids and surfactant sciences, colloidal stabilization, lubrication polymers, and compatibilization between organic and inorganic materials. In particular, PEG brushes can be used to repress protein adsorption, which is responsible for several unwanted effects, such as fouling of contact lenses, clotting on blood-containing devices, and triggering of inflammation around artificial organs [108]. Several studies were focused on the grafting of PEG in order to improve the cytocompatibility of substrate and creating the cell's attractive scaffolds [60]. Three PEGs with different molecular weights (molecular weight was 300, 6 000, and 20 000 g mol^{-1}) were grafted on the plasma-activated polymer surface [109]. Dependences of the CA on the aging time for plasma-modified PE and plasma-modified PE and subsequently grafted with PEG showed a pronounced decrease in CA after PEG grafting due to the binding of oxygen-rich compounds onto the activated polymer surface. Based on the biological experiments performed *in vitro* with the VSMCs, it was found that the cell's adhesion, proliferation, and viability is strongly dependent on PEG concentration on the polymer surface and on the length of its chain (Fig. 10).

Cells very well proliferated and were grown on the substrates grafted with PEG with very low or very high molecular weight. In addition, the viability of cells cultivated on these samples was very high. The best results (the higher number of the cells) was achieved on the samples treated in plasma for 50 s and subsequently grafted with 20 000 g mol^{-1} PEG. On the other hand, the samples grafted with 6 000 g mol^{-1} PEG showed a significant anti-adhesion character and were for the cell's undesirable cultivation (Fig. 11) [109].

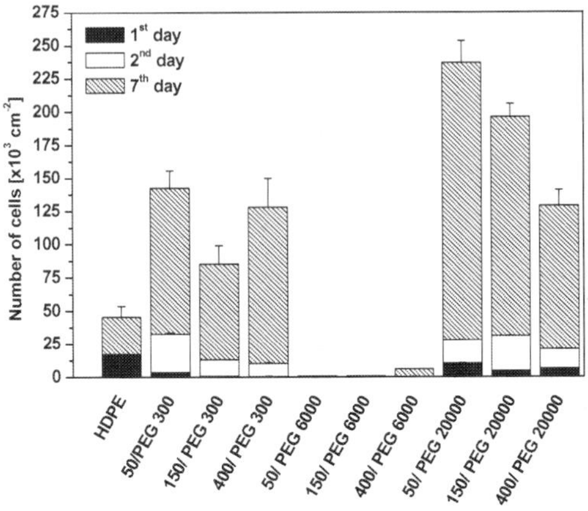

Figure 10. Number of adhered (first day) and proliferated (second and seventh days) VSMCs on pristine HDPE and HDPE treated in plasma discharge for 50, 150, or 400 s and subsequently grafted with PEG with a molecular weight of 300, 6 000, or 20 000 g mol^{-1}.

The influence of the presence of the gold nanoparticles on the surface wettability and biocompatibility of substrates was investigated in [110]. Golden, spherically shaped nanoparticles (AuNPs) with an average diameter of 12 nm were bound on the plasma pretreated polymer surface, which is concretely polyolefin, HDPE. The AuNPs were "grafted" from the citrate colloid solution. Surface wettability was measured by goniometry, i.e., the static (sessile) water drop CA method. The samples exhibited an increase in wettability after modification by plasma discharge and AuNPs. This increase was more pronounced for samples pretreated in plasma for a lower time. The cell–material interaction was studied using two different types of cells (VSMCs) and mouse embryonic fibroblasts (NIH 3T3)). Based on the test results, it was evident that the plasma treatment of the polymer and the presence of AuNPs had a positive influence on the spreading, proliferation activity, and homogeneity of the distribution of VSMCs and NIH 3T3 on the material surface [110].

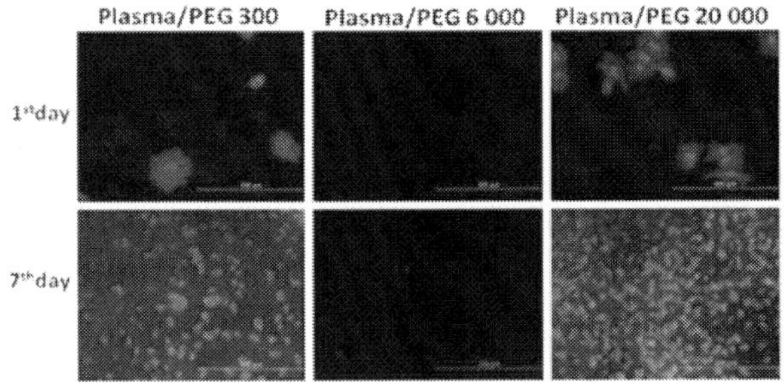

Figure 11. Photographs of the adhered (first day) and proliferated (seventh day) VSMCs on PE treated on plasma discharge and subsequent grafted by PEG with a molecular weight of 300, 6 000, or 20 000 g mol^{-1}.

http://dx.doi.org/10.5772/60824

Several vicinal compounds grafted on plasma-treated polymer foils were studied in [111] and [112]. First, surfaces of PET and PTFE were modified by plasma discharge and subsequently grafted with three different thiols: (i) 2-mercaptoethanol; (ii) 4,4′-biphenyldithiol; and (iii) cysteamine. The thiols are expected to be fixed through one of –OH, –SH, or $-NH_2$ groups to reactive places on the polymer surface created by the plasma treatment. ''Free'' –SH groups are allowed to interact (graft) with gold nanoparticles and nanorods [111]. During this research, we have found an interesting behavior of cysteamine molecule, which bonds to the surface through either the –SH group with the "free" $-NH_2$ group or reversely depending on the polarity of substrate surface.

Cysteamine was grafted on several polymer foils of different polarity (PLLA, PS, low-density polyethylenes (LDPE), HDPE, PET, PTFE, polyvinylfluoride (PVF) and polyvinylidenefluoride (PVDF)) previously activated in plasma discharge [112]. Surface properties including surface wettability of the sample before and after plasma treatment and cysteamine grafting were studied using various methods. Surface chemistry was studied with X-ray photoelectron spectroscopy, chemistry, and polarity by electrokinetic analysis and by goniometry, and roughness and morphology by atomic force microscopy. Representatives of unmodified and modified polymers were used for an *in vitro* study of the adhesion and proliferation of VSMCs [112].

Wettability was determined by measurement of WCA on all samples, pristine, plasma treated for different times, plasma treated, and subsequently cysteamine grafted. It was found that plasma treatment leads to a dramatic decrease in CA, indicating an increase in surface wettability of all polymers. For some polymers, this CA decrease strongly depended on the plasma exposure time (e.g., PTFE), but for others, the plasma exposure time is not of such importance (e.g., HDPE and LDPE) [112]. The most dramatic dependence of CA is observed on PTFE. The cysteamine grafting caused a dramatic increase in CA (i.e., decrease in surface wettability) on all polymers, and the increase was the highest on PTFE. These results are mainly a combination of the so-called aging process on polymer surface after plasma treatment, which has been earlier studied for different polymers [92, 112–114] and subsequent grafting of cysteamine. Aging of the plasma-treated surface leads to an increase in CA and hydrophobicity due to the rearrangement of polar groups created on the surface during plasma treatment [92, 112–114], whereas subsequent grafting of hydrophilic functional groups on the surface increased surface hydrophilicity [112]. Due to a small concentration of cysteamine bonded on the surface, the hydrophilicity only slightly increased in this case, and the combination of these steps resulted in hydrophobicity increase compared with the freshly plasma-treated surface.

Electrokinetic analysis and XPS measurement work provided information on surface chemistry and surface charge. Both of these are important for cell adhesion. Fig. 12 shows the results of electrokinetic analysis (zeta potential values) for pristine polymers, plasma-treated polymers, and plasma-treated polymers subsequently grafted with cysteamine. The zeta potential is known to depend on surface chemistry, polarity, charge, and surface morphology and roughness [115]. It is evident in Fig. 12 that the zeta potential depends on the polymer surface properties and it changes after plasma treatment and also after cysteamine grafting. Plasma treatment leads to changes in the surface charge, chemistry, and polarity and to a change in the zeta potential too. A significant increase in the zeta potential was observed on PTFE, HDPE,

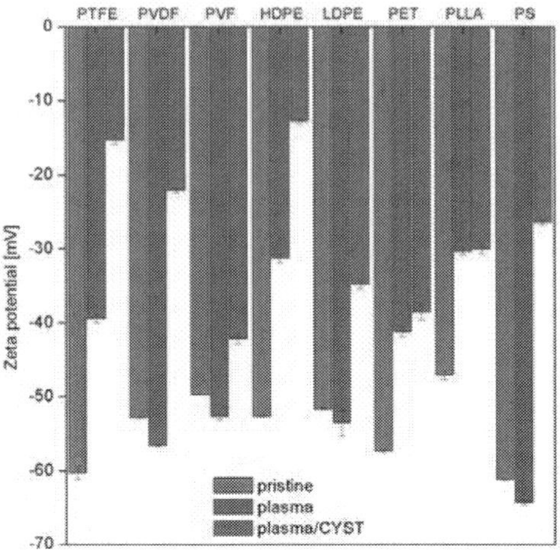

Figure 12. Zeta potentials of polymers: pristine (pristine), plasma treated for 120 s (plasma), and plasma treated and cysteamine grafted (plasma/CYST) determined by streaming the current method and calculated by the Helmholtze–Smoluchowski equation [112].

PET, and PLLA. Plasma treatment resulted in a cleavage of original bonds (e.g., C–H, C–C, and C–O) and the creation of new reactive places on the polymer surface ("free" radicals, double bonds, and new chemical groups, e.g., carbonyl, carboxyl, and ester groups) [116]. The plasma treatment led to the creation of excessive oxygen containing groups and subsequently to an increase in surface polarity on these four polymers mentioned above [112]. On PVF, PVDF, LDPE, and PS, the plasma treatment caused only a mild decrease in the zeta potential, indicating a decrease in surface polarity. The effect may also be due to an increase in surface roughness [112]. More pronounced results are obtained after cysteamine grafting. The most dramatic increase in the zeta potential was obtained on nonpolar polymers (PTFE and HDPE), whereas on more polar polymers (PET and PLLA), the zeta potential remained almost unchanged. The different behavior of these two groups of polymers may be due to (i) different amounts of grafted cysteamine and also (ii) different characters of the preferential bond of cysteamine (via the –SH or $-NH_2$ group) to the surface. Cysteamine grafting resulted in the creation of the new chemical functional groups on the polymer surface.

The results of ARXPS measurement of C, O, N, S, and F atomic concentrations on pristine polymers, plasma-treated polymers, and plasma-treated polymers subsequently grafted with cysteamine confirmed changes in the surface chemistry of all polymers after (i) the plasma treatment and (ii) subsequent cysteamine grafting [112]. After subsequent cysteamine grafting, the sulphur (S) and the nitrogen (N) containing groups appeared on the surface of all polymers, but their concentrations were different for individual polymers [112]. The highest concentration of N and S was detected at PTFE, PS, PVF, and PVDF, quite less at HDPE and LDPE, and minimal at PET and PLLA. This correlates well with the results of electrokinetic analysis (see Fig. 12), according to which the highest amount of cysteamine was grafted (the highest increase of zeta potential) at PTFE, PS, PVF, and PVDF, quite less at HDPE and LDPE, and almost none at PET and PLLA. Comparison of the XPS results obtained on grafted samples showed a higher amount of concentration of N on PTFE, PS, PVF, and PVDF compared with S [112]. This indicates that cysteamine is preferentially bonded at these surfaces through the –SH group to the surface and with the $-NH_2$ group remaining "free". On the other hand, the concentrations of N and S at PET and PLLA are almost the same, so no preferential bonding of cysteamine at these polymers was observed [112]. Cysteamine is bonded to these polar polymers without preference through the –SH and $-NH_2$ groups. Both electrokinetic analysis and XPS measurement confirmed the successful grafting of cysteamine. The schematic diagram of cysteamine grafting on different polymers is presented in Fig. 13. On PTFE, PS, PVF, and PVDF, cysteamine is bonded with a higher amount and even preferentially through the –SH group with the "free" $-NH_2$ group, which creates a positive charge on the surface in the presence of a KCl water solution. As a result, the zeta potential increases to more positive values. On the other hand, on polar polymers (PET and PLLA), cysteamine is grafted much less (the concentration of N and S is significantly lower), and none of the possible bonding mechanisms is preferred [112]. Probably, on these polymers, the binding of cysteamine is accomplished by both mechanisms, i.e., via (i) the –SH group with the "free" $-NH_2$ group, which creates a positive charge on surface in presence of a KCl water solution and also (ii) the $-NH_2$ group with the "free" –SH group, which creates a negative surface charge in the presence of a KCl water solution connected with a more negative zeta potential [112].

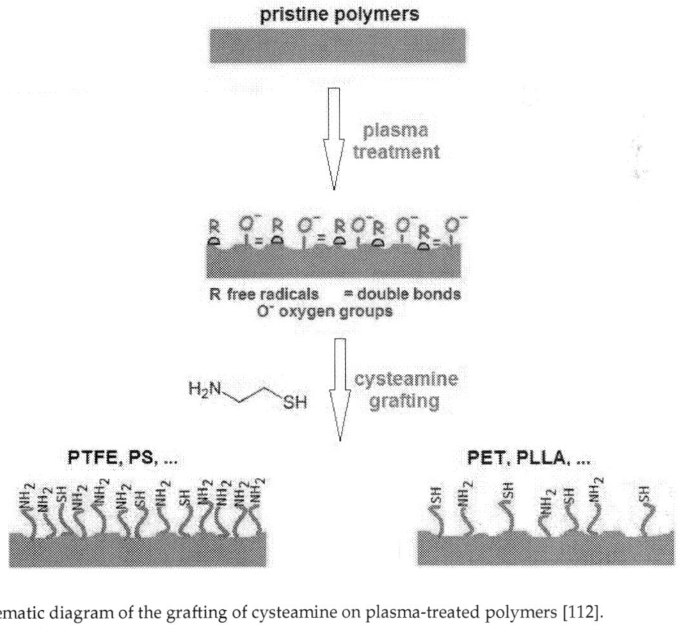

Figure 13. Schematic diagram of the grafting of cysteamine on plasma-treated polymers [112].

5. Conclusions

Surface properties of polymer substrates may be significantly enhanced by different treatment methods. Laser and plasma treatment or grafting processes can be applied to alter surface wettability, which can be also influenced by several other factors, including surface chemistry (introducing new functional groups on the solid substrate surface) or morphology (specific pattern and roughness change). The surface and interface properties of polymers play a key role in many technological processes, such as polymer processing, metallization, or their biocompatibility. All these processes can be controlled by wetting and adhesion phenomena. Based on the conclusions of the wettability measurement, we can estimate the surface chemistry with regard to polar or nonpolar functional groups as well as acidic or basic sites available at the modified surface region. The wettability changes in the laser-treated surface with respect to the type of laser treatment (excimer laser, fs/ns laser, etc.) and the dependence of the CA on the processing parameters of laser treatment (number of pulses and laser fluence) were described. The changes in wettability of polymers and biopolymers due to the treatment and during aging were described. Related properties (morphology and chemistry changes, and ablation) were discussed with respect to the CA and surface-free energy of the surface. The polymer generally has a low surface reactivity. Different types of plasma treatment were described in detail with respect to the change of the surface properties of exposed substrates. A broad range of functional groups can be introduced at the surface by variation of the gas that is used or by consequential grafting processes. The influence of grafting of either metal nanoparticles (e.g., Au) or chemical substances (e.g., PEG) on the wettability of modified polymers was described in detail. The properties of plasma treatment and grafting procedures that have a positive influence on the spreading, proliferation activity, and homogeneity of the distribution of several cell types were introduced.

Acknowledgements

This work has been supported by the Grant Agency of the Czech Republic under Project 108/12/G108.

Author details

Nikola Slepickova Kasalkova[1], Petr Slepicka[1], Zdenka Kolska[2] and Vaclav Svorcik[1]

*Address all correspondence to: nikola.kasalkova@vscht.cz

1 Department of Solid State Engineering, University of Chemistry and Technology Prague, Prague, Czech Republic

2 Faculty of Science, J. E. Purkyne University, Usti nad Labem, Czech Republic

References

[1] Mazzola L., Bemporad E., Carassiti F. Flame treatment on plastic: a new surface free energy statistical prediction model and characterization of treated surfaces. Applied Surface Science 2011; 257 2148–2158.

[2] Gauvin R., Guillemette M., Langer R., Khademhosseini A. Emerging trends in tissue engineering. In: Cui Z. (ed). Comprehensive Biotechnology. Burlington: Elsevier Ltd. Academic Press: 2011. p251–263.

[3] Dhandayuthapani B., Yoshida Y., Maekawa T., Kumar D. S. Polymeric scaffolds in tissue engineering application: a review. International Journal of Polymer Science 2011; 2011 1–19.

[4] Nair L.S., Laurencin C. T. Biodegradable polymers as biomaterials. Progress in Polymer Science 2007; 32 762–798.

[5] Bhat S. V. Biomaterials. First edition, Alpha Science International Ltd, UK: 2002.

[6] Clarke D. E., McCullen S.D., Chow A. G. Y, Steevens M. M. Functional biomaterials. In: Moo-Young M. (ed.). Comprehensive Biotechnology, 2nd edition, Academic Press: Burlington 2011: 5 p3–10.

[7] Abdul-Kader A. M., Turos A., Radwan R. M., Kelany A. M. Surface free energy of ultra-high molecular weight polyethylene modified by electron and gamma irradiation. Applied Surface Science 2009; 255 7786–7790.

[8] Sun T., Feng L., Gao X., Jiang L. Bioinspired surfaces with special wettability. Accounts of Chemical Research 2005; 38 644–652.

[9] Lee J. H., Lee S. J., Khang G., Lee H. B. The effect of fluid shear stress on endothelial cell adhesiveness to polymer surfaces with wettability gradient. Journal of Colloid and Interface Science 2000; 230 84–90.

[10] Fujinami A., Matsunaka D., Shibutani X. Water wettability/non-wettability of polymer materials by molecular orbital studies. Polymer 2009; 50 716–720.

[11] Pereir P. M. M., Moita A. S., Monteiro G. A., Prazeres D. M. F. Characterization of the topography and wettability of English weed leaves and biomimetic replicas. Journal of Bionic Engineering 2014; 11 346–359.

[12] Drake S. S., O'Carroll D. M., Gerhard J. I. Wettability contrasts between fresh and weathered diesel fuels. Journal of Contaminant Hydrology 2013; 144 46–57.

[13] Krawczyk J., Szymczyk K., Zdziennicka A., Jańczuk B. Wettability of polymers by aqueous solution of binary surfactants mixture with regard to adhesion in polymer–solution system I—correlation between the adsorption of surfactants mixture and contact angle. International Journal of Adhesion and Adhesives 2013; 45 98–105.

[14] Darmanin T., Guittard F. Wettability of conducting polymers: from superhydrophilicity to superoleophobicity. Progress in Polymer Science 2014; 39 656–682.

[15] Slepičková Kasálková N., Slepička P., Kolská Z., Hodačová P., Kučková Š., Švorčík V. Grafting of bovine serum albumin proteins on plasma-modified polymers for potential application in tissue engineering. Nanoscale Research Letters 2014; 9 1–7.

[16] Siegel J., Slepička P., Heitz J., Kolská Z., Sajdl P., Švorčík V. Gold nano-wires and nano-layers at laser-induced nano-ripples on PET. Applied Surface Science 2010; 256 2205–2209.

[17] Barb R. A., Hrelescu C., Dong L., Heitz J., Siegel J., Slepička P., Vosmanská V., Švorčík V., Magnus B., Marksteiner R., Schernthaner M., Groschner K. Laser-induced periodic surface structures on polymers for formation of gold nanowires and activation of human cells. Applied Physics A 2014; 117 295–300.

[18] Slepička P., Neděla O., Sajdl P., Kolská Z., Švorčík V. Polyethylene naphthalate as an excellent candidate for ripplenanopatterning. Applied Surface Science 2013; 285B 885–892.

[19] Slepička P., Chaloupka A., Sajdl P., Heitz J., Hnatowicz V., Švorčík V. Angle dependent laser nanopatterning of poly(ethylene terephthalate) surfaces. Applied Surface Science 2011; 257 6021–6025.

[20] Rebollar E., Frischauf I., Olbrich M., Peterbauer T., Hering S., Preiner J., Hinterdorfer P., Romanin Ch., Heitz J. Proliferation of aligned mammalian cells on laser-nanostructured polystyrene. Biomaterials 2008; 29 1796-1806.

[21] Zhao Ch., Xue J., Ran F., Sun S. Modification of polyethersulfone membranes—a review of methods. Progress in Materials Science 2013; 58 76–150.

[22] Soto R. R., Comesana R., Boutinguiza M., del Val J., Quintero F., Lusquinosa F., Pou J. Laser surface modification of PEEK. Applied Surface Science 2012; 258 9437–9442.

[23] Gotoh K., Nakata Y., Tagawa M., Tagawa M., Gotoh K., Nakata Y., Tagawa M., Tagawa M. Wettability of ultraviolet excimer-exposed PE, PI and PTFE films determined by the contact angle measurements. Colloids and Surfaces A 2003; 224 165-173.

[24] Akkan C. K., Hammadeh M., Brück S., Park H. W., Veith M., Abdul-Khaliq H., Aktas C. Plasma and short pulse laser treatment of medical grade PEEK surfacesfor controlled wetting. Materials Letters 2013; 109 261–264.

[25] Riveiro A., Soto R., del Val J., Comesana R., Boutinguiza M., Quintero F., Lusquinos F., Pou J. Laser surface modification of ultra-high-molecular-weightpolyethylene (UHMWPE) for biomedical applications. Applied Surface Science 2014; 302 236–242.

[26] Waugh D. G., Lawrence J. Wettability and osteoblast cell response modulation through UV laser processing of nylon 6,6. Applied Surface Science 2011; 257 8798–8812.

[27] Waugh D. G., Lawrence J., Walton C. D., Zakaria R. B. On the effects of using CO_2 and F_2 lasers to modify the wettability of a polymeric biomaterial. Optics & Laser Technology 2010; 42 347–356.

[28] He L., Chen J., Farson D. F., Lannutti J. J., Rokhlin S. I. Wettability modification of electrospun poly(e-caprolactone) fibers by femtosecond laser irradiation in different gas atmospheres. Applied Surface Science 2011; 257 3547–3553.

[29] Belaud V., Valette S., Stremsdoerfer G., Bigerelle M., Benayoun S. Wettability versus roughness: multi-scales approach. Tribology International 2015; 82 343–349.

[30] Waugh D. G., Lawrence J. On the use of CO_2 laser induced surface patterns to modify the wettability of poly(methyl methacrylate)(PMMA). Optics and Lasers in Engineering 2010; 48 707–715.

[31] Mirzadeh H., Bagheri Sh. Comparison of the effect of excimer laser irradiation and RF plasma treatment on polystyrene surface. Radiation Physics and Chemistry 2007; 76 1435–1440.

[32] Pfleging W., Bruns M., Welle A., Wilson S. Laser-assisted modification of polystyrene surfaces for cell culture applications. Applied Surface Science 2007; 253 9177–9184.

[33] Oliveira V., Nunes B., Vilar R. Wetting response of KrF laser ablated polyimide surfaces. Nuclear Instruments and Methods in Physics Research B 2010; 268 1626–1630.

[34] Petit S., Laurens P., Amouroux J., Arefi-Khonsari F. Excimer laser treatment of PET before plasma metallization. Applied Surface Science 2000; 168 300–303.

[35] Tiaw K. S., Goh S. W., Hong M., Wang Z., Lan B., Teoh S. H. Laser surface modification of poly(e-caprolactone) (PCL) membrane for tissue engineering applications. Biomaterials 2005; 26 763–769.

[36] Pfleging W., Kohler R., Torge M., Trouillet V., Danneil F., Stüber M. Control of wettability of hydrogenated amorphous carbon thin films by laser-assisted micro- and nanostructuring. Applied Surface Science 2011; 257 7907–7912.

[37] Grundke K., Pöschel K., Synytska A., Frenzel R., Drechsler A., Nitschke M., Cordeiro A. L., Uhlmann P., Welzel P. B. Experimental studies of contact angle hysteresis phenomena on polymer surfaces—toward the understanding and control of wettability for different applications. Advances in Colloid and Interface Science 2014, in press, DOI: 10.1016/j.cis.2014.10.012.

[38] Slepička P., Neděla O., Siegel J., Krajcar R., Kolská Z., Švorčík V. Ripple polystyrene nano-pattern induced by KrF laser. eXPRESS Polymer Letters 2014; 8 459–466.

[39] Rytlewski P., Zenkiewicz M. Laser-induced surface modification of polystyrene. Applied Surface Science 2009; 256 857–861.

[40] Slepička P., Neděla O., Sajdl P., Kolská Z., Švorčík V. Polyethylene naphthalate as an excellent candidate for ripplenanopatterning. Applied Surface Science 2013; 285B 885–892.

[41] Řezníčková A., Chaloupka A., Heitz J., Kolská Z., Švorčík V. Surface properties of polymers treated with F_2 laser. Surface and Interface Analysis 2012; 44 296–300.

[42] Slepička P., Michaljaničová I., Fitl P., Sajdl P., Švorčík V. Surface ablation of PLLA induced by KrF excimer laser. Applied Surface Science 2013; 283 438–444.

[43] Slepička P., Michaljaničová I., Švorčík V. Controlled biopolymer roughness induced by plasma and excimer laser treatment. eXPRESS Polymers Letters 2013; 7 950–959.

[44] Pérez S., Rebollar E., Oujja M., Martín M., Castillejo M. Laser-induced periodic surface structuring of biopolymers. Applied Physics A 2013; 110 683–690.

[45] Antonczak A. J., Stepak B. D., Szustakiewicz K., Wojcik M. R., Abramski K. M. Degradation of poly(L-lactide) under CO_2 laser treatment above the ablation threshold. Polymer Degradation and Stability 2014; 109 97-105.

[46] Bogaerts A., Neyts E., Gijbels R., van der Mullen J. Gas discharge plasmas and their applications. Spectrochimica Acta Part B 2002; 57 (4) 609–658.

[47] Friedrich J. F., Geng Sh., Unger W., Lippitz A., Erdmann J., Gorsler H. V., Wöll Ch., Schertel A., Bierbaum K. Plasma functionalization and reorientation of macromolecules at polymer surfaces. Surface and Coatings Technology 1995; 74–75 664–669.

[48] Riekering M. B. O. Structural and chemical modification of polymer surface by gas plasma etching. Ph.D. dissertation. University of Twente, Enschede, The Netherlands: 2001.

[49] Ferreira B. M. P., Pinheiro L. M. P., Nascente P. A. P., Duek E. A. R. Plasma surface treatments of poly(l-lactic acid) (PLLA) and poly(hydroxybutyrate-co-hydroxyvalerate) (PHBV). Materials Science and Engineering C 2009; 29 806–813.

[50] Goddard J. M., Hotchkiss J. H. Polymer surface modification for the attachment of bioactive compounds. Progress in Polymer Science 2007; 32 698–725.

[51] Morent R., De Geyter N., Desmet T., Dubruel P., Leys C. Plasma surface modification of biodegradable polymers: a review. Plasma Processes and Polymers 2011; 8 171–190.

[52] Shi M. K., Selmani A., Martinu L., Sacher E., Wertheimer M. R., Yelon A. Mittal K. L. (ed.) Polymer Surface Modification: Relevance to Adhesion. Utrecht: VSP BV: 1996. p73–85.

[53] Hudis M. Plasma treatment of solid materials. In: Hollahan J. R., Bell A. T. (eds.) Techniques and Applications of Plasma Chemistry. New York: John Wiley: 1974. p113–147.

[54] Triolo P. M., Andrade J. D. J. Surface modification and evaluation of some commonly used catheter materials I: surface properties. Journal of Biomedical Materials Research Part B: Applied Biomaterials 1983; 17 129–147.

[55] Amstein C. F., Hartman P. A. J. Adaptation of plastic surfaces for tissue culture by glow discharge. Journal of Clinical Microbiology 1975; 2 46–54.

[56] van Os M. T. Surface modification by plasma polymerization: film deposition, tailoring of surface properties and biocompatibility. PhD dissertation. University of Twente, Enschede, The Netherlands: 2000.

[57] Slepičková Kasálková N., Slepička P., Bačáková L., Sajdl P., Švorčík V. Biocompatibility of plasma nanostructured biopolymers. Nuclear Instruments and Methods in Physics Research B 2013; 307 642–646.

[58] Wilhelm R. ECR plasmas for thin-film deposition. In: Schlüter H., Shivarova A. (eds.), Advanced Technologies Based on Wave and Beam Generated Plasmas. Kluwer, Dordrecht: NATO Science Series: 1999. p111–122.

[59] Šorli I., Petasch W., Kegel B., Schmit H., Liebel G. Plasma processes part I: plasma basics, plasma generation. Informacije MIDEM 1996; 26 35–45.

[60] de Geyter N., Morent R., Leys C., Gengembre L., Payen E. Treatment of polymer films with a dielectric barrier discharge in air, helium and argon at medium pressure. Surface and Coatings Technology 2007; 201 7066–7075.

[61] Vesel A., Junkar I., Cvelbar U., Kovač J., Mozetič M. Surface modification of polyester by oxygen- and nitrogen-plasma treatment. Surface and Interface Analysis 2008; 40 1444–1453.

[62] Kim Y. J., Kang I. K., Huh M. W., Yoon S. C. Surface characterization and in vitro blood compatibility of poly(ethylene terephthalate) immobilized with insulin and/or heparin using plasma glow discharge. Biomaterials 2000; 21 121-130.

[63] Vesel A., Mozetic M., Zalar A. XPS study of oxygen plasma activated PET. Vacuum 2007; 82 248–251.

[64] Chappel P. J. C., Brown J. R., George G. A., Willis H. A. Surface modification of extended chain polyethylene fibres to improve adhesion to epoxy and unsaturated polyester resins. Surface and Interface Analysis 1991; 17 143–150.

[65] Nakayama Y., Takahagi T., Soeda F., Hatada K., Nagaoka S., Suzuki J., Ishitani A. XPS analysis of NH3 plasma-treated polystyrene films utilizing gas phase chemical modification. Journal of Polymer Science Part A: Polymer Chemistry 1988; 26 559–572.

[66] Clark D. T., Dilks A. ESCA applied to polymers XV: RF glow-discharge modification of polymers, studied by means of ESCA in terms of a direct and radiative energy-

transfer model. Journal of Polymer Science: Polymer Chemistry Edition 1977; 15 2321–2345.

[67] Inagaki N., Tasaka S, Kawai H. Improved adhesion of poly(tetrafluoroethylene) by NH3-plasma treatment. Journal of Adhesion Science and Technology 1989; 3 637–649.

[68] Chan C. M., Ko T. M., Hiraoka H. Polymer surface modification by plasmas and photons. Surface Science Reports 1996; 24 1–54.

[69] Hu S. G., Jou C. H., Yang M. C. Biocompatibility and antibacterial activity of chitosan and collagen immobilized poly(3-hydroxybutyric acid-co-3-hydroxyvaleric acid). Carbohydrate Polymers 2004; 58 173–179.

[70] Zhijiang C., Guoxiang Ch., Zhiyuan T. Surface modification of poly(3-hydroxybutyrate) (PHB) by photografting and its properties evaluation. Journal of Polymer Research 2004; 11 99–104.

[71] Demina T., Zaytseva-Zotova D., Yablokov M., Gilman A., Akopova T., Markvicheva E., Zelenetskii A. DC discharge plasma modification of chitosan/gelatin/PLLA films: surface properties, chemical structure and cell affinity. Surface and Coatings Technology 2012; 207 508–516.

[72] Peyvandi A., Abideen S. U., Huang Y., Lee I., Soroushian P., Lu J. Surface treatment of polymer microfibrillar structures for improved surface wettability and adhesion. Applied Surface Science 2014; 289 586–591.

[73] Vassallo E., Cremona A., Ghezzi F., Ricci A. Characterization by optical emission spectroscopy of an oxygen plasma used for improving PET wettability. Vacuum 2010; 84 902–906.

[74] Farhat S., Gilliam M., Rabago-Smith M., Baran C., Walter N., Zand A. Polymer coatings for biomedical applications using atmospheric pressure plasma. Surface and Coatings Technology 2014; 241 123–129.

[75] Akishev Y., Grushin M., Dyatko N., Kochetov I., Napartovich A., Trushkin N., Duc T. M, Descours S. Studies on cold plasma–polymer surface interaction by example of PP- and PET-films. Journal of Physics D: Applied Physics 2008; 41 235203 (13pp).

[76] Mirabedini S. M., Arabi H., Salem A., Asiaban S. Effect of low-pressure O_2 and Ar plasma treatments on the wettability and morphology of biaxial-oriented polypropylene (BOPP) film. Progress in Organic Coatings 2007; 60 105–111.

[77] Lehocký M., Drnovská H., Lapčíková H., Barros-Timmons A. M., Trindade T., Zembala M., Lapčí L. Jr. Plasma surface modification of polyethylene. Colloids and Surfaces A: Physicochemical and Engineering Aspects 2003; 222 125-131.

[78] Slepička P., Trostová S., Slepičková Kasálková N., Kolská Z., Sajdl P., Švorčík V. Surface modification of biopolymers by argon plasma and thermal treatment. Plasma Processes and Polymers 2012; 9 197–206.

[79] Slepička P., Stýblová Š., Slepičková Kasálková N., Rimpelová S., Švorčík V. Cytocompatibility of polyhydroxybutyrate modified by plasma discharge. Polymer Engineering and Science 2014; 54 1231–1238.

[80] Kolářová K., Kasálková N., Dvořánková B., Michálek J., Švorčík V. Surface modification of hydrogels and cell adhesion. Materials Science Forum 2008; 567–568 265–268.

[81] Švorčík V., Kotál V., Bláhová O., Špírková M., Sajdl P., Hnatowicz V. Modification of surface properties of polyethylene by Ar plasma discharge. Nuclear Instruments and Methods in Physics Research Section B: Beam Interactions with Materials and Atoms 2006; 224 365–372.

[82] Kim K. S., Ryu Ch. M., Park Ch. S., Sur G. S., Park Ch. E. Investigation of crystallinity effects on the surface of oxygen plasma treated low density polyethylene using X-ray photoelectron spectroscopy. Polymer 2003; 44 6287–6295.

[83] Hagiwara K., Hasebe T., Hotta A. Effects of plasma treatments on the controlled drug release from poly(ethylene-co-vinyl acetate). Surface and Coatings Technology 2013; 216 318–323.

[84] Morent R., De Geyter N., Leys C., Gengembre L., Payen E. Study of the aging behavior of polymer films treated with a dielectric barrier discharge in air, helium and argon at medium pressure. Surface and Coatings Technology 2007; 201 7847–7854.

[85] Bhattacharya A., Mistra B. N. Grafting: a versatile means to modify polymers: Techniques, factors and applications. Progress in Polymer Science 2004; 28: 767–814.

[86] Kang H. J., Cha E. J., Park H. D. Protein immobilization onto various surfaces using a polymer-bound isocyanate. Applied Surface Science 2015; 324 198–204.

[87] Iwata R., Satoh R., Iwasaki Y., Akiyoshi K. Covalent immobilization of antibody fragments on well-defined polymer brushes via site-directed method. Colloids and Surfaces B: Biointerfaces 2008; 62 288–298.

[88] Muguruma H. Plasma-polymerized films for biochip design. Plasma Processes and Polymers 2010; 7 151–162.

[89] Yin Y., Nosworthy N. J., Gong B., Bax D., Kondyurin A., McKenzie D. R., Bilek M. Plasma polymer surfaces compatible with a CMOS process for direct covalent enzyme immobilization. Plasma Polymer Surfaces 2009; 6 68–75.

[90] Heyse P., Van Hoeck A., Roeffaers M. B. J., Raffin J. P., Steinbüchel A., Stöveken T., Lammertyn J., Verboven P., Jacobs P. A., Hofkens J., Paulussen S., Sels B. F. Exploration of atmospheric pressure plasma nanofilm technology for straightforward bio-ac-

tive coating deposition: enzymes, plasmas and polymers: an elegant synergy. Plasma Processes and Polymers 2011; 8 965–974.

[91] Mahouche-Cherguia S., Guerrouachea M., Carbonniera B., Chehimib M. M. Polymer-immobilized nanoparticles. Biosensors and Bioelectronics 2013; 41 43–53.

[92] Slepičková Kasálková N., Slepička P., Kolská Z., Sajdl P., Bačáková L., Rimpelová S., Švorčík V. Cell adhesion and proliferation on polyethylene grafted with Au nanoparticles. Nuclear Instruments and Methods in Physics Research B 2012; 272 391–395.

[93] Boisseau P., Loubaton B. Nanomedicine, nanotechnology in medicine. Comptes Rendus Physique 2011; 12: 620–636.

[94] Lehner R., Wang X., Marsch S., Hunziker P. Intelligent nanomaterials for medicine: carrier platforms and targeting strategies in the context of clinical application. Nanomedicine: Nanotechnology, Biology, and Medicine 2013; 9 742–757.

[95] Bačáková L., Grausová L., Vacík J., Fraczek A., Blazewicz S., Kromka A., Vaneček M., Švorčík V. Improved adhesion and growth of human osteoblast-like MG 63 cells on biomaterials modified with carbon nanoparticles. Diamond and Related Materials 2007; 16 2133–2140.

[96] Foley S., Crowley C., Smaihi M., Bonfils C., Erlanger B. F., Seta P., Larroque C. Cellular localisation of a water-soluble fullerene derivative. Biochemical and Biophysical Research Communications. 2002; 294 116-119.

[97] Yamawaki H., Iwai N. Genotoxicity in cell lines induced by chronic exposure to water-soluble fullerenes using micronucleus test. American Journal of Physiology 2006; 290 1495–1502.

[98] Tang Y. J., Ashcroft J. M., Chen D., Min G., Kim C. H., Murkhejee B., Larabell C., Keasling J. D., Chen F. F. Charge-associated effects of fullerene derivatives on microbial structural integrity and central metabolism. NanoLetters 2007; 7 754–760.

[99] Katz E., Willner I. Integrated nanoparticle–biomolecule hybrid systems: synthesis, properties, and applications. Angewandte Chemie International Edition 2004; 43 6042–6108.

[100] Pissuwan D., Cortie C. H., Valenzuela S. M., Cortie M. B. Functionalised gold nanoparticles for controlling pathogenic bacteria. Trends in Biotechnology 2010; 28 207–213.

[101] Enache D. I., Edwards J. K., Landon P., Solsona-Espriu B., Carley A. F., Herzing A. A., Watanabe M., Kiely Ch. J., Knight D. W., Hutchings G. J. Solvent-free oxidation of primary alcohols to aldehydes using Au-Pd/TiO2 catalysts. Science 2006; 311 362–365.

[102] Ročková K., Švorčík V., Bačáková L., Dvořánková B., Heitz J. Bio-compatibility of ion beam-modified and RGD-grafted polyethylene. Nuclear Instruments and Methods in

Physics Research Section B: Beam Interactions with Materials and Atoms 2004; 225 275–279.

[103] Santos M. I., Tuzlakoglu K., Fuchs S., Gomes M. E., Peters K., Unger R. E., Piskin E., Reis R. L., Kirkpatrick J. Endothelial cell colonization and angiogenic potential of combined nano- and micro-fibrous scaffolds for bone tissue engineering. Biomaterials 2008; 29 4306–4313.

[104] Jain P. K., Huang X., El-Sayed I. H., El-Sayed M. A. Noble metals on the nanoscale: optical and photothermal properties and some applications in imaging, sensing, biology, and medicine. Accounts of Chemical Research 2008; 41 1578–1586.

[105] Oehr Ch. Plasma surface modification of polymers for biomedical use. Nuclear Instruments and Methods in Physics Research Section B: Beam Interactions with Materials and Atoms 2003; 208 40–47.

[106] Rimpleová S., Slepičková Kasálková N., Slepička P., Lemerová H., Švorčík V., Ruml T. Plasma treated polyethylene grafted with adhesive molecules for enhanced adhesion and growth of fibroblasts. Materials Science and Engineering C: Materials for Biological Applications 2013; 33 1116–1124.

[107] Pařízek M., Kasálková N., Bačáková L., Slepička P., Lisá V., Blažková M., Švorčík V. Improved adhesion, growth and maturation of vascular smooth muscle cells on polyethylene grafted with bioactive molecules and carbon particles. International Journal of Molecular Sciences 2009; 10 4352–4374.

[108] Brouette N., Sferrazza M. PEG polymer brushes via the Langmuir–Schaefer technique: The transfer ratio. Journal of Colloid and Interface Science 2013; 394 643–645.

[109] Kasálková N., Makajová Z., Pařízek M., Slepička P., Kolářová K., Bačáková L., Hnatowicz V., Švorčík V. Cell adhesion and proliferation on plasma-treated and poly(ethylene glycol)-grafted polyethylene. Journal of Adhesion Science and Technology 2010; 24 743–754.

[110] Švorčík V., Kasálková N., Slepička P., Záruba K., Král V., Bačáková L., Pařízek M., Lisá V., Ruml T., Gbelcová H., Rimpelová S., Macková A. Cytocompatibility of Ar+ plasma treated and Au nanoparticle-grafted PE. Nuclear Instruments and Methods in Physics Research B 2009; 267 1904–1910.

[111] Řezníčková A., Kolská Z., Siegel J., Švorčík V. Grafting of gold nanoparticles and nanorods on plasma-treated polymers by thiols. Journal of Material Science 2012; 47 6297–6304.

[112] Kolská Z., Řezníčková A., Nagyová M., Slepičková Kasálková N., Sajdl P., Slepička P., Švorčík V. Plasma activated polymers grafted with cysteamine improving surfaces cytocompatibility. Polymer Degradation and Stability 2014; 101 1-9.

[113] Bachurová M., Wiener J. Contact angle hysteresis on plasma treated polyethylene terephthalate. e-Polymers 2012; 041 1-9.

[114] Svorčík V., Kolářová K., Slepička P., Macková A., Novotná M., Hnatowicz V. Modification of surface properties of high and low density polyethylene by Ar plasma discharge. Polymer Degradation and Stability 2006; 91 1219–1225.

[115] Kirby B. J., Hasselbrink Jr. E. F. Zeta potential of microfluidic substrates 1: theory, experimental techniques, and effects on separations. Electrophoresis 2004; 25 187-202.

[116] Bačáková L., Filová E., Pařízek M., Ruml T., Švorčík V. Modulation of cell adhesion, proliferation and differentiation on materials designed for body implants. Biotechnology Advances 2011; 29 739–767.

Chapter 13

Wettability of Carbonaceous Materials with Molten Iron at 1550°C

R. Khanna, I. Mansuri and V. Sahajwalla

Additional information is available at the end of the chapter

http://dx.doi.org/10.5772/61198

Abstract

In the direct iron smelting process, interfacial reactions of carbonaceous materials with molten iron are among some of the key factors that dictate the rate of carbon transfer into molten iron and establish a carbon concentrated melt to reduce iron oxide in the slag phase. Detailed wettability investigations on a range of carbonaceous materials, e.g. synthetic graphite, natural graphite, coke, coal-chars, coke-polymer blends, waste plastics in contact with molten iron at 1550°C were carried out using the sessile drop method. Experimental results on dynamic contact angles are presented in this chapter and are discussed in terms of the basic characteristics of carbons and the changing composition of the interfacial region as a function of time.

The influence of carbon and sulfur content in Fe-C-S melts on the wettability of synthetic graphite was investigated. The initial melt carbon content was in the range of 0.13 to 2.24 wt %, and the melt sulfur content was in the range of 0.05 to 0.37 wt %. It was found that the contact of solid graphite with Fe-C-S melts resulted in a non-equilibrium reactive wetting. It involved the transfer of carbon from the solid to the liquid and iron transfer from the liquid to the solid. The Fe-C-S melts exhibited relatively poor wetting in the absence of such a material exchange.

The wetting of natural graphite, which contained 8.8 pct ash, by iron was investigated to establish a fundamental understanding of the influence of ash on interactions between graphite and iron. It was found that the formation of an ash interfacial layer between the carbonaceous material and liquid iron had a strong influence on the mass transfer and interfacial reaction.

Wetting investigations of pure liquid iron with three types of metallurgical cokes at 1550°C showed a non-wetting behaviour for all cokes with contact angles ranging from 123° – 126° initially, which marginally decreased to 109°-114° after 60 minutes of contact. The slight decrease in contact angles was attributed to the transfer of carbon from coke to liquid iron, which then led to a lower the interfacial tension.

The wettability of four non-graphitic coal-chars with electrolytic pure iron was investigated at 1550°C. The ash concentration in chars ranged between 9.04 and 12.61 wt %, with alumina and silica as predominant ash components. The initial contact angles ranged between 106° and 137°, the contact angles for all coal-chars were quite similar after 60 minutes of contact (105 - 110°). While no well-defined correlations could be observed between the initial char structure and ash concentration/composition and contact angles in the initial stages of contact, the contact angles over extended periods were significantly affected by the presence of reaction products and impurity deposits in the interfacial region.

Keywords: Carbonaceous materials, molten iron, wettability, contact angles

1. Introduction

Ironmaking is one of the largest consumers of carbon fuels in the world, with an annual energy consumption of 24 EJ (x 10^{18}J) and 69% of the world's total energy consumption in iron/steelmaking industry [1]. Conventional ironmaking involves temperatures of more than 1500°C, leading to high-energy consumption. It is also dependent upon high quality coking coal, and cannot use low-grade waste materials such as coal fines. To lower both the pollution and carbon emissions, the energy consumption levels need to be lowered and this can be achieved by lowering the temperature of ironmaking. Previous attempts to lower the temperatures have focused on macro-scale reactions. There have been attempts to develop a lower-temperature ironmaking process to reduce energy requirements and emissions by replacing coke with coal fines; but what have been predominantly iron oxide reduction and industry-oriented experiments, have encountered crucial issues of operational stability and low-temperature carburization.

'Mechanisms for liquid iron carburization were extensively investigated in a previous study from our group [2] and the results showed a close relationship with composition of inorganic materials, carbon structure, and temperature. However, solid state carburization is rather complex due to the combination of gas phase carburization reaction and solid state diffusion of carbon [3, 4]. Due to difficulties associated with measuring contact angles with fine powders, very limited information is currently available on the wettability of non-graphitic carbons, such as coal-chars, and the role of carbon type and ash composition is not well determined.

In the direct iron smelting process, interfacial reactions of carbonaceous materials with molten iron are among some of the key factors that dictate the rate of carbon transfer into molten iron and establish a carbon concentrated melt to reduce iron oxide in the slag phase. Basic parameters such as wettability, fixed carbon, ash content/composition, and crystalline order of the carburizing material influence the dissolution kinetics of carbon [5, 6]. Various metal oxides present as ash impurities can also form a viscous layer in the interfacial region affecting wettability and may act as a physical barrier for the transfer of carbon. An understanding of wettability and associated interfacial phenomena is required to establish carburizing fundamentals of various carbonaceous materials. The interfacial phenomena between carbonaceous materials and molten iron is very important due to the extensive use of these materials in iron and steelmaking.

Wu and Sahajwalla provided an alternative explanation for the initial decrease in the contact angle for Fe-C-S melts on graphite followed by a recovery to the equilibrium value [7]. They postulated that the initial decrease was due to the diffusion of carbon and sulfur atoms to the interfacial layer resulting in a reduction in the interfacial tension between the solid and the liquid. Once the diffusing species reduced the chemical potential difference between the solid and the liquid to negligible levels, the Van der Waals forces begin to dominate the wettability. This results in the contact angle approaching its equilibrium value.

The presence of surface-active elements at the interface can affect the interfacial energy. Surface activity is defined as the magnitude of change in surface tension brought about by the addition of a unit quantity of the species. The ultimate surface activity of a surface-active species in a binary system generally decreases with increasing temperature. Additions of more than one solute to a system complicate the determination of surface activity. The compounded results cannot be predicted from the individual contributions. For example carbon, a non-surface active element, increases the activity coefficient of the dissolved sulfur in liquid iron, lowers the surface tension of Fe-C-S solutions thereby enhancing the surface concentrations of sulfur. In an atomistic computer simulation study on the Fe-C-S system, S and C were found to mutually displace each other. Carbon-rich regions were depleted of sulfur and vice versa [8].

Detailed wettability investigations on a range of carbonaceous materials, e.g., synthetic graphite, natural graphite, coke, coal-chars, coke-polymer blends, and waste plastics, in contact with molten iron at 1550°C were carried out using the sessile drop method. Experimental results on dynamic contact angles are presented in this chapter and are discussed in terms of the basic characteristics of carbons and the changing composition of the interfacial region as a function of time. The systems of interest in this study is rather complex: along with a heterogeneous composition, the carbonaceous materials contain non-graphitic carbon and ash minerals; the liquid metal typically contains carbon, sulfur, and quite a few trace elements. In combination, these can result in a number of reactions occurring in the interfacial region, which could a have significant influence on the wettability of the system.

2. Experimental

The wettability of carbonaceous materials and Fe-C-S system was investigated in a laboratory-scale, horizontal tube resistance furnace using the sessile drop approach. The carbonaceous materials formed into cylindrical substrates (25 mm diameter, 3–4 mm height) and compressed under 7 tonne pressure (for approximately 1 minute) using a hydraulic press (ENERPAC, Australia®). The weight of the metal used was ~0.70 g of electrolytic-grade iron and Fe-C-S alloys, respectively. The static nature of the droplet in the sessile drop arrangement helps to preserve the interfacial layer that forms between the iron and carbonaceous materials, allowing a detailed investigation of the interface. Apart from Marangoni flow, there is minimal flow within the liquid iron droplet in the sessile drop method.

A schematic diagram of the experimental arrangement is shown in Fig. 1. The furnace tube had an inside diameter of 50 mm. Initially, the metal/carbon assembly was held on a specimen holder, which could be pushed to the center of the hot zone in the furnace with the help of a graphite rod. The metal/carbon assembly was held in the cold zone of the furnace until the desired temperature (1550°C) was attained. The assembly was then inserted into the hot zone; this eliminated any reaction that could occur at lower temperatures and possibly influence the phenomena to be studied at the temperature of interest. The melting of iron marked the beginning of contact time. The furnace tube was purged with argon throughout the duration of the experiment with a flow rate of 0.5 L/min.

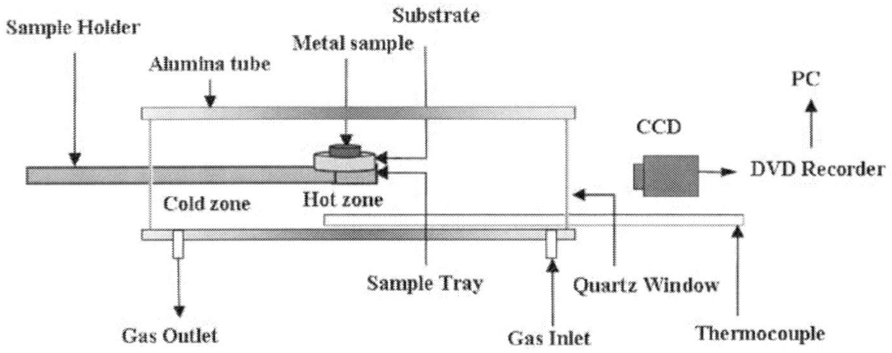

Figure 1. A schematic representation of experimental arrangement.

The wettability behavior of the metal/carbon system was investigated using a closely controlled and visually monitored sessile drop technique. A high-quality, high-resolution charge-coupled device (CCD) camera fitted with an IRIS lens was used to capture the live in-situ phenomena in the furnace. The output from the camera was channeled to a video cassette recorder (VCR) and a television (TV) monitor to record the entire process as a function of time. This allows specific images, displaying the contact between the metal and carbonaceous material, to be captured from the videotape, as a function of time, into a computer using a frame grabber. A time-date generator is used in the system to display the duration of the process. For a better understanding of reaction dynamics, contact angles were recorded up to 2 hours in most cases. Fig. 2 shows a representative image of molten iron droplet during melting in the horizontal tube furnace.

Specially designed computer software was used to determine the contact angle from the captured images, on the basis of a curve-fitting exercise. The contact angle between the iron

http://dx.doi.org/10.5772/61198

Figure 2. Representative image of molten iron droplet during melting in the horizontal tube furnace.

droplet and the substrate was measured by using computer image analysis software based on a curve-fitting algorithm. The working principle of this algorithm is detailed in the work conducted by [7] and is shown in Fig. 2. Typical accuracy of contact angle measurements was within ± 5%.

3. Wettability and interfacial phenomena

Wettability investigations were carried out on a number of carbonaceous materials in contact with molten iron at 1550°C. Detailed results are presented in this section.

3.1. Synthetic and natural graphite

A number of researchers have investigated the wettability between graphite and pure iron, graphite and Fe-C alloys at temperatures of 1300–1550°C under atmospheres of helium, argon and hydrogen gases and in vacuum. These studies have shown that pure iron showed good wetting with graphite at high temperatures under these conditions. During sessile drop experiments of iron with graphite, it was observed that in the initial stages, when the iron had melted completely, the carbon content had reached 4.23%; and in the final stages, it had reached the saturation carbon level of 5.62%. The contact angle changed from 64° at the initial state to 38° in final stages when the carbon content had reached saturation.

Lower contact angles in C/Fe interaction have been explained through the lowering of interfacial tension between the two phases due to the mass transport across the interface. The low value of the contact angle was maintained even when the carbon transport rate slowed down at higher carbon content in the liquid droplet during the later period of carbon had dissolution reaction [9, 10]. Thus for the sessile drop experiment, the reaction area was

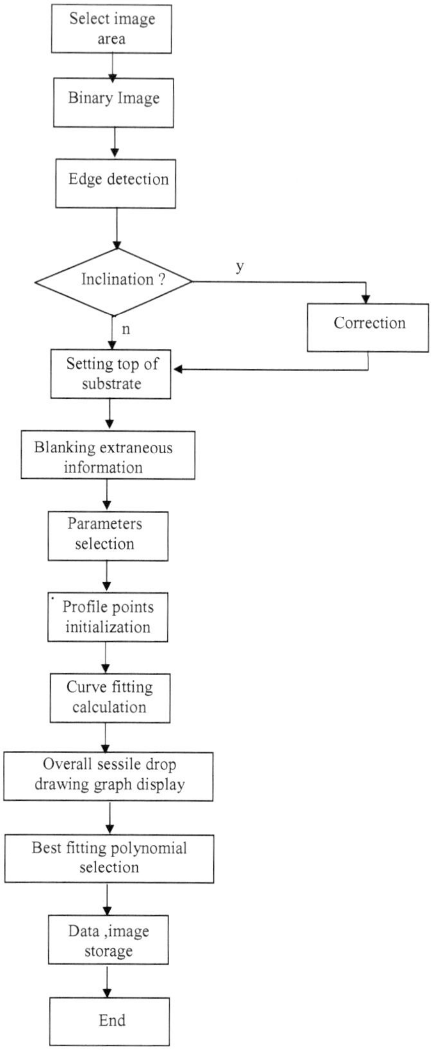

Figure 3. The flow sheet of "interface software" for analysis of the contact angle of a sessile drop image [7].

estimated reasonably well by the lower value of the contact angle. Some experimental results are listed in Table 1.

Sun et al. [13] found that the contact angle of iron on carbon with 0% initial carbon was 59° at 1500 °C, while that of a sample with 4.8 wt% carbon was 118°. It was considered that the contact angle for a low carbon sample is very small, because a lower interfacial tension exists during the rapid initial carbon pickup reaction occurring at the metal-graphite interface. This initial rapid carbon pickup is considered to occur during and immediately after melting. Once the interfacial area extends during the initial carbon pickup reaction, it cannot recover to the shape where a higher carbon content sample was used, even though carbon content reaches the saturation level. These studies showed that pure iron had good wetting with graphite at high temperatures under vacuum, helium and hydrogen and argon gases [7, 11].

Metal	Substrate	Atmosphere	T (°C)	θ (°)	Reference
Molten iron	Graphite	H_2	1550	37	[11]
Molten iron	Graphite	Helium	1550	51	[11]
Molten iron	Graphite	Vacuum	1550	37-51	[12]
Fe+3.77%C	Graphite	Vacuum	1550	107	[12]
Pure iron	Graphite	Argon	1300–1500	66	[13]
Fe+4.8%C	Graphite	Argon	1300–1500	110	[13]
Pure iron	Graphite	Argon	1600	97-55-62	[7]
Fe+0.13%C	Graphite	Argon	1600	72-54-61	[7]

Table 1. Some experimental results on the wettability of synthetic graphite.

Wu and Sahajwalla [7] investigated the dynamic wetting behavior of Fe-C-S melts with graphite substrate using a sessile drop technique by focusing on the influence of melt carbon and sulfur concentration on the wettability at 1600°C. The wettability between the Fe-C-S melt and solid graphite has been presented in terms of the changes in contact angles with time (Fig. 4).

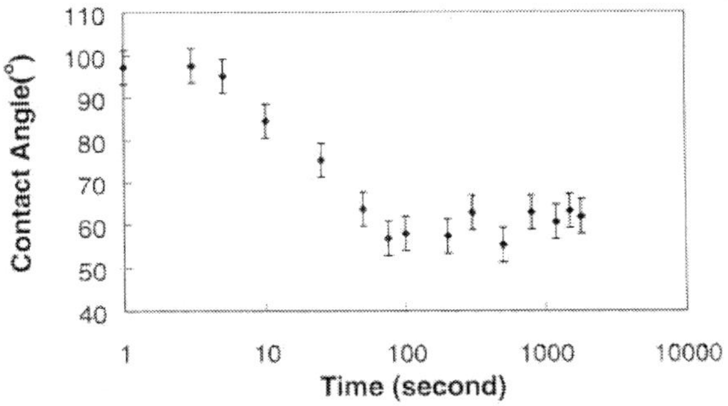

Figure 4. The variation of contact angles as a function of time for molten Fe-S-C and graphite system.

From Fig. 4, in the initial state (within ~8 seconds of contact), the system is observed to be non-wetting (~100°). Thereafter, this system becomes wetting (~60°) with a considerable decrease in the contact angles (in the time period ~8 to 90 seconds). The contact angles were observed to stabilize at ~60° for the rest of the experimental time. The changes in contact angles in this system were attributed to carbon and iron inter-diffusing into the solid graphite and the liquid iron.

Wu et al. [14] investigated the wetting behavior of natural graphite and liquid iron system using the sessile drop technique and the results are shown in Fig. 5. The contact angles of this system decreased from ~100° to become wetting within the first 10 seconds (~60°). Thereafter, it became non-wetting with the contact angles sharply increasing to stabilize at ~100° within 100 seconds. This dynamic wetting behavior of iron and natural graphite is very different from the behavior of molten Fe-C-S and graphite systems; the deposition of ash oxides at the interface played a strong role in this phenomena.

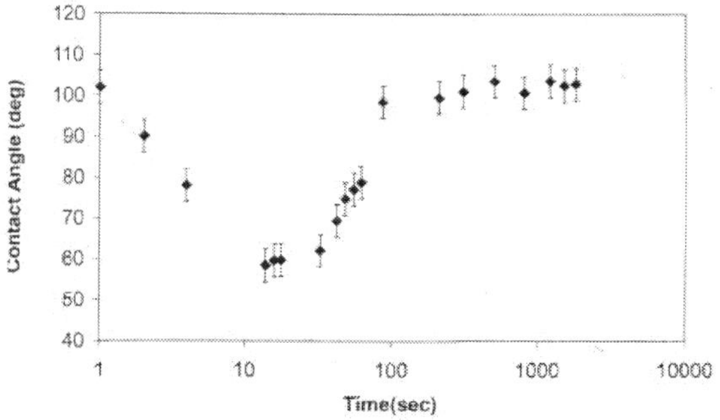

Figure 5. The variation in the contact angles as a function of time for molten iron on natural graphite substrate.

3.2. Coal chars

Raw coals were crushed in a jaw crusher and vibrating grinder, and wet sieved and the resulting size graded samples were dried in an oven prior to char making. The chars were produced from raw coals in a drop-tube furnace in an atmosphere of 23 % O_2 in N_2 at a temperature of 1200 °C.

The wettability for four chars was determined using the sessile drop approach and the contact angles were measured. Chars 1 and 2 had a very high initial contact angle of 145° and 140° respectively, which decreased to ~110° after 60 minutes of contact. While the contact angle for Char 1 showed a gradual decline with time, the contact angles for Char 2 showed a sharp reduction to 110° after 2 minutes of contact and then fluctuated around that value over long times of contact. Chars 3 and 4 had somewhat lower initial contact angles, 115° and 120° respectively, but these showed no improvement with time. Following a trend contrary to Chars 1 and 2, these chars did not show a sharp initial decline; there were only marginal fluctuations in contact angles as a function of time. After 60 minutes of contact, the contact angles for all four chars were quite similar ranging between 105–110°.

Only a marginal improvement in wettability (decrease in contact angles) was observed as a function of time. The authors identified the factors influencing the wettability as: transfer of carbon and sulfur across the metal/carbon interface, the formation of Al_2O_3 and CaS complexes at the interface and the reduction of reducible oxides, such as SiO_2 and Fe_2O_3, and the transfer of Si into the liquid metal.

3.3. Metallurgical coke

An experimental investigation of the wetting behavior of pure liquid iron with three types of metallurgical coke at 1550°C was conducted by Cham et al [15]. It was observed that all cokes showed a non-wetting behavior with contact angles ranging from 123°–126° initially, which marginally decreased to 109°–114° after 60 minutes of contact. It was concluded that the slight decrease in contact angles was due to the transfer of carbon from coke to liquid iron, which then decreased the interfacial tension. On the other hand, the effects of sulfur, ash content and composition and carbon structure on the wettability could not be observed in this study.

The wetting images of molten steel droplet on coke after 60 minutes of reactions are shown in Fig. 6. It can be seen that the liquid steel droplet showed a non-wetting behavior with the coke (contact angle 129°) initially and the value increased slightly within the first few minutes. A very small change was observed throughout the experimental run with the contact angle reaching 128° after 60 mins of contact.

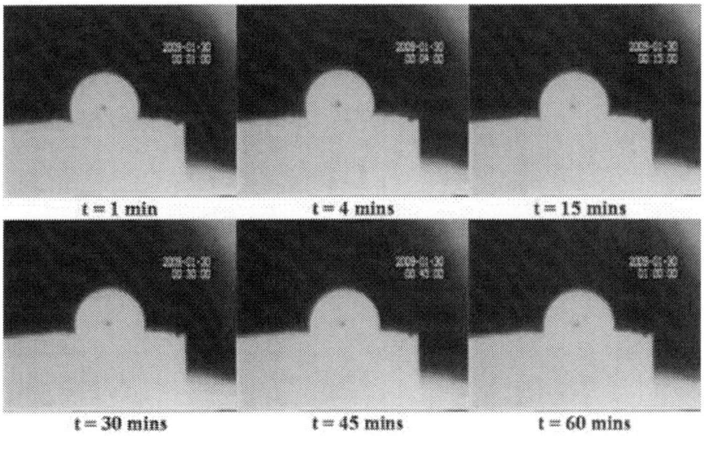

Figure 6. Wetting images of steel droplet after reaction with coke for 60 minutes at 1550°C.

3.4. Metallurgical coke/Bakelite blends

A number of polymers and metallurgical coke were blended in three different ratios: Blend 1, Blend 2, and Blend 3. These were labelled as BK1, BK2, BK3; H1, H2, H3; and P1, P2, P3 for

Blends 1, 2 and 3 of Bakelite/Coke, HDPE/Coke, and PET/Coke, respectively. Fig. 7 represents the blending ratios used in these blends.

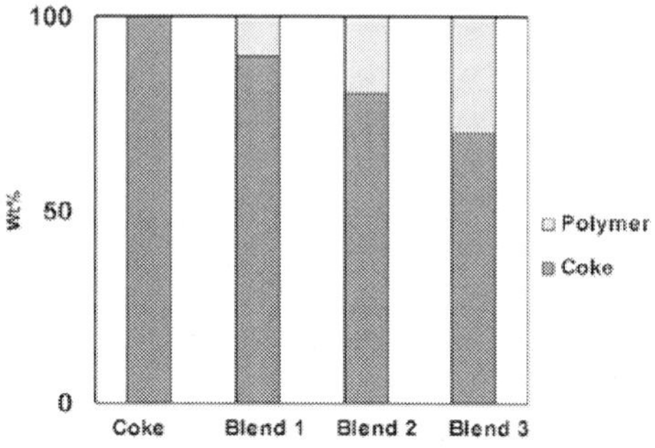

Figure 7. Relative proportions between the plastics and metallurgical coke in the polymer/coke blends.

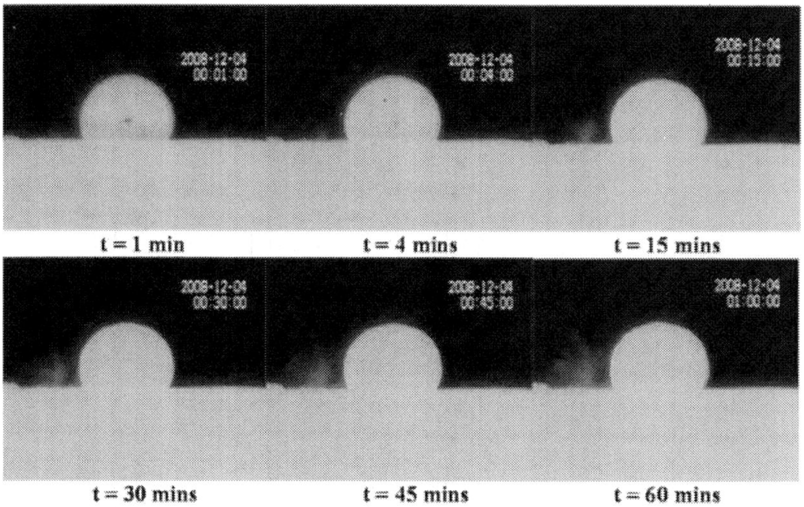

Figure 8. Wetting images of steel droplet after reaction with blend BK1 for 60 minutes at 1550°C.

The raw polymer/coke blends were combusted in a drop tube furnace (DTF) at 1200°C under an atmosphere of 80% nitrogen and 20% oxygen with a gas flow rate of 1.0 L/min, and the chars obtained were used for steel/carbon interaction investigations. The devolatilization was carried out to simulate the real process in Electric Arc Furnace (EAF) steelmaking. In EAF steelmaking process, the carbonaceous materials are injected into the furnace and these undergo partial combustion before coming into contact the molten steel.

http://dx.doi.org/10.5772/61198

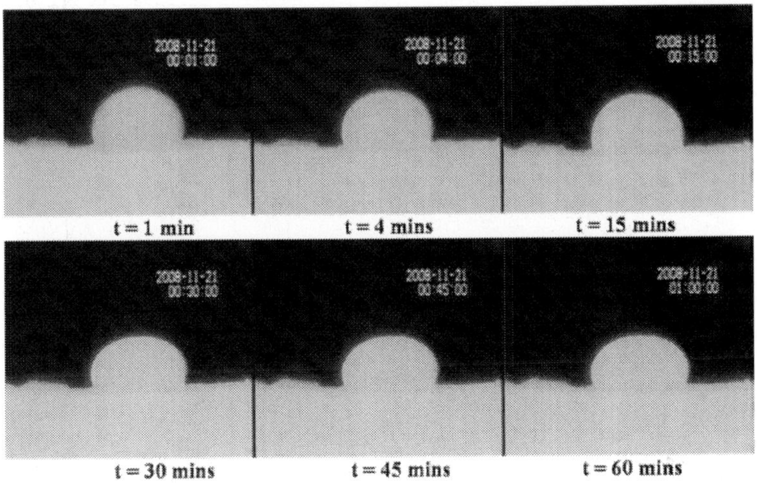

Figure 10. Wetting images of steel droplet after reaction with blend BK2 for 60 minutes at 1550°C.

Wetting results showed variations in contact angle between the Bakelite/coke blends. A general improvement in the wettability of the liquid steel with Bakelite/coke blends was observed as compared to wettability with coke alone.

The wetting images of molten steel droplet on blends BK1, BK2, and BK3 and corresponding contact angles after 60 minutes of reactions are shown in Figs. 8–13, respectively.

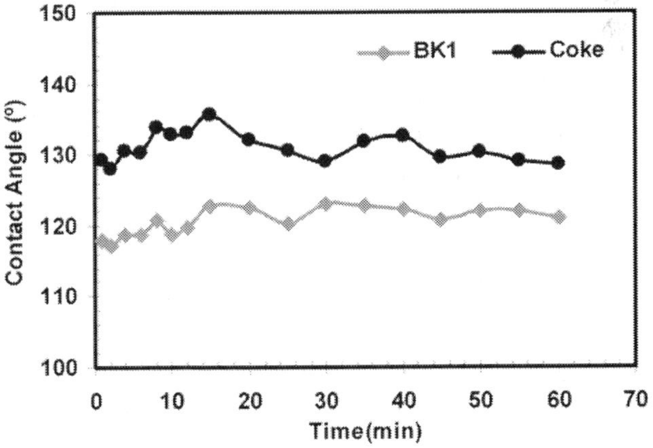

Figure 9. Variation of contact angles of steel droplet with blend BK1 and coke substrates at 1550°C.

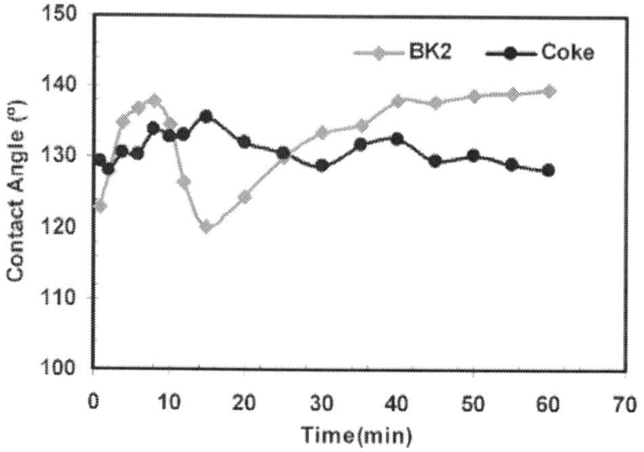

Figure 11. Variation of contact angles of steel droplet with blend BK2 and coke substrates at 1550°C.

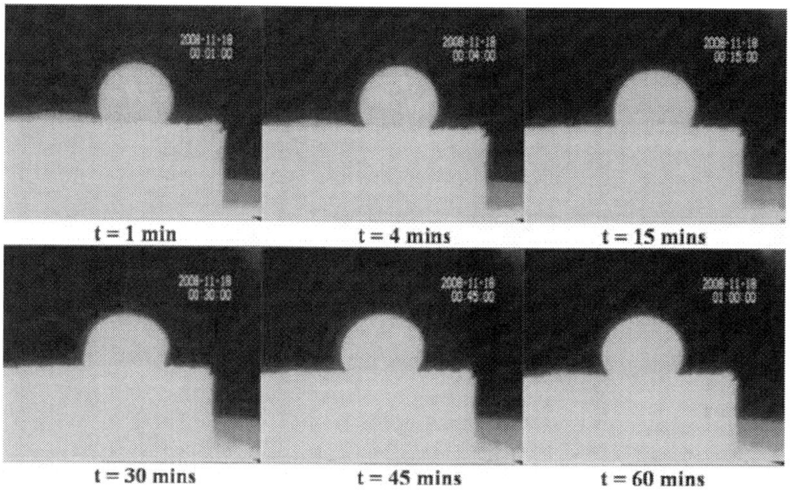

Figure 12. Wetting images of steel droplet after reaction with blend BK3 for 60 minutes at 1550°C.

From these figures, it can be seen that the liquid steel droplet showed a non-wetting behavior with the coke (contact angle 129°) initially and the value increased slightly within the first few minutes. Bakelite containing blends were also found to exhibit non-wetting behavior with liquid steel. The liquid steel droplet exhibited better wetting with blend BK1 substrates with the measured contact angles being approximately 10° lower than that seen with coke alone. The initial contact angle was 118° that increased to 121° after 60 minutes of reaction. In the case of blend BK2, the initial contact angle was 123° that decreased to 120° after 15 minutes and

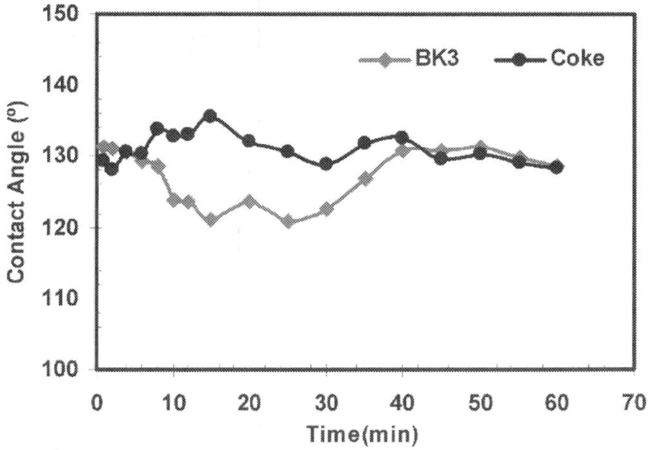

Figure 13. Variation of contact angles of steel droplet with blend BK3 and coke substrates at 1550°C.

then increased to almost 139° after 60 minutes. In the case of BK3, the initial contact angle was 131° that initially decreased to 120° after 25 minutes and then increased to 129° after 60 minutes.

The wetting behavior observed in the case of blends BK2 and BK3 was similar to that observed in natural graphite and liquid iron system by Wu et al. [14]. These authors investigated the wetting behavior of natural graphite with liquid iron at 1600°C using the sessile drop technique. They observed a decrease in contact angles (from ~100° to ~60°) along with a rapid increase in the carbon content in the metal droplet within the first 10 seconds of reaction. Thereafter, the melt carbon content was observed to slightly decrease and the contact angles were observed to increase from ~60° to ~100° within approximately 2 minutes of contact and then stabilize for the rest of the reaction times. Wu et al. [14] explained that the decrease in contact angles within the first 10 seconds was due to a decrease in the interfacial tension due to the dissolution of carbon into the melt. Thereafter, the contact angles increased because of a slowing down in carbon dissolution since the the formation of ash layer at the metal/carbon interface became significant (interfacial blockage and consumption of solute carbon by reducible oxides). The carbon dissolution into the metal droplet and the ash deposits in the interfacial region were also expected to influence the wetting behavior between molten steel and Bakelite/coke blends.

3.5. Bakelite char

Wettability investigation was carried out using Bakelite char substrate and molten iron. Video images and the measured contact angles between the substrate and molten iron is shown in Figs. 14 and 15. A non-wetting behavior (contact angle > 90°) was observed for Bakelite char with liquid iron. Initial contact angle (at 0 minutes) between Bakelite char substrate was found to be 125.7° and decreased to 122.4° after 1 minute of reaction time. Not much change in contact angle was observed at longer times and angles remained more or less constant later on.

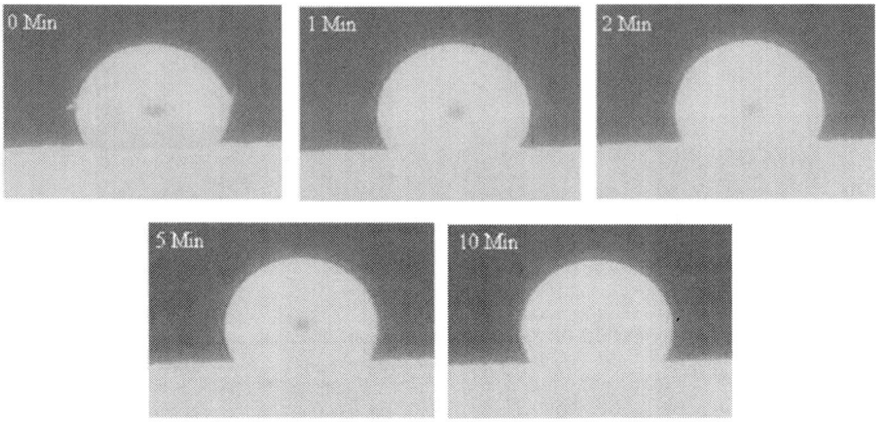

Figure 14. The wetting behavior of Bakelite char in contact with molten iron as function of time.

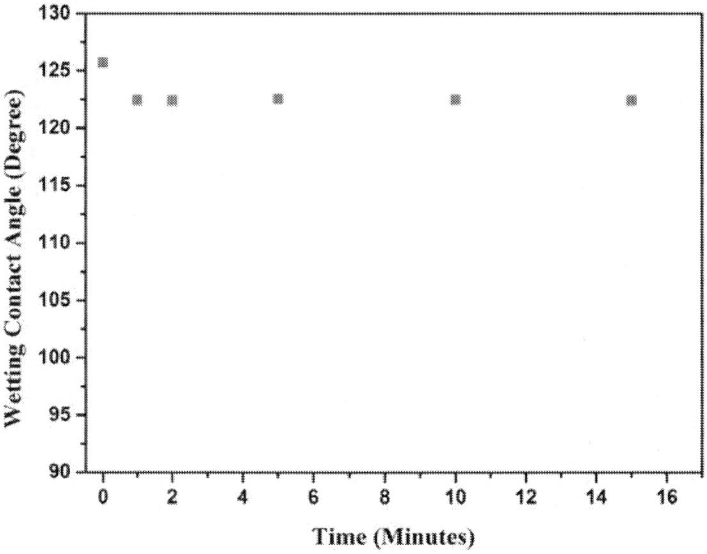

Figure 15. Contact angle as a function of time between iron and Bakelite char substrate

3.6. Waste CD char

A wettability study was carried out using waste CD char substrate and molten iron. The contact angle between the substrate and molten iron was measured and is shown in Figs. 16 and 17.

http://dx.doi.org/10.5772/61198

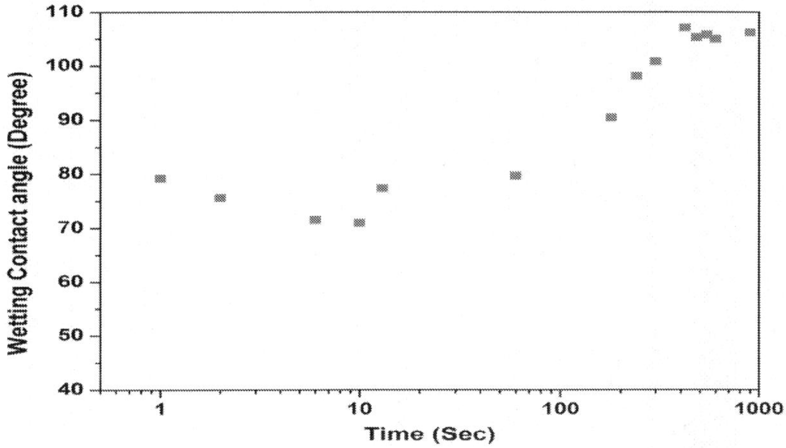

Figure 17. Contact angle as a function of time between iron and waste CD char substrate.

During initial contact from 0 to 10 seconds, a small decrease in contact angle was observed from 79° to 70°, possibly due to the transfer of carbon and also due to a change in interfacial energy due to interaction between char and molten iron. Wu et al. had observed a similar behavior while using graphite as a carbon source. During later periods, an increase was observed in contact angles that reached 100° after 3 minutes and stabilized at 105° after 15 minutes thereby indicating a much poorer wetting with the carbon level in molten iron approaching saturation. This result points to a much lower driving force for carbon dissolution.

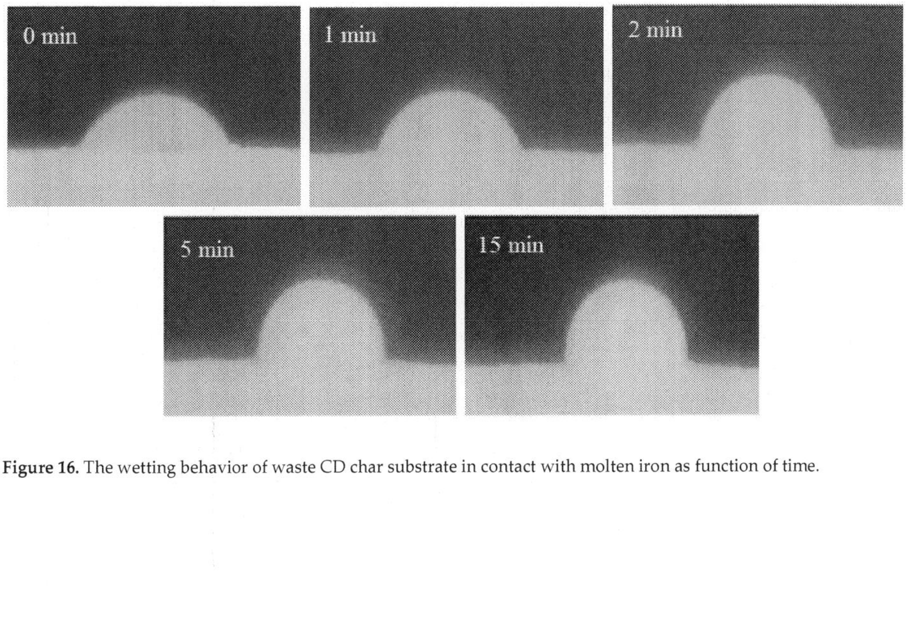

Figure 16. The wetting behavior of waste CD char substrate in contact with molten iron as function of time.

4. Concluding remarks

Detailed results have been reported on the wettability of a wide range of carbonaceous materials with liquid iron at 1550°C and important fundamental data on contact angles have been provided. This study has identified important drivers in the form of interfacial reactions; the evidence suggests that the deposition of reaction products in the interfacial region strongly affected the wettability and therefore the contact area between carbons and liquid iron. Main findings of this investigation are:

The effect of structure on the initial wettability of carbons were found to be marginal, with no well- defined trends observed with L_c values. Wide variations were observed in contact angles for chars possessing very similar L_c values. Due to the likelihood of structure evolution and changes in particle morphologies, no comments however can be made regarding the influence of structure on contact angles over prolonged periods of contact.

While no correlations were observed between initial contact angles and ash concentration/ composition, final contact angles were significantly affected by the presence of reaction products and impurity deposits in the interfacial region. Aluminum, calcium, sulfur, oxygen, and iron were detected in the interfacial layer. Small concentrations of magnesium and silicon were also detected indicating the partial presence of the parent ash. The presence of ash impurities and the nature of the interfacial layer play a very important role in affecting the wetting behavior, as it is this material that the liquid iron is in quasi-equilibrium with.

It is quite likely that combinations of several reactions occur simultaneously to dictate the wetting behavior. Reactions that could occur and subsequently influence the contact angles are: the formation of an enriched interfacial layer containing calcium, sulfur, and alumina; transfer of carbon and sulfur by mass transport across the interface; and reduction of reducible oxides such as silica and iron oxides, and possible transfer of these elements into liquid iron.

The mass transfer between solid carbons and iron melts was observed to strongly enhance the wetting phenomena and lowered contact angles. The reduction of system free energy corresponding to the mass transfer reactions was found to influence the formation of the interface region and resulted in the progressive spreading of the wetting line. The composition and thickness of the carbon/iron interfacial layer was dependent on the intensity of mass transfer across the interface. The resulting change in the interfacial energy γ_{ls} was a strong function of mass transfer, and was found to vary with the time of contact. The influence of carbon content on the wetting phenomena could only be seen in the initial stages, whereas the influence of sulfur on the wettability was found when the system approached equilibrium.

Acknowledgements

This work was supported by the Australian Research Council (ARC) funds. We gratefully acknowledge their financial contribution. We thank Ikram-ul-Haq and Somyote Kongkarat for some of the experimental work reported in this article.

Author details

R. Khanna*, I. Mansuri and V. Sahajwalla

*Address all correspondence to: ritakhanna@unsw.edu.au

Centre for Sustainable Materials Research and Technology, School of Materials Science and Engineering, The University of New South Wales, NSW Sydney, Australia

References

[1] Li K, Zhang J, Liu Z, Jiang X. Critical analyses about the development of iron-making process based on the principle of energy-saving and emission reduction. Chinese J Proc Eng. 2014;14(1):162-72.

[2] Khanna R, McCarthy F, Sahajwalla V. Investigation of wetting behavior of coal-chars with liquid iron by sessile drop method. Carbon - Science and Technology. 2008;1(2008):39-45.

[3] Young T. An Essay on the Cohesion of Fluids. Philosophical Transactions of the Royal Society of London. 1805;95:65-87.

[4] Aksay IA, Hoge CE, Pask JA. Wetting under chemical equilibrium and nonequilibrium conditions. Journal of Physical Chemistry. 1974;78(12):1178-83.

[5] Espie L, Drevet B, Eustathopoulos N. Experimental study of the influence of interfacial energies and reactivity on wetting in metal/oxide systems. Metallurgical and materials transactions A. 1994;25(3):599-605.

[6] Chatain D, Condurier L, Steinchen A, Eustathopoulos N. Interface in New Materials. Elsevier Applied Science, Amsterdam; 1991.

[7] Wu C, Sahajwalla V. Influence of melt carbon and sulfur on the wetting of solid graphite by Fe-CS melts. Metallurgical and Materials Transactions B. 1998;29(2): 471-7.

[8] Sahajwalla V, Khanna R. Effect of sulfur on the dissolution behaviour of graphite in Fe-C-S melts: A Monte Carlo simulation study, Scan. J. Metallurgy, 32(1), pp. 53-57.

[9] Khanna R, Sahajwalla V. An atomistic model for the graphite-alumina/liquid iron system: Monte-Carlo simulations on carbon dissolution. Acta materialia. 2005;53(4): 1205-14.

[10] Wright J, Baldock B. Dissolution kinetics of particulate graphite injected into iron/carbon melts. Metallurgical and Materials Transactions B. 1988;19(3):375-82.

[11] Humenik M, Kingery WD. Metal-Ceramic Interactions: III, Surface Tension and Wettability of Metal-Ceramic Systems. Journal of the American Ceramic Society. 1954;37(1):18-23.

[12] Naidich YV, Kolesnichenko GA. Investigation of the wetting of diamond and graphite by molten metals and alloys. Soviet Powder Metallurgy and Metal Ceramics. 1965;3(3):191-5.

[13] Sun H, Mori K, Sahajwalla V, Pehlke R. Carbon solution in liquid iron and iron alloys. High-temperature materials and processes. 1998;17(4):257-70.

[14] Wu C, Sahajwalla V. Dissolution rates of coals and graphite in Fe-C-S melts in direct ironmaking: Influence of melt carbon and sulfur on carbon dissolution. Metallurgical and Materials Transactions B. 2000 2000/04/01;31(2):243-51.

[15] Cham ST. Investigating factors that influence carbon dissolution from coke into molten iron. Sydney: The University of New South Wales; 2007.